Methods in Molecular Biology

Series Editor
John M. Walker
School of Life and Medical Sciences
University of Hertfordshire
Hatfield, Hertfordshire, UK

For further volumes:
http://www.springer.com/series/7651

For over 35 years, biological scientists have come to rely on the research protocols and methodologies in the critically acclaimed Methods in Molecular Biology series. The series was the first to introduce the step-by-step protocols approach that has become the standard in all biomedical protocol publishing. Each protocol is provided in readily-reproducible step-by step fashion, opening with an introductory overview, a list of the materials and reagents needed to complete the experiment, and followed by a detailed procedure that is supported with a helpful notes section offering tips and tricks of the trade as well as troubleshooting advice. These hallmark features were introduced by series editor Dr. John Walker and constitute the key ingredient in each and every volume of the Methods in Molecular Biology series. Tested and trusted, comprehensive and reliable, all protocols from the series are indexed in PubMed.

Manuscript Instructions https://drive.google.com/file/d/1K4Az5vpAFVzYadJ3jKx2-QSo7Dqa_SNA/view?usp=sharing

Protein Post-Translational Modifications

Methods and Protocols

Edited by

Xu Zhang

Biologic and Radiopharmaceutical Drugs Directorate, Health Products and Food Branch, Health Canada, Ottawa, ON, Canada

Rui Chen

Human Health Therapeutics Research Centre, National Research Council Canada, Ottawa, ON, Canada

Jianjun Li

Human Health Therapeutics Research Centre, National Research Council Canada, Ottawa, ON, Canada

Humana Press

Editors
Xu Zhang
Biologic and Radiopharmaceutical Drugs Directorate
Health Products and Food Branch
Health Canada
Ottawa, ON, Canada

Rui Chen
Human Health Therapeutics Research Centre
National Research Council Canada
Ottawa, ON, Canada

Jianjun Li
Human Health Therapeutics Research Centre
National Research Council Canada
Ottawa, ON, Canada

ISSN 1064-3745 ISSN 1940-6029 (electronic)
Methods in Molecular Biology
ISBN 978-1-0716-5165-0 ISBN 978-1-0716-5166-7 (eBook)
https://doi.org/10.1007/978-1-0716-5166-7

This Humana imprint is published by the registered company Springer Science+Business Media, LLC, part of Springer Nature.
The registered company address is: 1 New York Plaza, New York, NY 10004, U.S.A.

Preface

Post-translational modifications (PTMs) play a pivotal role in the regulation of protein function, stability, localization, and protein–protein interactions. These chemical modifications, ranging from simple types such as phosphorylation to more complex types such as glycosylation and ubiquitylation, dramatically expand the proteome diversity in cells beyond what is encoded by the genome and enable cells to respond more rapidly to environmental changes.

In recent years, the field of PTM research has grown rapidly, thanks to advances in enrichment methodologies, mass spectrometer instrumentations, proteomics/glycomics strategies, new technologies, such as nanopore-based protein sequencing, and more efficient computational tools. These developments have dramatically facilitated more comprehensive PTM studies, deepening our mechanistic understanding of disease and biology and enhancing the analytical characterization of protein-based therapeutics, where PTMs are considered critical quality attributes. In both basic and translational contexts, robust and reproducible analytical workflows for PTM analysis are essential, which is therefore the main purpose of this book.

This book in the *Methods in Molecular Biology* series brings together a collection of 20 chapters, structured into four parts. It begins with foundational reviews of the field (Part I), and then delves into detailed experimental and computational protocols covering both *traditional bottom-up strategies* (Part II) and *top-down proteoform analysis* (Part III). The final section (Part IV) is dedicated to the PTM analysis of *biotherapeutic proteins*, reflecting the growing emphasis on PTMs as critical quality attributes in biologic drug development and regulation. Across these sections, the detailed methods are applied to a wide range of sample types (including *cellular and animal models*, *clinical samples*, and *microbiomes*) showcasing how PTM studies are conducted in diverse research settings.

More specifically, the book opens with an overview of mass spectrometry methodologies (Chap. 1) and a review of PTM analysis in biotherapeutic drug development (Chap. 2). The subsequent protocol chapters explore a wide array of PTMs, ranging from common modifications like phosphorylation (Chaps. 6–8), glycosylation (Chaps. 5, 14, 18, and 19), methylation (Chaps. 9 and 16), sulfation (Chap. 17), deamidation (Chap. 20), acetylation (Chap. 10), and succinylation (Chap. 11) to more complex forms such as ubiquitylation (Chap. 4), SUMOylation (Chap. 3), and ADP-ribosylation (Chap. 15). The analytical techniques covered include both targeted and untargeted mass spectrometry when coupled with separation techniques like liquid chromatography and capillary electrophoresis (Chaps. 12, 13, and 20), as well as complementary approaches, including nuclear magnetic resonance (Chap. 16) and emerging nanopore-based methods (Chap. 17).

Collectively, this book offers modular and methodologically diverse toolkits for the researchers who study PTMs in basic, translational, or therapeutic contexts. Each chapter provides detailed, step-by-step protocols accompanied by troubleshooting advice, expected outcomes, and data analysis strategies. This book is intended for both early-career and advanced researchers in the field of proteomics, molecular biology, chemical biology, systems biology, bioinformatics, and biotherapeutic development. It is also useful for analytical scientists in industrial environments seeking robust methods for PTM characterization.

We hope this book serves as both a foundational and forward-looking resource to support and inspire continued exploration in this rapidly evolving field.

Ottawa, ON, Canada *Xu Zhang*
Ottawa, ON, Canada *Rui Chen*
Ottawa, ON, Canada *Jianjun Li*

Contents

Contributors

GHASEM AZEMI • *Computational Neurosurgery Lab (CNS), Macquarie Medical School, Faculty of Medicine, Health and Human Sciences, Macquarie University, Sydney, NSW, Australia*

MAXIM V. BEREZOVSKI • *John L. Holmes Mass Spectrometry Facility, Faculty of Science, Department of Chemistry and Biomolecular Sciences, University of Ottawa, Ottawa, ON, Canada*

KYLE K. BIGGAR • *Institute of Biochemistry, Carleton University, Ottawa, ON, Canada; Department of Biology, Carleton University, Ottawa, ON, Canada*

JOANNA BONS • *Buck Institute for Research on Aging, Novato, CA, USA*

RUI CHEN • *Human Health Therapeutics Research Centre, National Research Council Canada, Ottawa, ON, Canada*

ANAND CHOPRA • *The Shraga Segal Department of Microbiology, Immunology and Genetics, Ben-Gurion University of the Negev, Be'er-Sheva, Israel; National Institute for Biotechnology in the Negev, Ben-Gurion University of the Negev, Be'er-Sheva, Israel; Institute of Biochemistry, Carleton University, Ottawa, ON, Canada; Department of Biology, Carleton University, Ottawa, ON, Canada*

ROGER CHUNG • *Motor Neuron Disease Research Centre, Macquarie Medical School, Faculty of Medicine, Health and Human Sciences, Macquarie University, Sydney, NSW, Australia*

ROCHELLE D'MELLO • *John L. Holmes Mass Spectrometry Facility, Faculty of Science, Department of Chemistry and Biomolecular Sciences, University of Ottawa, Ottawa, ON, Canada*

JENNILEE M. DAVIDSON • *Motor Neuron Disease Research Centre, Macquarie Medical School, Faculty of Medicine, Health and Human Sciences, Macquarie University, Sydney, NSW, Australia; ALS Scholars in Therapeutics, Sean M. Healey and AMG Center for ALS at MassGeneral, Massachusetts General Hospital, Boston, MA, USA*

ANTONIO DI IEVA • *Computational Neurosurgery Lab (CNS), Macquarie Medical School, Faculty of Medicine, Health and Human Sciences, Macquarie University, Sydney, NSW, Australia*

FEI FANG • *Department of Chemistry, Michigan State University, East Lansing, MI, USA*

MIN FANG • *Molecular Analysis, Medicines and Healthcare Products Regulatory Agency, South Mimms, UK*

EMILY E. F. FEKETE • *Biologic and Radiopharmaceutical Drugs Directorate, Health Products and Food Branch, Health Canada, Ottawa, ON, Canada*

DANIEL FIGEYS • *Faculty of Medicine, University of Ottawa, Ottawa, ON, Canada; Quadram Institute Bioscience, Norwich Research Park, Norwich, UK*

LUCA FORNELLI • *School of Biological Sciences, University of Oklahoma, Oklahoma, USA; Department of Chemistry and Biochemistry, University of Oklahoma, Oklahoma, USA*

YAN GAO • *School of Chemistry and Chemical Engineering, Nanjing University, Nanjing, China*
HANSJÖRG HABISCH • *Research Unit Integrative Structural Biology, Medicinal Chemistry, Otto Loewi Research Center, Medical University of Graz, Graz, Austria; BioTechMed-Graz, Graz, Austria*
GRAYSON HATFIELD • *Regulatory Research Division, Centre for Oncology, Radiopharmaceuticals and Research, Biologics and Radiopharmaceutical Drugs Directorate, Health Canada, Ottawa, ON, Canada*
LYNDAL HENDEN • *Motor Neuron Disease Research Centre, Macquarie Medical School, Faculty of Medicine, Health and Human Sciences, Macquarie University, Sydney, NSW, Australia*
MATTHEW HOEKSTRA • *Institute of Biochemistry, Carleton University, Ottawa, ON, Canada; Department of Biology, Carleton University, Ottawa, ON, Canada*
TOBIAS HUBERTS • *Research Unit Integrative Structural Biology, Medicinal Chemistry, Otto Loewi Research Center, Medical University of Graz, Graz, Austria; BioTechMed-Graz, Graz, Austria*
JIE JIANG • *School of Chemistry and Chemical Engineering, Nanjing University, Nanjing, China*
JAKE T. KLINE • *School of Biological Sciences, University of Oklahoma, Oklahoma, USA*
ANGELA S. LAIRD • *Motor Neuron Disease Research Centre, Macquarie Medical School, Faculty of Medicine, Health and Human Sciences, Macquarie University, Sydney, NSW, Australia*
MARIE-FRANCE LANGELIER • *Department of Biochemistry and Molecular Medicine, Université de Montréal, Montréal, QC, Canada*
ALBERT LEE • *Motor Neuron Disease Research Centre, Macquarie Medical School, Faculty of Medicine, Health and Human Sciences, Macquarie University, Sydney, NSW, Australia*
CHONGYANG LI • *Institute for Research in Immunology and Cancer, Université de Montréal, Montréal, QC, Canada*
JIANJUN LI • *Human Health Therapeutics Research Centre, National Research Council Canada, Ottawa, ON, Canada*
MENG-YIN LI • *School of Chemistry and Chemical Engineering, Nanjing University, Nanjing, China*
YINGXI LI • *John L. Holmes Mass Spectrometry Facility, Faculty of Science, Department of Chemistry and Biomolecular Sciences, University of Ottawa, Ottawa, ON, Canada*
BING LIU • *School of Chemistry and Chemical Engineering, Liaoning Normal University, Dalian, China*
YI-TAO LONG • *School of Chemistry and Chemical Engineering, Nanjing University, Nanjing, China*
HUIXIN LU • *Biologic and Radiopharmaceutical Drugs Directorate, Health Products and Food Branch, Health Canada, Ottawa, ON, Canada*
JUNFENG MA • *Department of Oncology, Lombardi Comprehensive Cancer Center, Georgetown University Medical Center, Washington, DC, USA*
TOBIAS MADL • *Research Unit Integrative Structural Biology, Medicinal Chemistry, Otto Loewi Research Center, Medical University of Graz, Graz, Austria; BioTechMed-Graz, Graz, Austria*

ZORAN MINIC • *John L. Holmes Mass Spectrometry Facility, Faculty of Science, Department of Chemistry and Biomolecular Sciences, University of Ottawa, Ottawa, ON, Canada*
MANIJA MIRHASAN • *Department of Biochemistry and Molecular Medicine, Université de Montréal, Montréal, QC, Canada*
ZHIBIN NING • *Faculty of Medicine, University of Ottawa, Ottawa, ON, Canada*
JOHN M. PASCAL • *Department of Biochemistry and Molecular Medicine, Université de Montréal, Montréal, QC, Canada*
SUTTINEE POOLSUP • *John L. Holmes Mass Spectrometry Facility, Faculty of Science, Department of Chemistry and Biomolecular Sciences, University of Ottawa, Ottawa, ON, Canada*
ANASTASIYA POTAPENKO • *Motor Neuron Disease Research Centre, Macquarie Medical School, Faculty of Medicine, Health and Human Sciences, Macquarie University, Sydney, NSW, Australia*
TIMOTHY R. RUDD • *Molecular Analysis, Medicines and Healthcare Products Regulatory Agency, South Mimms, UK*
BIRGIT SCHILLING • *Buck Institute for Research on Aging, Novato, CA, USA*
ISABELLE SIROIS • *CHU Sainte-Justine Research Center, Université de Montréal, Montreal, QC, Canada*
LIANGLIANG SUN • *Department of Chemistry, Michigan State University, East Lansing, MI, USA*
AYAKO TAKEMORI • *Advanced Research Support Center, Ehime University, Toon, Ehime, Japan*
NOBUAKI TAKEMORI • *Advanced Research Support Center, Ehime University, Toon, Ehime, Japan*
ROGER Y. TAM • *Regulatory Research Division, Centre for Oncology, Radiopharmaceuticals and Research, Biologics and Radiopharmaceutical Drugs Directorate, Health Canada, Ottawa, ON, Canada; Department of Chemistry, University of Ottawa, Ottawa, ON, Canada*
MINCHU TANG • *Department of Analytical Chemistry, State Key Laboratory of Drug Research, Shanghai Institute of Materia Medica, Chinese Academy of Sciences, Shanghai, China; University of Chinese Academy of Sciences, Beijing, China*
LIOUDMILA TEPLIAKOVA • *Regulatory Research Division, Centre for Oncology, Radiopharmaceuticals and Research, Biologics and Radiopharmaceutical Drugs Directorate, Health Canada, Ottawa, ON, Canada*
PIERRE THIBAULT • *Institute for Research in Immunology and Cancer, Université de Montréal, Montréal, QC, Canada; Department of Chemistry, Université de Montréal, Montréal, QC, Canada*
YUN WANG • *Biologic and Radiopharmaceutical Drugs Directorate, Health Products and Food Branch, Health Canada, Ottawa, ON, Canada*
YUQIU WANG • *Department of Analytical Chemistry, State Key Laboratory of Drug Research, Shanghai Institute of Materia Medica, Chinese Academy of Sciences, Shanghai, China; Department of Otolaryngology, Eye & ENT Hospital, Fudan University, Shanghai, China*
GAIL C. WHITING • *Molecular Analysis, Medicines and Healthcare Products Regulatory Agency, South Mimms, UK*

WILLIAM G. WILLMORE • *Institute of Biochemistry, Carleton University, Ottawa, ON, Canada; Department of Biology, Carleton University, Ottawa, ON, Canada*
CI WU • *School of Chemistry and Chemical Engineering, Liaoning Normal University, Dalian, China; Department of Oncology, Lombardi Comprehensive Cancer Center, Georgetown University Medical Center, Washington, DC, USA*
ZHAOGUAN WU • *CHU Sainte-Justine Research Center, Université de Montréal, Montreal, QC, Canada*
WENFAN XIE • *School of Chinese Materia Medica, Nanjing University of Chinese Medicine, Nanjing, China*
ALICE ZHANG • *Buck Institute for Research on Aging, Novato, CA, USA*
XU ZHANG • *Biologic and Radiopharmaceutical Drugs Directorate, Health Products and Food Branch, Health Canada, Ottawa, ON, Canada; Faculty of Medicine, University of Ottawa, Ottawa, ON, Canada*

Part I

Introduction

Chapter 1

An Overview of Mass Spectrometry-Based Methods to Analyze Post-Translational Modifications

Rui Chen, Xu Zhang and Jianjun Li

Abstract

Post-translational modifications (PTM) are the covalent modifications of proteins after their biosynthesis. PTMs change the structure and functions of proteins, which eventually regulate various biological processes that are closely related to the onset and development of diseases. Hundreds of different forms of PTMs have been discovered so far and mass spectrometry (MS) has become one of the most important tools to analyze PTMs. With the development of new methods in sample preparation, advances in MS instrumentation, and inventions of new bioinformatic tools, MS-based methods can identify most forms of PTMs, precisely locate the site of modification, sensitively and accurately quantify the changes of modification levels from various of biological samples. This chapter will summarize generic procedures to analyze PTMs with MS methods and provide an update with the latest development in sample preparation, MS analysis, and database search that advanced the way for PTM analysis, improving our knowledge of PTMs and understanding of their roles in various diseases.

Keywords Post-translational modification, Mass spectrometry, Bottom-up, Top-down, Sample preparation, Database search

1 Introduction

Protein post-translational modifications (PTMs) are covalent changes made to proteins after their synthesis, which can occur on amino acid side chains or the protein backbone itself. These modifications can dramatically alter a protein's physiological and biological properties, vastly expanding the functional capacity of the proteome beyond what is encoded by the genome. To date, over 200 different types of PTMs have been discovered. These include the addition of small chemical groups to the amino acid residues, such as phosphorylation, methylation, and acetylation, changes of amino acids, such as deamidation, oxidation, and disulfide bond formation, changes to the protein backbone, such as proteolytic cleavage, and the attachment of more complex biomolecules, such as glycosylation, ubiquitylation, SUMOylation, and ADP-ribosylation. Accumulating evidence has

Xu Zhang et al. (eds.), *Protein Post-Translational Modifications*, Methods in Molecular Biology, vol. 3018,
https://doi.org/10.1007/978-1-0716-5166-7_1,

implicated the dysregulation of PTMs in numerous human diseases, including cancer, neurodegenerative disorders like Alzheimer's, metabolic diseases, such as diabetes, aging, and immune disorders [1]. Additionally, many of these PTMs are not exclusive to mammals; they are fundamental processes found across all domains of life, including bacteria, viruses, archaea, and fungal species. In microbial pathogens, PTMs have been demonstrated to be critical for their metabolism, pathogenic virulence, host–pathogen interactions, and antimicrobial resistance [2, 3]. PTMs has also been demonstrated to be prevalent in complex microbial communities or microbiomes from host-associated and environmental samples [4]. The PTMs of human gut microbiomes are implicated in disease onset, such as inflammatory bowel disease, and the PTMs in environmental microbiomes are associated with their adaption to extreme environmental conditions [5, 6]. Altogether, the analysis of PTMs in biological samples is of paramount importance in modern life science and biomedical research.

Currently, there are two ways to analyze PTMs with MS—bottom-up and top-down approaches (Fig. 1). Bottom-up approach

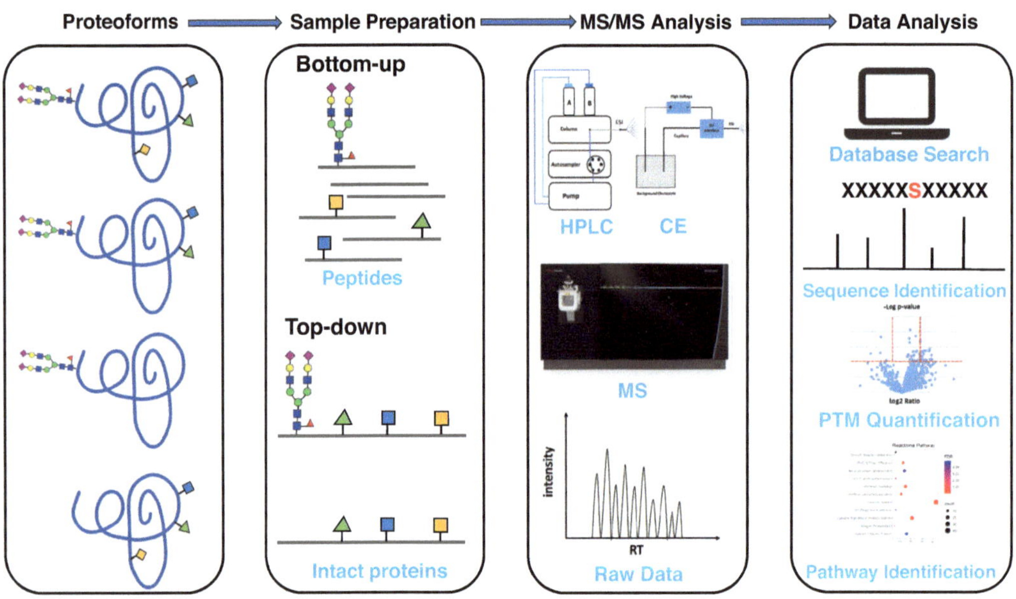

Fig. 1 Illustration of the procedures for two approaches in the analysis of protein post-translational modification with mass spectrometry. The majority of steps are similar for both top-down and bottom-up approach that include sample preparation, separation, MS analysis, and database search. The only and most significant difference is with top-down approach, the proteins are analyzed as intact form while with bottom-up methods, proteins are digested into peptides before enrichment of particular types of PTMs. Either intact proteins or peptides are separated by HPLC and CE before MS analysis to reduce the complexity of samples. MS/MS spectra of proteins/peptides are collected for database search to identify the sequence and location of PTMs. Further analysis can also quantify the difference in the level of PTMs between samples groups and potentially enable the identification of pathways to elucidate the function of PTMs. Both methods have advantages and drawbacks, and they are often used in combination to achieve higher depth in PTM characterization

is based on digestion of the proteins or full proteomes into short peptides prior to separation and analysis [7]. Bottom-up approach takes advantage of highly efficient chromatography separation and electrospray ionization, specific enrichment of peptides with PTMs and mature database search/site assignment, and has been applied in numerous large-scale PTM studies. However, digestion of proteins into peptides make it impossible to characterize all the PTMs on individual protein, which is termed as proteoform. Top-down approach was developed to solve this issue to perform both separation and analysis at the intact protein level. It has been proven to be a powerful tool for proteoform analysis and enables accurate protein identification and PTM localization, including possible sequence variants and the relative quantification of expression levels of various proteoforms. However, defining and pinpointing PTMs on intact proteins remains a challenge due to the complexity of spectra from multiple charged precursors, especially for highly modified proteins with molecular weight over 30 kDa. In many applications, both bottom-up and top-down are used to achieve more comprehensive coverage and characterization of PTMs. In addition to mass spectrometry, various other instrumentations have been used for native protein analysis, such as nanopore and nuclear magnetic resonance (NMR). Those methods are usually used along site MS to further analyze PTMs on certain proteins targets.

The analysis of many PTMs has always posed difficulties due primarily to the sub-stoichiometric nature of modified proteins within a proteome for both bottom-up or top-down approaches. MS has become the method of choice for qualitative and quantitative proteomic analyses of PTMs, which can be attributed to improvements in the three major steps in PTM analysis by mass spectrometry [8–10]. Firstly, new enrichment methods and antibodies are developed to specifically isolate protein or peptides with particular PTMs from complex proteome samples, which significantly enhanced the chance of modified peptides/proteins been picked up by MS [11]. Secondly, advances in the MS instrumentation, i.e., implementation of new mass analyzers like the Orbitrap [12], novel fragmentation methods like electron transfer dissociation (ETD) [13] and ultraviolet photodissociation (UVPD) [14], and data collection method like data-independent acquisition (DIA) [15]. These advances improved the accuracy, sensitivity, and capability in analyzing PTMs in high-throughput manner. Lastly, numerous computational and statistical methods have been invented to analyze large MS dataset and pinpoint the PTMs by connecting MS/MS spectra with peptide/protein sequences [16]. Compared to other methods, MS offered high sensitivity to detect PTMs with low stoichiometries from the background of complex proteome samples. With advances in fragmentation methods to generate various type of fragments ions during MS/MS analysis, it provides unique structural information to preciously location the site of modification, i.e., which amino acid the modification group is attached to. Thus, MS can both qualitatively and quantitative analyze PTMs from biological samples. Furthermore, most MS instruments are couples with high performance liquid chromatography or capillary electrophoresis that provide another dimension in terms of separation to reduce the complexity of samples. Recently, ion mobility technologies [17], which separate molecules by their cross-section area, offer complementary gas-phase separation and enable the differentiation of isobaric ions. All these technologies have been progressing rapidly in the last decades to make MS the most powerful in PTM analysis. This chapter will comprehensively summarize the methods that are currently used in each of three steps.

2 Sample Preparation

Regardless of whether bottom-up or top-down strategy is used for PTM analysis, the first and probably the most important step is the efficient extraction of proteins from biological samples, including cells, tissues, and various body fluids [18]. For both methods, protein extraction involves lysis with different buffers, which contains either detergents like sodium dodecyl sulfate (SDS), Triton X-100, sodium deoxycholate (DOC), 3-((3-cholamidopropyl) dimethylammonio)-1-propanesulfonate (CHAPS), etc., or chaotic reagents like urea and guanidine hydrochloride. Although different buffers have preferences in extracting certain types of proteins, most commonly used methods in protein extraction can provide decent coverage of proteome/proteoform and can be applied for PTM analysis [19, 20]. For a few PTMs, artifacts from sample preparation needs to be avoided when the mass difference introduced by artifacts cannot be distinguished with the genuine modifications. One example is carbamylation of lysine by urea during protein denaturation and digestion. Carbamylation has a mass shift of 43.008, and it could be recognized as acetylation (42, 106 Da) with heavier isotopes [21] and coenriched by lysine acetylation motif-specific antibody. Similarly, iodoacetamide, the alkylating agent commonly used in standard protein digestion protocols to block cysteine residues, can produce a 2-acetamidoacetamide covalent adduct to lysine residues (114,0429 Da), which has the same atomic composition as the diglycine adduct that comes from digestion of ubiquitinated proteins [22]. Thus, consideration should be given when selecting sample preparation method to eliminate potential artifacts. For PTM analysis, the enrichment of proteins/peptides with certain PTMs of interest after protein extraction is more important, which requires different and more sophisticated types of techniques [11, 23]. This section will discuss major techniques in PTM enrichment.

2.1 Ion Exchange Chromatography

Ion exchange chromatography (IEC), which is complementary to reversed phase liquid chromatography (RPLC) that based on the differences in hydrophobicity, separates peptides/proteins by the net charges they carry. Although not efficient nor directly compatible with mass spectrometer, IEC has been routinely used to fractionate peptides/proteins before RPLC-MS/MS analysis, which significantly reduces samples complexity and improve the coverage of proteome/proteoforms [24]. As chemical modification changes the net charge of peptides/proteins, IEC has also been used to sperate modified peptides/proteins from those without modifications. Strong cation exchange (SCX) chromatography is recommended for the enrichment of peptides carrying extra positive charges, e.g., acetylated peptide [25] and methylated peptides [26]. It can also be used to negatively select peptides carrying less positive charge like phosphopeptides, in which phosphopeptides are eluted from the flow-through and earlier fractions [27]. On the other hand, strong anion exchange (SAX) chromatography has a positive charge and bind to peptides carrying net negative charges. As a result, it can be used to enrich phosphopeptides directly [28]. When operated with high organic solvent, SAX or weak anion exchange (WAX) can be turned into electrostatic repulsion-hydrophilic interaction chromatography (ERLIC) to enrich hydrophilic peptides [29], including glycopeptides and phosphopeptides [30, 31]. It can also be used in combination with other methods as prefractionation to improve the coverage of phosphoproteome [32, 33]. In another application, the operation mode of WAX can be switched from ERLIC to HILIC by changing the eluting buffer to facilitate simultaneous and separate enrichment of both glycopeptides and phosphopeptides with a single solid phase extraction (SPE) [34]. These developments suggest IEX is a versatile method for sample preparation with broad applications.

2.2 Immobilized Metal Ion Chromatography/ Metal Oxide

Immobilized metal ion chromatography (IMAC) and metal oxide (MO) have been commonly exploited to enrich phosphorylated proteins and peptides [35, 36]. IMAC utilizes multivalent metal ions immobilized on a stationary phase to selectively chelate negatively charged phosphate groups of phosphoproteins or phosphopeptides. Currently, a variety of various metal ions are available for phosphorylated peptide enrichment, such as Fe^{3+}, Zr^{4+}, Ga^{3+}, and Ti^{4+} [37–40]. IMAC enrichment of phosphopeptides is characterized by high selectivity and sensitivity. However, this enrichment technique has the disadvantage as highly acidic non-phosphorylated peptides also have an affinity for metal ions and can interact with IMAC resin. While low pH can reduce the binding of non-phosphorylated peptides, it can also reduce the binding of specific phosphorylated peptides. Therefore, enormous effort has been made to improve the efficiency of IMAC-based phosphopeptide enrichment with the development new resins and chelating chemistry. MOs bind to negatively charged phosphate groups with the positive charge on their surface and are also widely used for phosphopeptide enrichment [41]. MOs are more tolerant to low pH and the specificity of the enrichment can be further improved with the adding of competing acid like lactic acid, glycolic acid, or dihydroxybenzoic acid to reduce the binding of C-terminal carboxylic group and acidic amino acids (glutamic and aspartic acid). There are several MOs utilized for this technique, including Al_2O_3, ZrO_2, and TiO_2, etc. Of these, TiO_2 is likely to be the most commonly used MO substrate with the strengths of high enrichment efficiency and specificity. However, different MOs have different affinities and specificities for phosphopeptides, MOs can be combined with either other MOs or IMACs for enrichment to increase the recovery of phosphopeptide. In some cases, fractionation with either RPLC or IEC is conducted before the enrichment to maximize the coverage of phosphoproteome [42–44].

2.3 Hydrophilic Interaction Chromatography

Hydrophilic interaction chromatography (HILIC) was developed as variation of normal phase chromatography with water compatible organic solvent to separate polar and hydrophilic compounds [29]. It employs semi-aqueous mobile phases to create a "water-enriched" layer within a hydrophilic stationary phase and separates analytes based on their differential partitioning between the aqueous layer adsorbed on the stationary phase and the organic-rich mobile phase. Thus, hydrophilic peptides, like phosphopeptides and glycopeptides, would have higher retention on the column and could be separated with other peptides. Although it has been used for phosphopeptides enrichment and fractionation, the main application of HILIC in PTM analysis is to enrichment glycopeptides that followed by deglycosylation for high-throughput quantification of glycoproteins for membrane and surface proteome analysis. More importantly, it enables the MS analysis of intact glycopeptides to achieve both glycan composition and peptide sequence identification. HILIC enriches glycopeptides via the hydrophilic properties imparted by the glycan, which contains several hydroxyl groups on each monosaccharide unit. Since its invention in 1990, it has become the most efficient method to enrich glycopeptides and the addition of ion paring reagents significantly improved its specificity [45]. The performance of HILIC was further enhanced with the synthesis of new materials [46]. A few HILIC materials and columns are commercially available to facilitate either the off-line SPE of glycopeptides from protein digest [47] or direct coupling of HILIC column to MS [48]. Recently, the application of glycopeptides has been expanded from *N*-glycopeptides to *O*-glycopeptides [49], which has fewer monosaccharides and is more difficult to be enriched [50, 51]. It is still challenging to comprehensive characterize the glycosylation from complex samples and more development in HILIC is needed for better enrichment.

2.4 Protein/Antibody-based Affinity Enrichment

Unlike chromatographic methods, protein/antibody-based affinity enrichment is dependent on the specific recognition of modified amino acid to develop a pan-antibody to enrich all modified peptides. It is more effective to develop PTM-specific antibody that only recognize the particular chemical structure like tyrosine phosphorylation (pY) [52], *N*-acetylglucosamine on serine/threonine (O-GlcNAc) [53], lysine/arginine methylation [54], ubiquitylation (K-ε-GG) [55], and series of metabolism-coupled modifications on lysine, including acetylation [56], succinylation [57], malonylation [58], glutarylation [59], lactylation [60], isobutyrylation [61], acetoacetylation [62], itaconylation [63], methacrylation [64], etc. Most antibodies are immobilized to either agarose or magnetic beads for the enrichment of modified peptides and various solid phase matrix are developed to reduce non-specific binding of peptides without modifications. Some of those modified peptides have very low stoichiometry of modification and it requires two steps of enrichment in some applications, i.e., enrichment at protein level first and followed by second enrichment at peptide level after digestion. Another approach to improve the recovery and specificity of enrichment is using chromatography-based fractionation prior to reduce the complexity of samples. One advantage of antibody-based affinity enrichment is that it allows the tandem enrichment of several modifications within one experiment [65], which significantly reduces the requirement of biological samples. However, for some modifications, it is challenging to develop pan-antibodies that recognize only the modification, but not its surrounding sequence, which is used to assess the overall change in the protein PTMs of interest.

In addition to antibodies, proteins specifically recognize modifications are also discovered. One example is the Af1521 macrodomain from Archaebacteria [66] that binds to both mono-ADP-ribosylation (MARylation) and poly-ADP-ribosylation (PARylation) on multiple amino acids (lysine, arginine, serine, threonine, aspartic acid, and glutamic acid) [67], and it has become the most common macrodomain employed to enrich MARylated/PARylated proteins for large-scale proteomics studies [68, 69]. The discovery of those antibodies and proteins have greatly improved the performance of MS based PTM analysis in both discovery and quantitative analysis.

3 MS Analysis

Once samples are prepared, the next step is the analysis by tandem mass spectrometry. Modern mass spectrometers are either coupled to high performance liquid chromatography (HPLC) or capillary electrophoresis (CE). Although HPLC is more popular in the field of proteomics or protein analysis, CE offers alternative mechanism of separation and sometimes has better separation and higher sensitivity in PTM analysis than HPLC [70–73]. Particularly in top-down analysis when intact proteins are analyzed, CE is capable of separating naïve form of proteins and its modified form to improve the identification of proteoforms of histone [74], tau [75], and cancer cell lines [76]. However, the application of CE-MS is limited by its stability and reproducibility issue, while LC-MS has become a routine technique in large-scale PTM studies.

3.1 Data Acquisition

For either CE-MS or LC-MS, two different data acquisition methods are generally used to acquire MS/MS data for identification and quantification of PTMs. In Data-dependent acquisition (DDA) mode, the mass spectrometer selects the most intense peptide ions in a first stage of tandem mass spectrometry, and then, they are fragmented and analyzed in a second stage of tandem mass spectrometry to collect the spectrum of a peptide, i.e., the mass over charge (m/z) ratio of its precursor and fragments. In DDA mode, only a few precursors are selected due to number of ions and the limit of cycle time, which results in the neglect of peptides with lower abundance. Thus, for the analysis of PTMs, peptides/proteins with modifications need to be separated from the complex proteome background. Another disadvantage of DDA mode is its reproducibility as its sampling is dependent on the complexity of samples. One common issue of DDA, especially for quantitative analysis, is the missing value of one peptide from some samples. The missing value causes problem for statistical and functional analysis. The alternative acquisition method, named data-independent acquisition (DIA), was developed to solve this problem [15]. In DIA, the instrument focuses on a narrow mass window of precursors and acquires MS/MS data from all precursors detected within that window. This mass window is then stepped across the entire mass range, systematically collecting MS/MS data from all detected precursors, thus, all precursor/fragment pairs are theoretically detectable. The advantage of DIA over DDA is that it offers higher reproducibility in quantification of PTMs as demonstrated in a few benchmarking studies for phosphoproteomics [77]. Although less studied, it also demonstrated promising result for intact glycopeptides analysis [78]. However, DIA usually requires a prebuilt spectral library, especially for PTM analysis. It is a common approach today to firstly generate a large spectral library by fractionation and DDA with pooled samples for PTM identification, and then quantitatively analyze level of PTMs through individual samples by DIA. This strategy has been used in several large-scale phosphoproteomics studies and showed higher accuracy and lower false discovery rate (FDR) than using in-silico generated library. Adding modified precursor and fragment significantly increase the possible combinations of precursor/fragment pairs for spectra library building and will need more stringent control of FDR. It is also more difficult to predict the retention time and MS/MS fragmentation pattern with modified peptides using in-silico approach. Thus, the application of DIA methods for a broader range of PTMs still requires further exploration.

3.2 Fragmentation Methods

Another advantage of DDA is its compatibility with various kinds of fragmentation techniques, which is crucial to identify and characterize PTMs, especially for peptides are labile with conventional collision induced dissociation (CID) or recent high-energy collision dissociation (HCD) [79]. Both techniques break peptides at the amide bond and generate b- and y-types of ions. However, chemical bonds with lower bond strength, like glycosidic bond between two sugars are more favored to break during both CID and HCD. As a result, for modification with labile chemical groups, most fragments are generated either due to the loss of modification group (e.g., phosphate) or the breakage within the modification group (glycosylation, ADP-ribosylation). Although those fragmentations are not helpful for protein/peptide sequence identification, they produce distinct patterns of MS/MS spectrum and oxonium ions at low m/z range, which can be used as diagnostic ions to verify the existence of PTMs. For example, phosphotyrosine immonium at ion m/z 216.043 has been used to identify tyrosine phosphorylation in peptides and proteins [80]. A series of oxonium ions of glycans, m/z 163 (hexose, Hex), m/z 204 (acetyl hexosamine, HexNAc), m/z 292 (acetyl neuraminic acid, Neu5Ac), m/z 366 (HexHexNAc) have been used as diagnostic ions for glycopeptides [81]. Unique diagnostic ions are also used to detect uncommon PTMs, including ADP-ribosylation. Four low mass marker ions, adenine+ (m/z 136.06), (adenosine-H_2O)+(m/z 250.09), AMP^+ (m/z 348.07), and ADP^+ (m/z 428.03) are generated by the fragmentation of the ADP-ribose moiety attached to a peptide. These ions provided with a unique signature to confirm that the peptide of interest is ADP-ribosylated [69].

Alternative fragmentation technique has been developed in recent years to provide complementary types of ions to boost the confidence in characterization of PTMs. Among them, electron transfer dissociation (ETD) become one of the most popular techniques, partly due to its implementation with Orbitrap mass spectrometer [82]. ETD relies on the ion–ion reaction between multiply charged precursor peptide (or protein) cations and singly charged radical reagent anions. Once an electron from the reagent anion is transferred to the cation, an odd-electron cation is formed and undergoes free-radical-driven cleavage, which dissociates N–Cα bonds along the peptide backbone to generate even electron c-type and odd electron z-type product ions. The fragmentation is complementary to the b- and y-types of ions generated from CID/HCD. More importantly, it preserves most of the modification groups on side change of peptide/proteins and helps the localization of the site of modification.

However, the efficiency of ETD is dependent on the charge state as it favors peptides/proteins with higher charge density. This disadvantage limits the application of ETD in routine bottom-up proteomics analysis of PTMs as doubly charges peptides are the main products of tryptic digest and their fragmentation efficiency is poor. In the earlier days, this problem was solved by a strategy name "data-dependent tree decision" [83]. During a cycle of MS/MS analysis, the mass spectrometer firstly determines the charge state of a precursor, if the peptide is doubly charged, CID/HCD will be used to generate MS/MS spectrum while ETD is applied if the peptide is triply charged (or higher). This method has been proven valuable for increasing identifications and improving confidence of peptide spectral matches in a wide variety of analyses. Recently, supplemental activation was developed to increase efficiency of ETD by introducing extra energy into the reaction, including EThcD [84], which activate all ETD products with high-energy collisional dissociation. EThcD generates both c/z- and b/y-types product ion for peptides sequence identification while diagnostic ions can also be found from low m/z range. Since its invention, EThcD has been used in various applications of PTM analysis, including glycosylation [85], phosphorylation [86], ADP-ribosylation [69], etc.

Another fragmentation technique that gained attention in the last decade is ultraviolet photodissociation (UVPD), which relies on the absorption of one or more high-energy UV photons as peptide bond in proteins exhibits strong adsorption around 190 nm. Photoactivation using 193 or 213 nm photons have been used for UVPD of proteins. UVPD prominently generates a/x ions from the C-Cα backbone bonds that are complementary to the C-N bond from CID/HCD and N-Cα from ETD [14]. UVPD also preserves the labile PTMs like ETD, but the fragmentation is not dependent on charge state, which makes it ideal for MS analysis of PTMs. Currently, most applications of UVPD are focused on top-down analysis and its demonstrated higher sequence coverage and protein identification score than ETD and HCD [87, 88]. A recent study demonstrated that it could also be applied in bottom-up analysis of the *O*-glycosylation diversity of mucin [89] and similar advantage over HCD and EThcD was reported. Nevertheless, no single fragmentation technique is perfect, and all those fragmentation techniques are complementary to each other. Selective use of fragmentation that suits the PTMs of interest will be helpful to MS analysis.

3.3 Ion Mobility Mass Spectrometry

One challenge in the MS analysis of PTM is the isobaric or positional isomers introduced by the chemical group. Although both LC and CE offer some extent of separation, it is not sufficient to resolve those isomers. Ion mobility (IM) technique was added as another dimension of separation that based on the collision cross sections in a gas-filled chamber under the influence of an electric field [90]. As the collision cross sections are determined by the size, shape, and charge a molecule, the combination of ion mobility and mass spectrometer provides a powerful tool for characterizing molecules, including isomers, isobars, and protein complexes [91]. One example of application of IM-MS is in the studying of protein phosphorylation, it enabled separation of phosphorylated and non-phosphorylated peptides with the same mass to charge ratio as phosphorylation introduces changes in peptide conformation [92] and recent development of cyclic ion mobility is capable of separating of isomeric Tau phosphopeptides where three isomers with adjacent phosphosites (T212, S214, and T217) were identified. IM-MS has also been applied in protein glycosylation analysis. High field asymmetric wave ion mobility spectrometry (FAIMS) was able to separate isomeric O-linked glycopeptides that have identical sequences but differing sites of glycosylation [93]. In this study, two glycopeptides with O-linked *N*-acetylgalactosamine at different position from the glycoprotein mucin 5AC, were separated with different compensation voltages while co-eluted from reversed phase chromatography. IM-MS is also able to separate more complex N-linked glycopeptides with isomeric α2,3 and α2,6 NeuAc linkages. It was revealed that the α2,6 fragment has a considerably shorter drift time when compared to the α2,3 equivalent [94]. A recent study analyzed over 4000 pairs of matching modified and unmodified peptide ion species with 22 common PTMs by trapped ion mobility spectrometry (TIMS), and it revealed most PTMs changed the drift time of peptides except methylation [95]. This suggested IM-MS has the potential to be applied in other PTMs. However, the capability of IM-MS in large-scale study has not been investigated yet, with limited application in proteoforms analysis [96]. Adding ion mobility also increases the difficulty in data analysis as most data processing tools have not adapted such function module, particularly for structure elucidation.

4 Database Search and Data Processing

The final step in PTMs analysis is searching the MS data with various of bioinformatic tools to extract spectra from modified peptides and pinpoint the nature of modification and which amino acid its attaches to. The choice of software is dependent on the strategy used (bottom-up or top-down) and types of PTMs.

4.1 Database Search for Bottom-Up Approach

For bottom-up approach, database search methods for PTMs analysis are similar to those used for unmodified peptides when modifications will remain intact on the peptide during fragmentation, which can be done by most of the commonly used search engines. In these cases, the masses of the modification groups are added as variable modifications to a few preset amino acids and a complete list of modifications and masses can be found at Unimod [97]. For example, most acyl modifications on lysine, including acetyl, succinyl, and malonyl, etc., are stable during fragmentation and these modifications can be identified and quantified as any unmodified peptides with proper enrichment. Labile modifications, such as phosphorylation, glycosylation, sulfation, ADP-ribosylation, ubiquintylation, and SUMOlation, fragment themselves in addition to the peptide backbone fragmentation in tandem MS with collision activation/dissociation methods. For those peptides, additional consideration is given to allow specific neutral loss and diagnostic ions to improve the search, especially for those modifications that are partially labile, such as phosphorylation. This feature has been implemented in many traditional search engines to improve search result. A few tools even introduced labile mode that allows filtering of spectra for diagnostic ions with customizable parameters [98]. Another emerging technique to discover new forms of PTMs is named open search by searching MS/MS spectra without setting parameters for specific PTMs [99]. This is achieved by using larger mass tolerances or by searching for mass shifts without specifying the modification. Several tools, including MSFragger [100], pFind3 [101], MetaMorpheus [16], TagGraph [102], etc., have been developed with different algorithms to facilitate open search with large proteomic dataset and discover unknown modifications, such as RNA crosslinking. It was found that adding fragment ions shifting by unknown modifications could further improved the sensitivity in open search [103]. With increasing scanning speed and sensitivity of high-resolution mass spectrometer, open search strategy could play more important roles in PTM studies, especially in discovering novel and unknown forms.

4.2 Top-Down Proteomics

Unlike bottom-up proteomics, the software packages available for top-down proteoform identification are less developed due to complexity of top-down high-resolution mass spectra, as protein tends to have multiple charge states during electrospray ionization [104, 105]. Extensive deconvolution is required to convert the fragment ions from each charge state as singly charged or neutral masses and to combine it with the intensity [106]. Although different algorithms were developed, the deconvolution results often contain a considerable amount of misassigned peaks due to the consequence of the complexity of the high-resolution MS and MS/MS data generated in top-down proteomics experiments [107]. Database search tools for top-down proteomics analysis include Mascot Top-down [108], ProSight PTM 2.0 [109], pTop [110], MS-Align+ [111], MASH Suite Pro [112], TopPIC [113], MSPathFinder [114], etc. However, there is lack of benchmarking and cross-platform comparison of the performance from those tools to build a standard procedure in studying the PTMs with top-down proteomics. The identification of PTMs still replies on the closed search method and is limited to a few well studied modifications.

To solve this problem, new tools are developed to identify ultramodified proteins by top-down and bottom-up mass spectrometry in conjunction with annotations [115] and features like de novo sequencing are introduced [116]. Nevertheless, bioinformatic tools are still needed to resolve the complexity of top-down spectra and improve the performance in both identification and quantification of proteoforms.

4.3 Glycosylation Analysis

Special tools are needed for glycosylation analysis as it is one of the most complex PTMs with numerous combinations of several monosaccharides. Glycosylation produces complex fragmentation patterns that exhibit multiple modes of fragmentation, i.e., breakage of both peptide backbone and glycosidic bonds. In most collision-based fragmentation methods, this generated predominately fragments from glycans but not from peptides, making it difficult to elucidate the peptides sequence. Thus, for glycosylation analysis, the focus of developing search algorithm is to locate those low abundant ions from peptide fragmentation [117]. Byonic is one of the first tools for efficient intact glycopeptides identification and are still popular in many applications [118, 119]. It searches glycopeptides as an integral unit that considers all possible peptides and glycans, which increases search spaces and requires stringent control of false assignment. Various algorithms have been developed to search for fragments contain peptide sequence (i.e., the Y1 ion with peptide backbone and one *N*-acetyl glucosamine) and identify peptide sequence by either MS^3[120] or correlate with deglycosylated peptide that can be identified through regular database search [121]. This strategy become the basis of many other tools like MAGIC [122], pGlyco 2.0 [123], pGlyco3.0 [124], and GPSeeker [125]. A few tools implemented de novo identification of glycans to expand glycan identification beyond glycan database [126–128] or directly identify glycan without the need of database [129]. Unlike those tools, MSfragger-glyco uses a peptide-first strategy by matching the b and y ions of peptides in the spectrum directly. It significantly improved the sensitivity in glycopeptide identification and outperformed those tools using glycan first strategy [130]. Except identification, many tools are also capable of performing *label-free* quantification of intact glycopeptides and have been applied in biomarker discovery to find specific glycomarker for cancer diagnosis [131]. It is foreseeable that the advancement in the bioinformatic tools for glycosylation analysis will help understand the biological role of these complex modifications.

5 Conclusions

Tremendous developments in techniques for PTM analysis by mass spectrometry have been made in the last two decades and this chapter only provides a glimpse of many new methods. There are still lots of challenges in this field to comprehensively and quantitatively characterize diverse forms of PTMs from biological samples as current technologies can only study a few forms of PTM at one time or focus on certain proteins. Complete solution to this problem needs invention of novel methods for sample preparation and separation, new mass spectrometer with enhanced sensitivity, scanning speed and better fragmentation and improved bioinformatic tools to interpret the complex MS/MS data. Emerging technologies like machine learning and artificial intelligence have demonstrated great potential in predicting technical characteristics of a proteomic experiment, and we can expect they will play more important roles in the analysis of PTMs through deep interpreting of existing experiments and data to design better experiments and data analyses.

References

1. Zhong Q, Xiao X, Qiu Y, Xu Z, Chen C, Chong B et al (2023) Protein posttranslational modifications in health and diseases: functions, regulatory mechanisms, and therapeutic implications. MedComm 4(3):e261
2. Retanal C, Ball B, Geddes-McAlister J (2021) Post-translational modifications drive success and failure of fungal–host interactions. J Fungi 7(2):124
3. Macek B, Forchhammer K, Hardouin J, Weber-Ban E, Grangeasse C, Mijakovic I (2019) Protein post-translational modifications in bacteria. Nat Rev Microbiol 17(11):651–664
4. Duan H, Zhang X, Figeys D (2023) An emerging field: post-translational modification in microbiome. Proteomics 23(3–4):2100389
5. Zhang W, Sun J, Cao H, Tian R, Cai L, Ding W et al (2016) Post-translational modifications are enriched within protein functional groups important to bacterial adaptation within a deep-sea hydrothermal vent environment. Microbiome 4(1):49
6. Zhang X, Ning Z, Mayne J, Yang Y, Deeke SA, Walker K et al (2020) Widespread protein lysine acetylation in gut microbiome and its alterations in patients with Crohn's disease. Nat Commun 11(1):4120
7. Jiang Y, Rex DAB, Schuster D, Neely BA, Rosano GL, Volkmar N et al (2024) Comprehensive overview of bottom-up proteomics using mass spectrometry. ACS Meas Sci Au 4(4):338–417
8. Witze ES, Old WM, Resing KA, Ahn NG (2007) Mapping protein post-translational modifications with mass spectrometry. Nat Methods 4(10):798–806
9. Jensen ON (2004) Modification-specific proteomics: characterization of post-translational modifications by mass spectrometry. Curr Opin Chem Biol 8(1):33–41
10. Mann M, Jensen ON (2003) Proteomic analysis of post-translational modifications. Nat Biotechnol 21(3):255–261
11. Brandi J, Noberini R, Bonaldi T, Cecconi D (2022) Advances in enrichment methods for mass spectrometry-based proteomics analysis of post-translational modifications. J Chromatogr A 1678:463352
12. Hu Q, Noll RJ, Li H, Makarov A, Hardman M, Graham Cooks R (2005) The Orbitrap: a new mass spectrometer. J Mass Spectrom 40(4):430–443
13. Syka JE, Coon JJ, Schroeder MJ, Shabanowitz J, Hunt DF (2004) Peptide and protein sequence analysis by electron transfer dissociation mass spectrometry. Proc Natl Acad Sci 101(26):9528–9533
14. Brodbelt JS, Morrison LJ, Santos I (2019) Ultraviolet photodissociation mass spectrometry for analysis of biological molecules. Chem Rev 120(7):3328–3380
15. Röst HL, Rosenberger G, Navarro P, Gillet L, Miladinović SM, Schubert OT et al (2014) OpenSWATH enables automated, targeted analysis of data-independent acquisition MS data. Nat Biotechnol 32(3):219–223
16. Solntsev SK, Shortreed MR, Frey BL, Smith LM (2018) Enhanced global post-translational modification discovery with MetaMorpheus. J Proteome Res 17(5):1844–1851
17. Orsburn BC, Yuan Y, Bumpus NN (2022) Insights into protein post-translational modification landscapes of individual human cells by trapped ion mobility time-of-flight mass spectrometry. Nat Commun 13(1):7246
18. Kaulich PT, Jeong K, Kohlbacher O, Tholey A (2024) Influence of different sample preparation approaches on proteoform identification by top-down proteomics. Nat Methods 21(12):2397–2407

19. Yang Z, Shen X, Chen D, Sun L (2020) Toward a universal sample preparation method for denaturing top-down proteomics of complex proteomes. J Proteome Res 19(8):3315–3325
20. Varnavides G, Madern M, Anrather D, Hartl N, Reiter W, Hartl M (2022) In search of a universal method: a comparative survey of bottom-up proteomics sample preparation methods. J Proteome Res 21(10):2397–2411
21. Martinez-Val A, Garcia F, Ximénez-Embún P, Martinez Teresa-Calleja A, Ibarz N, Ruppen I et al (2017) Urea artifacts interfere with immuno-purification of lysine acetylation. J Proteome Res 16(2):1061–1068
22. Nielsen ML, Vermeulen M, Bonaldi T, Cox J, Moroder L, Mann M (2008) Iodoacetamide-induced artifact mimics ubiquitination in mass spectrometry. Nat Methods 5(6):459–460
23. Virág D, Dalmadi-Kiss B, Vékey K, Drahos L, Klebovich I, Antal I et al (2020) Current trends in the analysis of post-translational modifications. Chromatographia 83(1):1–10
24. Zhai Z, Mavridou D, Damian M, Mutti FG, Schoenmakers PJ, Gargano AF (2024) Characterization of complex proteoform mixtures by online nanoflow ion-exchange chromatography-native mass spectrometry. Anal Chem 96(22):8880–8885
25. Chang C-H, Chang H-Y, Rappsilber J, Ishihama Y (2021) Isolation of acetylated and unmodified protein N-terminal peptides by strong cation exchange chromatographic separation of TrypN-digested peptides. Mol Cell Proteomics 20:100003
26. Li Z, Wang Q, Wang Y, Wang K, Liu Z, Zhang W et al (2021) An efficient approach based on basic strong cation exchange chromatography for enriching methylated peptides with high specificity for methylproteomics analysis. Anal Chim Acta 1161:338467
27. Mohammed S, Heck AJ (2011) Strong cation exchange (SCX) based analytical methods for the targeted analysis of protein post-translational modifications. Curr Opin Biotechnol 22(1):9–16
28. Hardman G, Perkins S, Brownridge PJ, Clarke CJ, Byrne DP, Campbell AE et al (2019) Strong anion exchange-mediated phosphoproteomics reveals extensive human non-canonical phosphorylation. EMBO J 38(21):e100847
29. Alpert AJ (1990) Hydrophilic-interaction chromatography for the separation of peptides, nucleic acids and other polar compounds. J Chromatogr A 499:177–196
30. Cui Y, Yang K, Tabang DN, Huang J, Tang W, Li L (2019) Finding the sweet spot in ERLIC mobile phase for simultaneous enrichment of N-glyco and phosphopeptides. J Am Soc Mass Spectrom 30(12):2491–2501
31. Cao L, Yu L, Guo Z, Li X, Xue X, Liang X (2013) Application of a strong anion exchange material in electrostatic repulsion–hydrophilic interaction chromatography for selective enrichment of glycopeptides. J Chromatogr A 1299:18–24
32. Zarei M, Sprenger A, Rackiewicz M, Dengjel J (2016) Fast and easy phosphopeptide fractionation by combinatorial ERLIC-SCX solid-phase extraction for in-depth phosphoproteome analysis. Nat Protoc 11(1):37–45
33. Chien K-y, Liu H-C, Goshe MB (2011) Development and application of a phosphoproteomic method using electrostatic repulsion-hydrophilic interaction chromatography (ERLIC), IMAC, and LC–MS/MS analysis to study Marek's disease virus infection. J Proteome Res 10(9):4041–4053
34. Chen R, Williamson S, Fulton KM, Twine SM, Li J (2019) Simultaneous analysis of phosphopeptides and intact glycopeptides from secretome with mode switchable solid phase extraction. Anal Methods 11(41):5243–5249
35. Leitner A (2010) Phosphopeptide enrichment using metal oxide affinity chromatography. TrAC Trends Anal Chem 29(2):177–185
36. Dunn JD, Reid GE, Bruening ML (2010) Techniques for phosphopeptide enrichment prior to analysis by mass spectrometry. Mass Spectrom Rev 29(1):29–54
37. Feng S, Ye M, Zhou H, Jiang X, Jiang X, Zou H et al (2007) Immobilized zirconium ion affinity chromatography for specific enrichment of phosphopeptides in phosphoproteome analysis. Mol Cell Proteomics 6(9):1656–1665
38. Ruprecht B, Koch H, Medard G, Mundt M, Kuster B, Lemeer S (2015) Comprehensive and reproducible phosphopeptide enrichment using iron immobilized metal ion affinity chromatography (Fe-IMAC) columns. Mol Cell Proteomics 14(1):205–215
39. Posewitz MC, Tempst P (1999) Immobilized gallium (III) affinity chromatography of phosphopeptides. Anal Chem 71(14):2883–2892
40. Zhou H, Ye M, Dong J, Han G, Jiang X, Wu R et al (2008) Specific phosphopeptide enrichment with immobilized titanium ion affinity chromatography adsorbent for phosphoproteome analysis. J Proteome Res 7(9):3957–3967
41. Wu J, Shakey Q, Liu W, Schuller A, Follettie MT (2007) Global profiling of phosphopeptides by titania affinity enrichment. J Proteome Res 6(12):4684–4689
42. Zarei M, Sprenger A, Metzger F, Gretzmeier C, Dengjel J (2011) Comparison of

ERLIC–TiO2, HILIC–TiO2, and SCX–TiO2 for global phosphoproteomics approaches. J Proteome Res 10(8):3474–3483
43. Zhou H, Low TY, Hennrich ML, van der Toorn H, Schwend T, Zou H et al (2011) Enhancing the identification of phosphopeptides from putative basophilic kinase substrates using Ti (IV) based IMAC enrichment. Mol Cell Proteomics 10(10):M110.006452
44. Kettenbach AN, Gerber SA (2011) Rapid and reproducible single-stage phosphopeptide enrichment of complex peptide mixtures: application to general and phosphotyrosine-specific phosphoproteomics experiments. Anal Chem 83(20):7635–7644
45. Mysling S, Palmisano G, Højrup P, Thaysen-Andersen M (2010) Utilizing ion-pairing hydrophilic interaction chromatography solid phase extraction for efficient glycopeptide enrichment in glycoproteomics. Anal Chem 82(13):5598–5609
46. Qing G, Yan J, He X, Li X, Liang X (2020) Recent advances in hydrophilic interaction liquid interaction chromatography materials for glycopeptide enrichment and glycan separation. TrAC Trends Anal Chem 124:115570
47. Chen R, Seebun D, Ye M, Zou H, Figeys D (2014) Site-specific characterization of cell membrane N-glycosylation with integrated hydrophilic interaction chromatography solid phase extraction and LC–MS/MS. J Proteomics 103:194–203
48. Ding W, Nothaft H, Szymanski CM, Kelly J (2009) Identification and quantification of glycoproteins using ion-pairing normal-phase liquid chromatography and mass spectrometry. Mol Cell Proteomics 8(9):2170–2185
49. Yang W, Shah P, Hu Y, Toghi Eshghi S, Sun S, Liu Y et al (2017) Comparison of enrichment methods for intact N-and O-linked glycopeptides using strong anion exchange and hydrophilic interaction liquid chromatography. Anal Chem 89(21):11193–11197
50. Yang W, Ao M, Hu Y, Li QK, Zhang H (2018) Mapping the O-glycoproteome using site-specific extraction of O-linked glycopeptides (EXoO). Mol Syst Biol 14(11):e8486
51. Iwashkiw JA, Seper A, Weber BS, Scott NE, Vinogradov E, Stratilo C et al (2012) Identification of a general O-linked protein glycosylation system in *Acinetobacter baumannii* and its role in virulence and biofilm formation. PLoS Pathog 8(6):e1002758
52. Rush J, Moritz A, Lee KA, Guo A, Goss VL, Spek EJ et al (2005) Immunoaffinity profiling of tyrosine phosphorylation in cancer cells. Nat Biotechnol 23(1):94–101
53. Burt RA, Dejanovic B, Peckham HJ, Lee KA, Li X, Ounadjela JR et al (2021) Novel antibodies for the simple and efficient enrichment of native O-GlcNAc modified peptides. Mol Cell Proteomics 20:100167
54. Ong S-E, Mittler G, Mann M (2004) Identifying and quantifying in vivo methylation sites by heavy methyl SILAC. Nat Methods 1(2):119–126
55. Peng J, Schwartz D, Elias JE, Thoreen CC, Cheng D, Marsischky G et al (2003) A proteomics approach to understanding protein ubiquitination. Nat Biotechnol 21(8):921–926
56. Choudhary C, Kumar C, Gnad F, Nielsen ML, Rehman M, Walther TC et al (2009) Lysine acetylation targets protein complexes and co-regulates major cellular functions. Science 325(5942):834–840
57. Zhang Z, Tan M, Xie Z, Dai L, Chen Y, Zhao Y (2011) Identification of lysine succinylation as a new post-translational modification. Nat Chem Biol 7(1):58–63
58. Peng C, Lu Z, Xie Z, Cheng Z, Chen Y, Tan M et al (2011) The first identification of lysine malonylation substrates and its regulatory enzyme. Mol Cell Proteomics 10(12):M111.012658
59. Tan M, Peng C, Anderson KA, Chhoy P, Xie Z, Dai L et al (2014) Lysine glutarylation is a protein posttranslational modification regulated by SIRT5. Cell Metab 19(4):605–617
60. Zhang D, Tang Z, Huang H, Zhou G, Cui C, Weng Y et al (2019) Metabolic regulation of gene expression by histone lactylation. Nature 574(7779):575–580
61. Zhu Z, Han Z, Halabelian L, Yang X, Ding J, Zhang N et al (2021) Identification of lysine isobutyrylation as a new histone modification mark. Nucleic Acids Res 49(1):177–189
62. Gao Y, Sheng X, Tan D, Kim S, Choi S, Paudel S et al (2023) Identification of histone lysine acetoacetylation as a dynamic post-translational modification regulated by HBO1. Adv Sci 10(25):2300032
63. Liu D, Xiao W, Li H, Zhang Y, Yuan S, Li C et al (2023) Discovery of itaconate-mediated lysine acylation. J Am Chem Soc 145(23):12673–12681
64. Delaney K, Tan M, Zhu Z, Gao J, Dai L, Kim S et al (2021) Histone lysine methacrylation is a dynamic post-translational modification regulated by HAT1 and SIRT2. Cell Discov 7(1):122
65. Mertins P, Qiao JW, Patel J, Udeshi ND, Clauser KR, Mani D et al (2013) Integrated proteomic analysis of post-translational modifications by serial enrichment. Nat Methods 10(7):634–637
66. Allen MD, Buckle AM, Cordell SC, Löwe J, Bycroft M (2003) The crystal structure of

AF1521 a protein from Archaeoglobus fulgidus with homology to the non-histone domain of macroH2A. J Mol Biol 330(3):503–511
67. Karras GI, Kustatscher G, Buhecha HR, Allen MD, Pugieux C, Sait F et al (2005) The macro domain is an ADP-ribose binding module. EMBO J 24(11):1911–1920
68. Nowak K, Rosenthal F, Karlberg T, Bütepage M, Thorsell A-G, Dreier B et al (2020) Engineering Af1521 improves ADP-ribose binding and identification of ADP-ribosylated proteins. Nat Commun 11(1):5199
69. Hendriks IA, Larsen SC, Nielsen ML (2019) An advanced strategy for comprehensive profiling of ADP-ribosylation sites using mass spectrometry-based proteomics. Mol Cell Proteomics 18(5):1010a
70. Faserl K, Sarg B, Gruber P, Lindner HH (2018) Investigating capillary electrophoresis-mass spectrometry for the analysis of common post-translational modifications. Electrophoresis 39(9–10):1208–1215
71. Chen D, Ludwig KR, Krokhin OV, Spicer V, Yang Z, Shen X et al (2019) Capillary zone electrophoresis–tandem mass spectrometry for large-scale phosphoproteomics with the production of over 11,000 phosphopeptides from the colon carcinoma HCT116 cell line. Anal Chem 91(3):2201–2208
72. Johnson KR, Gregus M, Kostas JC, Ivanov AR (2022) Capillary electrophoresis coupled to electrospray ionization tandem mass spectrometry for ultra-sensitive proteomic analysis of limited samples. Anal Chem 94(2):704–713
73. Serrano LR, Mellors JS, Thompson JW, Lancaster NM, Robinson ML, Overmyer KA et al (2025) SPE-CZE-MS quantifies zeptomole amounts of phosphorylated peptides. J Proteome Res 24(6):3049–3061
74. Faserl K, Sarg B, Lindner HH (2020) Application of CE-MS for the analysis of histones and histone modifications. Methods 184:125–134
75. Fang F, Xu T, Chien Hagar H-T, Hovde S, Kuo M-H, Sun L (2024) Pilot study for deciphering post-translational modifications and proteoforms of tau protein by capillary electrophoresis-mass spectrometry. J Proteome Res 23(11):5085–5095
76. McCool EN, Xu T, Chen W, Beller NC, Nolan SM, Hummon AB et al (2022) Deep top-down proteomics revealed significant proteoform-level differences between metastatic and nonmetastatic colorectal cancer cells. Sci Adv 8(51):eabq6348
77. Hogrebe A, von Stechow L, Bekker-Jensen DB, Weinert BT, Kelstrup CD, Olsen JV (2018) Benchmarking common quantification strategies for large-scale phosphoproteomics. Nat Commun 9(1):1045
78. Yang Y, Yan G, Kong S, Wu M, Yang P, Cao W et al (2021) GproDIA enables data-independent acquisition glycoproteomics with comprehensive statistical control. Nat Commun 12(1):6073
79. Olsen JV, Macek B, Lange O, Makarov A, Horning S, Mann M (2007) Higher-energy C-trap dissociation for peptide modification analysis. Nat Methods 4(9):709–712
80. Steen H, Küster B, Fernandez M, Pandey A, Mann M (2001) Detection of tyrosine phosphorylated peptides by precursor ion scanning quadrupole TOF mass spectrometry in positive ion mode. Anal Chem 73(7):1440–1448
81. Dalpathado DS, Desaire H (2008) Glycopeptide analysis by mass spectrometry. Analyst 133(6):731–738
82. Riley NM, Coon JJ (2018) The role of electron transfer dissociation in modern proteomics. Anal Chem 90(1):40–64
83. Swaney DL, McAlister GC, Coon JJ (2008) Decision tree–driven tandem mass spectrometry for shotgun proteomics. Nat Methods 5(11):959–964
84. Frese CK, Altelaar AM, van den Toorn H, Nolting D, Griep-Raming J, Heck AJ et al (2012) Toward full peptide sequence coverage by dual fragmentation combining electron-transfer and higher-energy collision dissociation tandem mass spectrometry. Anal Chem 84(22):9668–9673
85. Zhang Y, Xie X, Zhao X, Tian F, Lv J, Ying W et al (2018) Systems analysis of singly and multiply O-glycosylated peptides in the human serum glycoproteome via EThcD and HCD mass spectrometry. J Proteomics 170:14–27
86. Frese CK, Zhou H, Taus T, Altelaar AM, Mechtler K, Heck AJ et al (2013) Unambiguous phosphosite localization using electron-transfer/higher-energy collision dissociation (EThcD). J Proteome Res 12(3):1520–1525
87. Greer SM, Brodbelt JS (2018) Top down characterization of heavily modified histones using 193 nm ultraviolet photodissociation mass spectrometry. J Proteome Res 17(3):1138–1145
88. McCool EN, Chen D, Li W, Liu Y, Sun L (2019) Capillary zone electrophoresis-tandem mass spectrometry with ultraviolet photodissociation (213 nm) for large-scale top–down proteomics. Anal Methods 11(22):2855–2861
89. Helms A, Chang V, Malaker SA, Brodbelt JS (2024) Unraveling O-glycan diversity of mucins: insights from SmE mucinase and ultraviolet photodissociation mass spectrometry. Anal Chem 96(49):19230–19237
90. Wu C, Siems WF, Klasmeier J, Hill HH (2000) Separation of isomeric peptides

using electrospray ionization/high-resolution ion mobility spectrometry. Anal Chem 72(2):391–395
91. Christofi E, Barran P (2023) Ion mobility mass spectrometry (IM-MS) for structural biology: insights gained by measuring mass, charge, and collision cross section. Chem Rev 123(6):2902–2949
92. Thalassinos K, Grabenauer M, Slade SE, Hilton GR, Bowers MT, Scrivens JH (2009) Characterization of phosphorylated peptides using traveling wave-based and drift cell ion mobility mass spectrometry. Anal Chem 81(1):248–254
93. Creese AJ, Cooper HJ (2012) Separation and identification of isomeric glycopeptides by high field asymmetric waveform ion mobility spectrometry. Anal Chem 84(5):2597–2601
94. Hinneburg H, Hofmann J, Struwe W, Thader A, Altmann F, Silva DV et al (2016) Distinguishing N-acetylneuraminic acid linkage isomers on glycopeptides by ion mobility-mass spectrometry. Chem Commun 52(23):4381–4384
95. Will A, Oliinyk D, Bleiholder C, Meier F (2023) Peptide collision cross sections of 22 post-translational modifications. Anal Bioanal Chem 415(27):6633–6645
96. Xu T, Wang Q, Wang Q, Sun L (2023) Coupling high-field asymmetric waveform ion mobility spectrometry with capillary zone electrophoresis-tandem mass spectrometry for top-down proteomics. Anal Chem 95(25):9497–9504
97. Creasy DM, Cottrell JS (2004) Unimod: protein modifications for mass spectrometry. Proteomics 4(6):1534–1536
98. Polasky DA, Geiszler DJ, Yu F, Li K, Teo GC, Nesvizhskii AI (2023) MSFragger-labile: a flexible method to improve labile PTM analysis in proteomics. Mol Cell Proteomics 22:5
99. Chick JM, Kolippakkam D, Nusinow DP, Zhai B, Rad R, Huttlin EL et al (2015) A mass-tolerant database search identifies a large proportion of unassigned spectra in shotgun proteomics as modified peptides. Nat Biotechnol 33(7):743–749
100. Kong AT, Leprevost FV, Avtonomov DM, Mellacheruvu D, Nesvizhskii AI (2017) MSFragger: ultrafast and comprehensive peptide identification in mass spectrometry–based proteomics. Nat Methods 14(5):513–520
101. Chi H, Liu C, Yang H, Zeng W-F, Wu L, Zhou W-J et al (2018) Comprehensive identification of peptides in tandem mass spectra using an efficient open search engine. Nat Biotechnol 36(11):1059–1061
102. Devabhaktuni A, Lin S, Zhang L, Swaminathan K, Gonzalez CG, Olsson N et al (2019) TagGraph reveals vast protein modification landscapes from large tandem mass spectrometry datasets. Nat Biotechnol 37(4):469–479
103. Yu F, Teo G, Kong A, Haynes S, Avtonomov D, Geiszler D et al (2020) Identification of modified peptides using localization-aware open search. Nat Commun 11: 4065
104. Brown KA, Melby JA, Roberts DS, Ge Y (2020) Top-down proteomics: challenges, innovations, and applications in basic and clinical research. Expert Rev Proteomics 17(10):719–733
105. Po A, Eyers CE (2023) Top-down proteomics and the challenges of true proteoform characterization. J Proteome Res 22(12):3663–3675
106. Liu X, Inbar Y, Dorrestein PC, Wynne C, Edwards N, Souda P et al (2010) Deconvolution and database search of complex tandem mass spectra of intact proteins. Mol Cell Proteomics 9(12):2772–2782
107. Tabb DL, Jeong K, Druart K, Gant MS, Brown KA, Nicora C et al (2023) Comparing top-down proteoform identification: deconvolution, PrSM overlap, and PTM detection. J Proteome Res 22(7):2199–2217
108. Karabacak NM, Li L, Tiwari A, Hayward LJ, Hong P, Easterling ML et al (2009) Sensitive and specific identification of wild type and variant proteins from 8 to 669 kDa using top-down mass spectrometry* S. Mol Cell Proteomics 8(4):846–856
109. Zamdborg L, LeDuc RD, Glowacz KJ, Kim Y-B, Viswanathan V, Spaulding IT et al (2007) ProSight PTM 2.0: improved protein identification and characterization for top down mass spectrometry. Nucleic Acids Res 35(2):W701–W706
110. Sun R-X, Luo L, Wu L, Wang R-M, Zeng W-F, Chi H et al (2016) pTop 1.0: a high-accuracy and high-efficiency search engine for intact protein identification. Anal Chem 88(6):3082–3090
111. Liu X, Sirotkin Y, Shen Y, Anderson G, Tsai YS, Ting YS et al (2012) Protein identification using top-down spectra. Mol Cell Proteomics 11(6):M111.008524
112. Cai W, Guner H, Gregorich ZR, Chen AJ, Ayaz-Guner S, Peng Y et al (2016) MASH suite pro: a comprehensive software tool for top-down proteomics. Mol Cell Proteomics 15(2):703–714
113. Kou Q, Xun L, Liu X (2016) TopPIC: a software tool for top-down mass spectrometry-based proteoform identification and characterization. Bioinformatics 32(22):3495–3497
114. Park J, Piehowski PD, Wilkins C, Zhou M, Mendoza J, Fujimoto GM et al (2017) Informed-proteomics: open-source software package for top-down proteomics. Nat Methods 14(9):909–914

115. Chen W, Ding Z, Zang Y, Liu X (2023) Characterization of proteoform post-translational modifications by top-down and bottom-up mass spectrometry in conjunction with annotations. J Proteome Res 22(10):3178–3189
116. Vyatkina K, Wu S, Dekker LJ, VanDuijn MM, Liu X, Tolić N et al (2016) Top-down analysis of protein samples by de novo sequencing techniques. Bioinformatics 32(18):2753–2759
117. Polasky DA, Nesvizhskii AI (2023) Recent advances in computational algorithms and software for large-scale glycoproteomics. Curr Opin Chem Biol 72:102238
118. Bern M, Kil YJ, Becker C (2012) Byonic: advanced peptide and protein identification software. Curr Protoc Bioinformatics 40(1):13.20.1–13.20.14
119. Roushan A, Wilson GM, Kletter D, Sen KI, Tang W, Kil YJ et al (2021) Peak filtering, peak annotation, and wildcard search for glycoproteomics. Mol Cell Proteomics 20:100011
120. Zeng W-F, Liu M-Q, Zhang Y, Wu J-Q, Fang P, Peng C et al (2016) pGlyco: a pipeline for the identification of intact N-glycopeptides by using HCD-and CID-MS/MS and MS3. Sci Rep 6(1):25102
121. Cheng K, Chen R, Seebun D, Ye M, Figeys D, Zou H (2014) Large-scale characterization of intact N-glycopeptides using an automated glycoproteomic method. J Proteomics 110:145–154
122. Lynn K-S, Chen C-C, Lih TM, Cheng C-W, Su W-C, Chang C-H et al (2015) MAGIC: an automated N-linked glycoprotein identification tool using a Y1-ion pattern matching algorithm and in silico MS2 approach. Anal Chem 87(4):2466–2473
123. Liu M-Q, Zeng W-F, Fang P, Cao W-Q, Liu C, Yan G-Q et al (2017) pGlyco 2.0 enables precision N-glycoproteomics with comprehensive quality control and one-step mass spectrometry for intact glycopeptide identification. Nat Commun 8(1):438
124. Zeng W-F, Cao W-Q, Liu M-Q, He S-M, Yang P-Y (2021) Precise, fast and comprehensive analysis of intact glycopeptides and modified glycans with pGlyco3. Nat Methods 18(12):1515–1523
125. Xiao K, Tian Z (2019) GPSeeker enables quantitative structural N-glycoproteomics for site-and structure-specific characterization of differentially expressed N-glycosylation in hepatocellular carcinoma. J Proteome Res 18(7):2885–2895
126. Shen J, Jia L, Dang L, Su Y, Zhang J, Xu Y et al (2021) StrucGP: de novo structural sequencing of site-specific N-glycan on glycoproteins using a modularization strategy. Nat Methods 18(8):921–929
127. Sun W, Zhang Q, Zhang X, Tran NH, Ziaur Rahman M, Chen Z et al (2023) Glycopeptide database search and de novo sequencing with PEAKS GlycanFinder enable highly sensitive glycoproteomics. Nat Commun 14(1):4046
128. Zeng W-F, Yan G, Zhao H-H, Liu C, Cao W (2024) Uncovering missing glycans and unexpected fragments with pGlycoNovo for site-specific glycosylation analysis across species. Nat Commun 15(1):8055
129. Fang Z, Qin H, Mao J, Wang Z, Zhang N, Wang Y et al (2022) Glyco-Decipher enables glycan database-independent peptide matching and in-depth characterization of site-specific N-glycosylation. Nat Commun 13(1):1900
130. Polasky DA, Yu F, Teo GC, Nesvizhskii AI (2020) Fast and comprehensive N-and O-glycoproteomics analysis with MSFragger-Glyco. Nat Methods 17(11):1125–1132
131. Liu L, Liu L, Wang Y, Fang Z, Bian Y, Zhang W et al (2024) Robust glycoproteomics platform reveals a tetra-antennary site-specific glycan capping with sialyl-Lewis antigen for early detection of gastric cancer. Adv Sci 11(9):2306955

Chapter 2

Post-Translational Modifications in Protein Therapeutics: Importance and Analytical Tools

Emily E. F. Fekete, Huixin Lu and Xu Zhang

Abstract

Post-translational modifications (PTMs) play a pivotal role in the activity and stability of proteins and can, therefore, significantly affect the quality of protein- or cell-based biological therapeutic products. Accordingly, PTMs are often evaluated as critical quality attributes, which require thorough characterization and evaluation throughout product development. Continued progress in understanding and characterizing protein PTMs is important to enable the distribution of safe and effective biologic medicines. Currently, a wide range of analytical techniques are employed for the detection and characterization of PTMs in the biopharmaceutical context. This chapter provides an overview of PTMs relevant to therapeutic proteins and commonly used or emerging analytical tools to analyze them.

Keywords Biotherapeutics, Mass spectrometry, Quality attributes, Monoclonal antibody

1 Introduction

A therapeutic protein is a protein-based drug designed to treat disease by replacing, supplementing, or modifying biological functions in the body [1]. These drugs have revolutionized modern medicine by providing targeted treatments for a wide variety of conditions, such as cancer, autoimmune disorders, and genetic diseases. The development and use of therapeutic proteins began in 1922 with the treatment of a young boy with type 1 diabetes in Toronto, Canada, using insulin isolated from animal pancreases [2, 3]. The success of animal-derived insulin in treating diabetes paved the way for the development of recombinant human insulin, which became the first FDA-approved recombinant protein therapeutic in 1983. Since then, advances in biotechnology have facilitated the development of numerous highly targeted and effective protein-based therapeutics for a wide array of diseases.

Therapeutic proteins are broadly categorized into groups based on their mechanism of action and therapeutic application, which

Xu Zhang et al. (eds.), *Protein Post-Translational Modifications*, Methods in Molecular Biology, vol. 3018,
https://doi.org/10.1007/978-1-0716-5166-7_2,

includes enzymatic or regulatory functions, specific targeting activity, protein vaccines, and diagnostic applications (Fig. 1) [1, 4]. Common types of protein therapeutics include, but are not limited to, monoclonal antibodies or mAbs (e.g., Rituximab, Adalimumab), antimicrobial peptides, recombinant cytokines (e.g., interferon), enzyme-based products (e.g., imiglucerase and dornase alfa), fusion proteins (e.g., Etanercept), hormone-based products (e.g., insulin), recombinant blood factors (e.g., factor VIII, factor IX), and protein vaccines (e.g., antirhesus immunoglobulin G, OspA in LYMErix, and Arexvy). Unlike small molecule pharmaceuticals, therapeutic proteins are larger in size, more structurally complex, and prone to degradation or other chemical modifications that can alter conformation and activity. Monitoring these physicochemical characteristics is important for characterizing and maintaining drug quality over the product life cycle, from development to commercial lot release, with each attribute requiring a specific analytical methodology. Given their complexity, protein therapeutics need to be well-characterized and manufactured with stringent control of variability. This requires a comprehensive assessment of critical quality attributes (CQAs), such as molecular weight (MW), amino acid sequence, and post-translational modifications (PTMs), using methodologies fit for the intended purpose.

In this chapter, we focus on providing a brief overview of the various types of PTMs present on therapeutic proteins, and their critical roles in stability, activity, interactions, localization, and immunogenicity. To support the characterization of these PTMs, we will

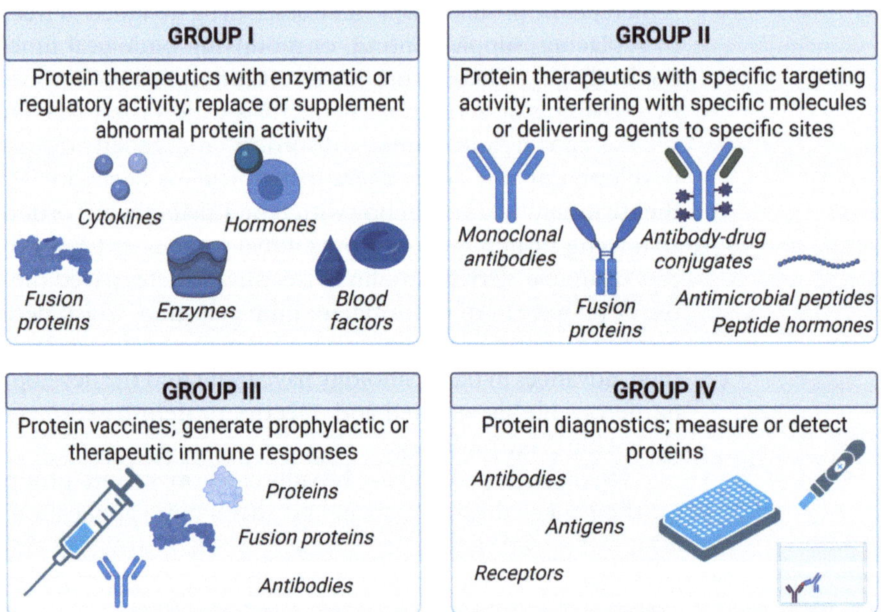

Fig. 1 Overview of therapeutic protein classifications with some examples

also discuss widely adopted methods with established applications in the quality assessments of therapeutic proteins.

It is also worth noting that the field of biological therapeutics is rapidly evolving, with increasingly complex products, such as cell, gene, phage, and microbiome therapies. While these are not the focus of this chapter, the analysis of proteins or proteomes, including PTMs, remains applicable to the characterization of novel advanced therapeutic products (ATPs). Accordingly, several chapters in this book describe methods that can be used for the large-scale characterization of PTMs in more complex systems, such as cells, extracellular vesicles, and microbiomes.

2 PTMs Present on Therapeutic Proteins

There are currently over 400 types of known PTMs [5], with some forms especially important to drug quality assessment due to their impact on product efficacy, consistency, and safety. An individual protein can carry several types of modifications, as well as multiple occurrences of the same type of modification [5]. Here, we will focus only on a few select modifications commonly highlighted for their roles in various protein therapeutics.

2.1 Glycosylation

Protein glycosylation is often viewed as one of the most important PTMs in therapeutic protein characterization, particularly for mAbs. Nearly all full-length IgG mAb therapeutics have Fragment crystallizable (Fc) glycosylation. Some also have glycosylation on the fragment antigen-binding (Fab) domain (e.g., Cetuximab). Glycosylation is the covalent addition of a glycan (carbohydrate) chain to proteins, typically at asparagine or serine/threonine residues, resulting in either N-linked or O-linked glycosylation, respectively [6–8]. This linkage, depending on the site and structure, can significantly influence the biological and physical properties of proteins. During drug development, strategies to manipulate glycan structures and glycosylation sites can be used to both mitigate and exploit the impacts of this modification. For example, site-directed mutagenesis to introduce an *N*-glycosylation site into the Fab region of adalimumab has been reported to increase the product's resistance to aggregation, reduce half-life in vivo, and decrease antigenicity/immunogenicity [9].

The process of glycan assembly and attachment to proteins is not template-driven but, instead, dynamically regulated by factors, such as availability of nucleotide sugar precursors, levels of glycosyltransferases, and other cellular signals [6, 7]. Thus, glycosylation patterns can vary greatly in length, structure, and composition depending on the intracellular environment. Variability in glycosylation can be particularly critical in large-scale therapeutic protein production, where factors like the passage number of master and working cell banks (MCB/WCB) and cell cultures conditions—such as pH, culture medium and duration—can strongly influence the glycosylation profile without tight control [10]. These variations can result in irregular glycan structures and localization sites that fluctuate between batches, ultimately affecting protein drug stability and activity and consequently impacting shelf-life, potency, and safety [6, 7]. As a result, glycosylation is considered a CQA for glycoprotein therapeutics, which must be well-controlled in the production process to ensure consistent product quality and efficacy.

2.2 Deamidation and Aggregation

Deamidation is a nonenzymatic chemical modification in which the side-chain amide group of asparagine or glutamine residues is removed [11, 12]. This loss of amide converts a neutral residue at physiological pH into an acidic or a negatively charged residue, changing the overall charge of the protein. This chemical modification has the potential to impact protein structural stability and immunogenicity, depending on the site of deamidation [11–13]. Deamidation within the complementarity determining region (CDR) of trastuzumab and pertuzumab has been shown to impair receptor binding, resulting in the loss of their pharmacological activity/therapeutic efficacy [11]. Since deamidation is considered a common degradation mechanism of protein therapeutics that can occur spontaneously in-solution, it is important to characterize the impact of this modification on stability and shelf-life [13, 14].

Deamidation, along with other nonenzymatic PTMs such as oxidation, can promote protein aggregation during product storage by altering physical properties and reducing structural stability [15]. Aggregation, the self-association of proteins, can occur through various mechanisms due to factors, such as the intrinsic physical properties of the therapeutic protein itself, production conditions, storage environment, or handling procedures [15, 16]. Aggregation often has a negative impact on protein activity, either through the loss of structure, solubility, or the induction of adverse immune reactions [15, 16]. Monitoring protein aggregates remain challenging due to the diverse forms of aggregation, which show varying attributes, with some being reversible and others irreversible [17]. Additionally, the mechanisms of aggregation and the effects of specific aggregate forms are not fully elucidated, which complicates the assessment of aggregation's impact on therapeutic proteins [17].

2.3 Oxidation and Disulfide Bond Formation

Oxidation for proteins is a chemical modification of amino acids that results in the loss of electrons or the acquisition of an oxygen atom. As a common degradation pathway, this process can occur at any stage during the manufacturing of a protein therapeutic [18]. Oxidation can lead to structural changes, reduced stability, aggregation, disrupted function/activity, and increased immunogenicity [18]. In one systematic evaluation of oxidation effects on two formulated mAb drug substances, Alam et al. revealed that methionine and tryptophan oxidation have significant but distinct impacts on drug stability and aggregation [19].

Disulfide bond formation, resulting from the oxidation of two cysteine residues, is an important process that can occur either spontaneously due to environmental oxidants or enzymatically through oxidases. The formation of the correct disulfide bond configuration of a protein is critical for proper folding and structure, which strongly impacts its stability and function. Missing or shuffled (i.e., incorrectly arranged) disulfide bonds in a protein therapeutic can result in misfolding, increased aggregation, degradation, or loss of activity from structural changes and diminished target binding [20]. Accordingly, disulfide bond engineering has been leveraged to enhance the stability of some protein therapeutics [9, 21]. For example, introducing specific disulfide bonds into insulin has been shown to increase stability, as well as receptor affinity [21]. In IgG mAbs, disulfide bond addition into the Fab constant domain can improve thermal stability with minimal impact on binding activity, offering a promising strategy for increasing mAb stability [9].

2.4 Amidation

Amidation is a reaction where an amine group reacts with a carboxylic acid group to form an amide. In the pharmaceutical context, amidation primarily refers to the conversion of a protein's C-terminal carboxyl group to an amide group [4]. Although the biological functions of this process are not fully elucidated, protein amidation may influence the function and stability of protein therapeutics, with particular importance demonstrated for the activity of peptide hormone therapies [22]. In humans, many bioactive peptides are amidated at their C-termini (e.g., neuropeptides, such as oxytocin and peptide hormones, such as gastrin and calcitonin). In peptide hormone drugs (e.g., forcaltonin), amidation has been used to improve product stability and activity [4]. Amidation has also been commonly observed in IgG4 and IgG1 mAb drug products; however, studies have not shown direct impacts on stability or function [23].

2.5 Sulfation

Sulfation is the addition of a sulfate group, typically to a tyrosine residue, mediated by sulfotransferases in eukaryotes. This modification alters the charge state of a protein, which can affect protein–protein interactions, such as receptor binding, consequently impacting therapeutic activity and immunogenicity [24, 25]. Blood factors VIII and IX are usually sulfated in cells, and some approved recombinant Factor VIII products contain tyrosine sulfation, which is critical for their clotting activity [26]. Accumulating evidence also demonstrates the impact of sulfation on mAb activity and stability. Lietz et al. found that, in a CHO-derived trivalent heterodimeric bispecific mAb called mAb-X, tyrosine sulfation modestly decreased antigen binding activity by increasing the rate of dissociation and aggregation, leading to reduced product stability [24]. Similarly, Luo et al. demonstrated that tyrosine sulfation within mAb CDRs can negatively impact antigen-receptor binding, suggesting that sulfation may be a CQA that warrants monitoring if present [25]. Additionally, sulfation can present on other PTMs, such as N-glycans, which could further impact the structure and activity of therapeutic proteins [27].

2.6 Acylation

Acylation involves the addition of either short-chain fatty acids (SCFAs) or long-chain fatty acids (LCFAs) to amino acids, with specific forms of acylations occurring preferentially on particular amino acids [28]. SCFA modifications, such as acetylation, butyrylation, and succinylation, usually occur on the protein N-terminus or a lysine side chain, while long chain acylations (e.g., myristoylation and palmitoylation) can occur on various amino acid residues [29]. Due to their long hydrocarbon chains, fatty acid addition can significantly alter a protein's physical properties, such as charge, size, and polarity, thereby affecting its function, interactions, stability, and localization within the cell [28]. Since fatty acid chains are hydrophobic, the acylation of protein therapeutics promotes membrane interactions. This feature is leveraged by lipopeptide antibiotics, such as daptomycin, colistin, and polymyxin B, where fatty acid additions are required for bacterial targeting and therapeutic activity [30]. Acylation can also be used for extending the half-life of therapeutics in the body [31]. Insulin detemir is an acylated version of human insulin with an added fatty acid chain which can reversibly bind to albumin, extending its circulation time in the body [32].

2.7 Proteolysis

Proteolysis is the enzymatic or chemical cleavage of the peptide bonds that link amino acids to form peptides and proteins. Proteolysis can be responsible for both the activation of protein precursors into their mature form, or the degradation of proteins, which renders proteins less efficient or non-functional. The impact of proteolysis on protein therapeutics depends on the nature of the protein itself. For protease-activated pro-drugs, cleavage at the site of interest is required for drug activation, activity, and efficacy [33, 34]. Protease-activated drugs are employed across various protein therapeutic formats and disease areas, with benefits, such as targeted or improved delivery, enhanced stability, increased activity of the activated therapeutic, and reduced off-target toxicity, thereby improving safety [33, 34]. For protein therapeutics that are not designed to undergo proteolytic activation, uncontrolled proteolysis can lead to protein/drug degradation, shortening its shelf-life and half-life in patients [35]. As a result, MW is a CQA and low MW product-related impurities, such as truncations, should be characterized and evaluated for their impact on drug quality [36].

3 Analytical Methods for PTM Analysis on Therapeutic Proteins

The accurate, reliable, and sensitive characterization of therapeutic protein quality attributes, such as MW, amino acid sequences, high order structures, PTMs, and process-related modifications (in-process intermediates), are critical for ensuring safety and efficacy for human use. Extensive analytical characterization data, along with clinical and safety data, are among the most critical components of a drug submission package and demonstrate that the product is well-understood and of high quality, with impurities that can be tightly controlled. Typically, national regulatory agencies establish their own requirements for therapeutic product review, regulation, and approval. However, international organizations often draft guidelines in consultation with Expert Working Groups composed of members from regulatory agencies and industry worldwide. These guidelines can be used as a foundation for drug submission and review. These guidelines help harmonize quality standards globally while also providing regulators and industry the flexibility to modify requirements as needed, given the complexity of each context. Such organizations include the International Council for Harmonisation of Technical Requirements for Pharmaceuticals for Human Use (ICH), the World Health Organization (WHO), as well as institutes that produce widely recognized pharmacopoeia, such as the European Pharmacopoeia (Ph. Eur.) and United States Pharmacopoeia (USP). These guidelines and standards serve as valuable references for both regulatory authorities and industry to streamline the development of innovative biotherapeutics.

PTMs are considered CQAs due to their impacts on the safety, efficacy, and stability of therapeutic proteins. Thus, for biotherapeutics, one focus of pharmacopoeial monographs is the development of methods to characterize and quantitate product-related impurities [37, 38]. While methodology and approaches for PTM characterization can vary by guideline or monograph, special focus is placed on glycosylation, oxidation, deamidation, and any PTMs considered critical to drug quality based on their mechanisms of action. In this section, we will discuss some commonly used analytical methods for PTM analysis, which can be used for therapeutic protein characterization.

3.1 Electrophoresis

Electrophoresis is a separation technique that resolves proteins based on size and overall charge by applying an electric field to move charged molecules through a matrix. Electrophoretic approaches are discussed in the ICH guidelines, Q5C and Q6B, for the analysis of protein PTM profiles, as well as degradation and aggregation impurities that may result from aberrant PTMs. The most widely known electrophoretic method is sodium dodecyl sulfate polyacrylamide gel electrophoresis (SDS-PAGE), where total protein content can be directly detected by dyes. SDS-PAGE can also be followed by membrane transfer and antibody staining to identify specific proteins including PTM-specific proteoforms (western blotting). Isoelectric focusing is another common electrophoretic technique that separates proteins by their isoelectric point (pI). This technique is used in one dimension to determine pI values or combined with conventional SDS-PAGE for two-dimensional gel electrophoresis, which provides higher resolution of protein separation and detection. Electrophoretic separation by charge allows for the detection of charge variants, which is of particularly importance for PTM analysis since many PTMs, such as deamidation and glycosylation, alter the charge states of proteins.

Capillary electrophoresis (CE) is another key technique in the biopharmaceutical industry that combines electrophoretic migration and electroosmotic flow to separate charged molecules within a capillary. CE offers more quantitative results and higher resolution than conventional gel electrophoresis, which has accelerated its use in PTM analysis for protein therapeutics. Additional advantages of CE include lower sample requirements, the capacity for automation, as well as compatibility with various downstream detection methods for further characterization of separated components [39, 40]. CE is routinely used in both drug development and quality control, especially for evaluating glycosylation and charge variant profiles in antibody therapeutics and biosimilars. Multiple CE-based analytical methods exist, which vary by the nature of the capillary and buffers used in the method, the principles of separation, as well as instrumentation configuration, such as detector coupling (e.g., absorbance, laser-induced fluorescence, or mass spectrometry) [39, 40]. When combined with mass spectrometry (CE-MS), charge-based separation provides significant advantages in both separation and identification of protein isoforms or PTMs. Some commonly used CE-based methods include capillary isoelectric focusing electrophoresis (cIEF), capillary zone electrophoresis (CZE), and CE-SDS. CE-based methods are often used in pharmacopoeial monographs for product-related impurity tests, particularly for charged variant profiling (e.g., Somatropin monograph 0951. Ph. Eur. Suppl. 11.8) and glycan/glycosylation analysis (e.g., Erythropoietin monograph 1316. Ph. Eur. Suppl. 11.8).

3.2 High-Resolution Chromatography

High-resolution chromatographic approaches are among the most widely used analytical methodologies for characterizing and monitoring therapeutics, including biological products. Broadly, high-resolution chromatographic techniques use specialized stationary phases, typically in a column format, to resolve individual components from a mixture. Chromatography is often coupled with different detectors (e.g., absorbance, fluorescence, or mass spectrometry) for the identification and quantification of specific components present in the sample. Biological samples are typically analyzed by high-performance liquid chromatography (HPLC), where the sample is carried by the mobile phase through the stationary phase packed inside a column. Individual sample components are then separated according to the strength of their interactions with the stationary and mobile phases, eluting either faster or slower through the column. Some types of high resolution chromatography include reversed-phase chromatography, ion-exchange chromatography, gel filtration chromatography (also known as size-exclusion chromatography), and affinity chromatography. While all of these chromatographic techniques have been used in both analytical characterization and process purification, this section will focus on their analytical application to PTM assessment.

Reverse-phased HPLC (RP-HPLC) is one of the most commonly used methods for the quality assessment of drugs, including protein therapeutics. In this technique, samples pass through a hydrophobic nonpolar stationary phase and a gradient of increasing polar organic solvent is applied to separate analytes within the sample, such as peptides and/or proteins, based on hydrophobicity. RP-HPLC is routinely applied to quantify changes in protein PTMs, such as oxidation, deamidation, sulfation, degradation, truncation, or aggregation, as well as other chemical modifications or impurities of therapeutic drug substances or drug products.

Ion exchange chromatography (IEC) uses a charged stationary phase and gradient of buffer pH or salt concentration to separate peptides or proteins by their charge states. Large biotherapeutic products, such as mAbs, are prone to charge variant impurities resulting from PTMs, such as deamidation and glycosylation. IEC can be used to effectively separate and quantify these charge variants, thereby, revealing PTM changes that impact the overall charge of a protein [41]. As a result, IEC has been widely used to evaluate product consistency and stability, as well as isolate components of interest for further characterization.

Gel filtration or size-exclusion chromatography (SEC) employs a porous and inert stationary phase, typically as beads, to separate samples by their approximate size and shape. Larger molecules elute earlier because they are excluded from the stationary phase pores when traveling through the column. Smaller molecules enter the pores and travel for longer distances, resulting in a later elution. SEC is commonly used to quantify size variants of biotherapeutics, primarily protein aggregates, which may arise from PTM changes and are key indicators of product stability [42]. Advantages of SEC include native conditions for analysis and minimal sample preparation, making it compatible with various downstream applications, such as the purification of samples for potency analysis.

Affinity chromatography separates proteins based on their highly selective and reversible binding to ligands immobilized on a solid support within a column. Samples are bound to the column and only eluted when a mobile phase containing a competing ligand or a shift in pH passes through to alter the binding capacity. This allows for the highly specific purification of proteins by separating them from non-binding components. Depending on the application, ligands can be metal ions, lectins, antibodies, enzyme substrates, antibody-binding proteins (A/G/L), or synthetic polymers. Affinity chromatography has been used to enrich and analyze numerous protein PTMs, such as phosphorylation using immobilized metal affinity chromatography (IMAC), glycosylation using lectin as ligands, and acylation using immune-affinity chromatography. This can be useful for evaluating PTM changes or proteoform impurities that are known to impact product bioactivity or target interaction, especially when combined with ligand-binding assays [43]. Affinity chromatography has also been explored for quantifying the titer of mAb products [44].

3.3 ELISA and Functionality Assay

Enzyme-linked immunosorbent assays (ELISA) are a widely used laboratory technique in both research, clinical, and pharmaceutical settings to detect or quantify specific substances, including proteins and mAb drug substances. In sandwich ELISAs, one of the most common approaches, an antigen-specific capture antibody is immobilized on a microplate and used to capture target antigens present in the sample. For detection, a second enzyme-linked detection antibody is, then, added and allowed to bind the captured antigen, forming a sandwich complex. Upon addition of a substrate, the enzyme catalyzes a reaction that generates a measurable signal ideally proportional to the antigen concentration. Since PTMs can affect antibody–antigen interactions, ELISA can be used to assess the presence of PTMs on therapeutic proteins. When a PTM-specific antibody is used as the capture antibody, the assay can selectively and accurately quantify PTM-specific proteoforms in a sample.

Similarly, specialized functionality assays can help infer or confirm the presence of desired or undesired PTMs in biological therapeutics. ICH guidelines discuss the use of binding assays, such as ELISA or surface plasmon resonance (SPR) along with appropriate cell-based assays for this purpose. A well-characterized example is the antibody-dependent cellular cytotoxicity (ADCC) assay for the analysis of mAb drugs. Glycosylation patterns on the Fc region of mAbs are key for binding to FcγRIIIa receptors on effector cells, which impacts mAb ADCC activity. Reduced fucosylation and increased galactosylation at specific mAb *N*-glycosylation sites can enhance ADCC activity [45]. SPR can be used to directly monitor binding of the mAb Fc region to FcγRIIIa, providing functional correlations to glycosylation patterns. In SPR, a target ligand (such as FcγRIIIa) is immobilized onto a sensor surface, followed by the introduction of the biologic drug or binding partner to the sensor surface. When bound, changes in the refractive index of the sensor surface are detected by the SPR system. These binding or functionality assays can also be used to compare novel biological products or biosimilars to accepted reference standards, providing an indirect assessment of glycan profiles that can be further characterized using analytical methods, such as mass spectrometry.

3.4 Mass Spectrometry and Peptide Mapping

Mass spectrometry (MS) measures the exact mass of analytes, including proteins, with high resolution (<10 ppm for high-resolution MS) to provide accurate detection of mass shifts caused by various PTMs, which can be highly variable. For example, the mass shift of phosphorylation is 79.96633 Da, while deamidation would add 0.98402 Da from the conversion of an amide to a carboxyl group, and glycosylation could add anywhere from 160 Da to over 3000 Da depending on the glycan chain added. Protein samples can be either directly infused into the mass spectrometer or separated prior to MS analysis with technologies, such as chromatography or CE. Dedicated bioinformatic tools, such as BioPharma Finder, can be used to deconvolute the MS spectra and calculate masses of different analytes in the sample.

Peptide mapping is an MS-based technique, often discussed in pharmacopoeia and ICH guidelines for identification and physicochemical characterization of therapeutic proteins. This technique utilizes proteolytic digestion, peptide separation, tandem mass spectrometry (MS/MS) fragmentation, and detection to determine the amino acid sequence of peptides. Peptides can then be mapped to the protein sequence of interest (Fig. 2). Peptide mapping is typically used to verify the identity of a protein, assess integrity, and localize PTM sites on a protein. Briefly, proteins are first digested into smaller peptides using proteolytic enzymes, such as trypsin or LysC. The resulting peptides are desalted, separated by chromatography or CE (as discussed above), and subject to the mass spectrometer for fragmentation and measurement. This generates mass-to-charge ratios (m/z), as well as intensity data for both parent (MS1) and fragment ion (MS2) mass spectra, which can then be used to infer the amino acid sequence of the peptides. Peptide identification can be achieved either through manual interpretation of the mass spectrum or a database search (peptide-spectrum match, PSM) [46]. For PSM, protein sequences will be digested in silico to generate a database of predicted peptides and expected fragment ions for each potential peptide. The database search engines look for how many fragment peaks match between the theoretical and experimental MS/MS spectra, as well as the mass accuracy of matching fragment peaks to assign best PSMs. A target-decoy strategy is usually applied to calculate and filter out false identifications by controlling false discovery rate (an FDR <1% is a commonly used threshold) with reverse sequences added in the predicted database [47]. To identify and locate the site of PTMs, expected PTM types can be added as variable modifications in the database search, or through open database search where the search space is expanded to allow for identifying any precursor mass shift

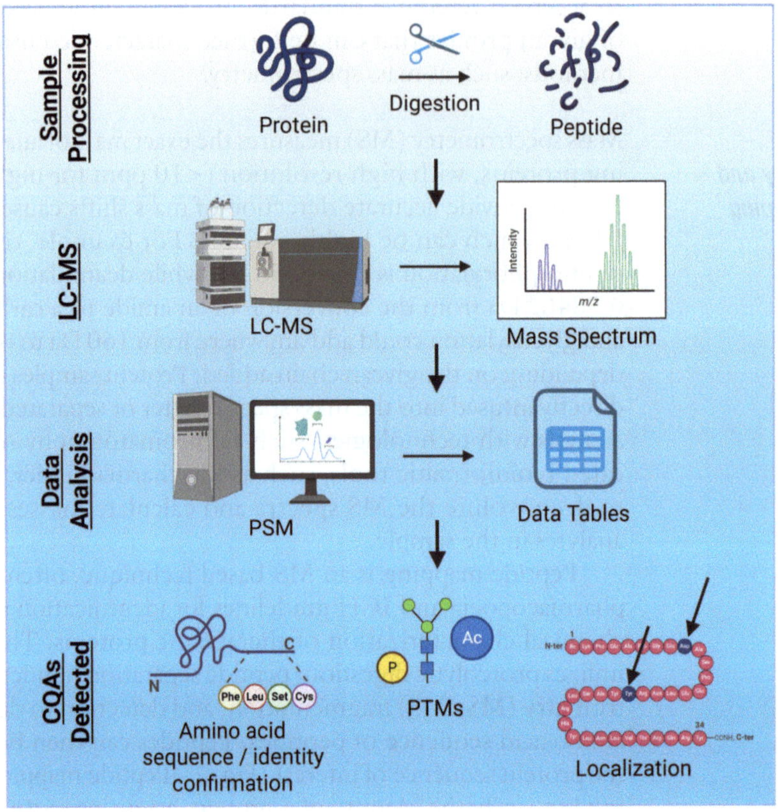

Fig. 2 Mass spectrometry-based peptide mapping and multi-attribute method (MAM) workflow for therapeutic protein quality assessment

introduced by PTMs. Identified peptides or modified peptides can then be mapped to protein sequences to determine the sequence coverage, PTM types, sites of modifications, and other quantitative information using algorithms, such as maxLFQ [48]. The PSM and protein inference/mapping can be performed automatically through some search engines, such as Proteome Discoverer [49], MaxQuant [50], and Mascot [51].

3.5 Mass Spectrometry-Based Multi-Attribute Method

During the quality assessment of biological therapeutics, a combination of analytical and functionality assays is needed to assess distinct CQAs. Moreover, given the complexity of biologics and the potential variations that can be introduced by manufacturing and storage, detecting both known and novel product-related impurities, including those related to PTMs, are becoming increasingly important. As technology evolves, so do the established methods for quality assessment. Increasingly, current discussions center on the Multi-Attribute Method (MAM) as a new standards of excellence for the quality assessment of biologicals.

Broadly, the MAM is a method that provides information on more than one quality attribute at the same time. The discussion of MAM in quality control environments started with the development of peptide mapping (discussed above) as it could identify multiple key elements of a biological therapeutic including intact mass, amino acid sequences, and PTM modifications. Developments in mass spectrometry and MS-based proteomics strategies have further expanded what is possible to examine in a single MS experiment, such as host cell protein impurities, quantitation, glycan composition and structure [52–54]. Although the MAM has not yet been widely adopted for routine use, its role in drug development is being actively proposed and discussed. For example, recently, the US Pharmacopoeia drafted General Chapter 1060 (Mass Spectrometry-Based Multi-Attribute Method for Therapeutic Proteins).

The most commonly explored MAM is LC-MS which, like peptide mapping, simultaneously determines multiple CQAs of protein type biologics, including MW, amino acid sequences, mutations, fragments, truncated products, and various types of PTMs including glycosylation structures [54–56]. Interlaboratory studies have been performed to show consistency and robustness of LC-MS as an assessment tool across various facilities and instrument platforms [55]. The usefulness of the MAM is also reinforced by its incorporation into various stages of drug development, such as research to characterize drug substances or manufacturing and quality control to monitor relevant CQAs, to ensure consistency for safety and efficacy [55]. Its use has been reinforced especially for mAbs to scan for protein mutations, identify PTMs, and assess impurities [54, 56, 57].

It is worth noting that there is no single method or MAM that can provide all required information for product characterization. For example, one challenging quality attribute for LC-MS-based MAMs is protein aggregates. Orthogonal methods are often used to validate MS results, although the robustness and integrity of MAM data have been established and reinforced. Nevertheless, the LC-MS-based MAM represents a promising analytical strategy, whose implementation could facilitate efficient PTM analysis and monitoring for a wide array of attributes relevant to both regulators and industry.

4 Conclusions

There is a plethora of PTMs that can be present on biological therapeutics, with potentially critical impacts on drug stability, function, efficacy, and safety profiles. A variety of analytical techniques can be leveraged to detect and characterize PTMs, with some emerging strategies taking advantage of recent technological advancements, such as advanced MS-based MAMs. These new strategies support future directions for detailed characterization of critical quality attributes in complex biological therapeutics. Understanding, monitoring, and controlling the presence, localization, and structure of PTMs on proteins in therapeutics is essential to ensure patients receive safe and efficient biotherapeutic treatments.

Acknowledgments

This work was supported by the Government of Canada through Health Canada and the funding from Genomics Research and Development Initiative (GRDI) of Canada (Phase VIII). This publication should not be taken to represent Health Canada's views of policies.

References

1. Leader B, Baca QJ, Golan DE (2008) Protein therapeutics: a summary and pharmacological classification. Nat Rev Drug Discov 7(1):21–39
2. Vecchio I, Tornali C, Bragazzi NL, Martini M (2018) The discovery of insulin: an important milestone in the history of medicine. Front Endocrinol (Lausanne) 9:613
3. White Jr. JR (2014) A brief history of the development of diabetes medications. Diabetes Spectr 27(2):82–86
4. Walsh G, Jefferis R (2006) Post-translational modifications in the context of therapeutic proteins. Nat Biotechnol 24(10):1241–1252
5. Khoury GA, Baliban RC, Floudas CA (2011) Proteome-wide post-translational modification statistics: frequency analysis and curation of the swiss-prot database. Sci Rep 1:90
6. Thompson N, Wakarchuk W (2022) O-glycosylation and its role in therapeutic proteins. Biosci Rep 42(10):BSR20220094
7. Chen B, Liu W, Li Y, Ma B, Shang S, Tan Z (2022) Impact of N-linked glycosylation on therapeutic proteins. Molecules 27(24):8859
8. Kohler J, Patrie SM (ed) (2013) Mass spectrometry of glycoproteins: methods and protocols. In: Methods in molecular biology, vol 951. Springer, Heidelberg
9. Nakamura H, Yoshikawa M, Oda-Ueda N, Ueda T, Ohkuri T (2021) A comprehensive analysis of novel disulfide bond introduction site into the constant domain of human Fab. Sci Rep 11(1):12937
10. Delobel A (2021) Glycosylation of therapeutic proteins: a critical quality attribute. Methods Mol Biol 2271:1–21
11. Bults P, van der Voort A, Meijer C, Sonke GS, Bischoff R, van de Merbel NC (2022) Analytical and pharmacological consequences of the in vivo deamidation of trastuzumab and pertuzumab. Anal Bioanal Chem 414(4):1513–1524
12. Taraj K (2025) Review on characterization of deamidation in therapeutic proteins by mass spectrometric methods. Technium BioChemMed 13:13–26
13. Haberger M, Bomans K, Diepold K, Hook M, Gassner J, Schlothauer T, Zwick A, Spick C, Kepert JF, Hienz B, Wiedmann M, Beck H, Metzger P, Molhoj M, Knoblich C, Grauschopf U, Reusch D, Bulau P (2014) Assessment of chemical modifications of sites in the CDRs of recombinant antibodies: susceptibility vs. functionality of critical quality attributes. Mabs 6(2):327–339
14. Delmar JA, Wang J, Choi SW, Martins JA, Mikhail JP (2019) Machine learning enables accurate prediction of asparagine deamidation probability and rate. Mol Ther Methods Clin Dev 15:264–274
15. Lundahl MLE, Fogli S, Colavita PE, Scanlan EM (2021) Aggregation of protein therapeutics enhances their immunogenicity: causes and mitigation strategies. RSC Chem Biol 2(4):1004–1020
16. Eon-Duval A, Broly H, Gleixner R (2012) Quality attributes of recombinant therapeutic proteins: an assessment of impact on safety and efficacy as part of a quality by design development approach. Biotechnol Prog 28(3):608–622
17. Singh SK, Afonina N, Awwad M, Bechtold-Peters K, Blue JT, Chou D, Cromwell M, Krause HJ, Mahler HC, Meyer BK, Narhi L, Nesta DP, Spitznagel T (2010) An industry perspective on the monitoring of subvisible particles as a quality attribute for protein therapeutics. J Pharm Sci 99(8):3302–3321
18. Torosantucci R, Schoneich C, Jiskoot W (2014) Oxidation of therapeutic proteins and peptides: structural and biological consequences. Pharm Res 31(3):541–553

19. Alam ME, Slaney TR, Wu L, Das TK, Kar S, Barnett GV, Leone A, Tessier PM (2020) Unique impacts of methionine oxidation, tryptophan oxidation, and asparagine deamidation on antibody stability and aggregation. J Pharm Sci 109(1):656–669
20. Coghlan J, Benet A, Kumaran P, Ford M, Veale L, Skilton SJ, Saveliev S, Schwendeman AA (2022) Streamlining the characterization of disulfide bond shuffling and protein degradation in IgG1 biopharmaceuticals under native and stressed conditions. Front Bioeng Biotechnol 10:862456
21. Das A, Shah M, Saraogi I (2022) Molecular aspects of insulin aggregation and various therapeutic interventions. ACS Bio Med Chem Au 2(3):205–221
22. Merkler DJ (1994) C-terminal amidated peptides: production by the in vitro enzymatic amidation of glycine-extended peptides and the importance of the amide to bioactivity. Enzyme Microb Technol 16(6):450–456
23. Xu CA, Feng AZ, Ramineni CK, Wallace MR, Culyba EK, Guay KP, Mehta K, Mabry R, Farrand S, Xu J, Feng J (2019) L(445)P mutation on heavy chain stabilizes IgG(4) under acidic conditions. Mabs 11(7):1289–1299
24. Lietz CB, Deyanova E, Cho Y, Cordia J, Franc S, Kabro S, Wang S, Mikolon D, Banks DD (2023) Identification of tyrosine sulfation in the variable region of a bispecific antibody and its effect on stability and biological activity. Mabs 15(1):2259289
25. Luo H, Mahon D, Wong P, Madayiputhiya N, Chen Y, Stauffer T, Tao L, Zeng M (2024) Structure-function relationship study for sulfated protein therapeutics using hydrophobic interaction chromatography and mass spectrometry. J Chromatogr B Analyt Technol Biomed Life Sci 1233:123981
26. Kannicht C, Ramstrom M, Kohla G, Tiemeyer M, Casademunt E, Walter O, Sandberg H (2013) Characterisation of the post-translational modifications of a novel, human cell line-derived recombinant human factor VIII. Thromb Res 131(1):78–88
27. She YM, Dai S, Tam RY (2022) Highly sensitive characterization of non-human glycan structures of monoclonal antibody drugs utilizing tandem mass spectrometry. Sci Rep 12(1):15109
28. Shang S, Liu J, Hua F (2022) Protein acylation: mechanisms, biological functions and therapeutic targets. Signal Transduct Target Ther 7(1):396
29. Zhong Q, Xiao X, Qiu Y, Xu Z, Chen C, Chong B, Zhao X, Hai S, Li S, An Z, Dai L (2023) Protein posttranslational modifications in health and diseases: functions, regulatory mechanisms, and therapeutic implications. MedComm (2020) 4(3):e261
30. Ledger EVK, Sabnis A, Edwards AM (2022) Polymyxin and lipopeptide antibiotics: membrane-targeting drugs of last resort. Microbiology 168(2):001136
31. van Witteloostuijn SB, Pedersen SL, Jensen KJ (2016) Half-life extension of biopharmaceuticals using chemical methods: alternatives to PEGylation. ChemMedChem 11(22):2474–2495
32. Tibaldi JM (2014) Evolution of insulin: from human to analog. Am J Med 127(10):S25–38
33. Bleuez C, Koch WF, Urbach C, Hollfelder F, Jermutus L (2022) Exploiting protease activation for therapy. Drug Discov Today 27(6):1743–1754
34. Poreba M (2020) Protease-activated prodrugs: strategies, challenges, and future directions. FEBS J 287(10):1936–1969
35. Shi M, McHugh KJ (2023) Strategies for overcoming protein and peptide instability in biodegradable drug delivery systems. Adv Drug Deliv Rev 199:114904
36. Wang S, Liu AP, Yan Y, Daly TJ, Li N (2018) Characterization of product-related low molecular weight impurities in therapeutic monoclonal antibodies using hydrophilic interaction chromatography coupled with mass spectrometry. J Pharm Biomed Anal 154:468–475
37. ICH (1995) Quality of biotechnological products: stability testing of biotechnological/biological products [Q5C]. In: ICH Quality Guidelines. https://www.ich.org/page/quality-guidelines. Accessed 01 May, 2025
38. ICH (1995) Specifications: test procedures and acceptance criteria for biotechnological/biological products [Q5B]. In: ICH Quality Guidelines. https://www.ich.org/page/quality-guidelines. Accessed 01 May, 2025
39. Cageling R, Carillo S, Boumeester AJ, Lubbers-Geuijen K, Bones J, Jooss K, Somsen GW (2024) Microfluidic capillary electrophoresis – mass spectrometry for rapid charge-variant and glycoform assessment of monoclonal antibody biosimilar candidates. J Pharm Biomed Anal 248:116301
40. Cernosek T, Dalphin M, Sakowski J, Behrens S, Wunderli P (2024) Application of analytical quality by design to the development and validation of reduced and non-reduced capillary electrophoresis analytical procedures for mAb purity determination. J Pharm Biomed Anal 249:116386
41. Haberger M, Heidenreich AK, Hook M, Fichtl J, Lang R, Cymer F, Adibzadeh M, Kuhne F, Wegele H, Reusch D, Bonnington L, Bulau P (2021) Multiattribute monitoring of antibody charge variants by cation-exchange chromatography coupled to native mass spectrometry. J Am Soc Mass Spectrom 32(8):2062–2071

42. Winter C, Springer A, Descamps JL, Hoefinghoff J, Mohammad-Sadegh S, Paudel A, Stankovic-Brandl M (2025) Unraveling the effects of filtration, process interruptions, and post-process agitation on protein aggregation. AAPS PharmSciTech 26(3):85
43. Zhang Z, Yan Y, Wang S, Li N (2022) A competitive binding-mass spectrometry strategy for high-throughput evaluation of potential critical quality attributes of therapeutic monoclonal antibodies. Mabs 14(1):2133674
44. Cernosek T, Dalphin M, Jain N, Lor J, Richter N, Mellal M, Behrens S, Wunderli P (2024) A case study application of AQbD to the re-development and validation of an affinity chromatography analytical procedure for mAb titer quantitation. J Chromatogr B Analyt Technol Biomed Life Sci 1242:124205
45. Hatfield G, Tepliakova L, Gingras G, Stalker A, Li X, Aubin Y, Tam RY (2022) Specific location of galactosylation in an afucosylated antiviral monoclonal antibody affects its FcγRIIIA binding affinity. Front Immunol 13:972168
46. Fernandez-de-Cossio J, Gonzalez J, Betancourt L, Besada V, Padron G, Shimonishi Y, Takao T (1998) Automated interpretation of high-energy collision-induced dissociation spectra of singly protonated peptides by 'SeqMS', a software aid for de novo sequencing by tandem mass spectrometry. Rapid Commun Mass Spectrom 12(23):1867–1878
47. Elias JE, Gygi SP (2007) Target-decoy search strategy for increased confidence in large-scale protein identifications by mass spectrometry. Nat Methods 4(3):207–214
48. Cox J, Hein MY, Luber CA, Paron I, Nagaraj N, Mann M (2014) Accurate proteome-wide label-free quantification by delayed normalization and maximal peptide ratio extraction, termed MaxLFQ. Mol Cell Proteomics 13(9):2513–2526
49. Orsburn BC (2021) Proteome discoverer-a community enhanced data processing suite for protein informatics. Proteomes 9(1):15
50. Tyanova S, Temu T, Cox J (2016) The MaxQuant computational platform for mass spectrometry-based shotgun proteomics. Nat Protoc 11(12):2301–2319
51. Brosch M, Yu L, Hubbard T, Choudhary J (2009) Accurate and sensitive peptide identification with Mascot Percolator. J Proteome Res 8(6):3176–3181
52. Rogstad S, Yan H, Wang X, Powers D, Brorson K, Damdinsuren B, Lee S (2019) Multi-attribute method for quality control of therapeutic proteins. Anal Chem 91(22):14170–14177
53. Rogers RS, Nightlinger NS, Livingston B, Campbell P, Bailey R, Balland A (2015) Development of a quantitative mass spectrometry multi-attribute method for characterization, quality control testing and disposition of biologics. Mabs 7(5):881–890
54. Yang F, Zhang J, Buettner A, Vosika E, Sadek M, Hao Z, Reusch D, Koenig M, Chan W, Bathke A, Pallat H, Lundin V, Kepert JF, Bulau P, Deperalta G, Yu C, Beardsley R, Camilli T, Harris R, Stults J (2023) Mass spectrometry-based multi-attribute method in protein therapeutics product quality monitoring and quality control. Mabs 15(1):2197668
55. Millan-Martin S, Jakes C, Carillo S, Gallagher L, Scheffler K, Broster K, Bones J (2024) Multi-attribute method (MAM): an emerging analytical workflow for biopharmaceutical characterization, batch release and cGMP purity testing at the peptide and intact protein level. Crit Rev Anal Chem 54(8):3234–3251
56. Carillo S, Criscuolo A, Fussl F, Cook K, Bones J (2022) Intact multi-attribute method (iMAM): a flexible tool for the analysis of monoclonal antibodies. Eur J Pharm Biopharm 177:241–248
57. Alhazmi HA, Albratty M (2023) Analytical techniques for the characterization and quantification of monoclonal antibodies. Pharmaceuticals (Basel) 16(2):291

Part II

Bottom-Up Approaches

Check for updates

Chapter 3

Decoding the SUMO Proteome: A Mass Spectrometry Workflow for Site-Specific Identification

Chongyang Li and Pierre Thibault

Abstract

Protein SUMOylation is a dynamic post-translational modification that regulates numerous cellular processes, including DNA repair, transcription, and proteostasis. SUMO modifiers are conjugated to lysine residues on substrate proteins via a conserved enzymatic cascade and can form diverse chain architectures that encode specific cellular outcomes. The identification of SUMOylated proteins and their modification sites has historically been challenging due to the low abundance of SUMOylation and the complexity of SUMO remnants after proteolysis. Recent advances in proteomics have led to the development of enrichment strategies and mass spectrometry (MS)-based methods that now enable the site-specific mapping of SUMO modifications. This chapter provides an overview of the biological roles and structural diversity of SUMOylation, and presents an MS-based workflow designed to identify SUMOylation sites with high specificity and depth. These tools offer new opportunities to dissect the SUMO-modified proteome in health and disease.

Keywords SUMOylation, Ubiquitin-like proteins (UBLs), Posttranslational modifications, Mass spectrometry, Proteomics, Immunoaffinity enrichment

1 Introduction

Posttranslational modifications (PTMs) fine-tune nearly every aspect of protein function, enabling cells to respond to dynamic intracellular and environmental cues. Among these, the conjugation of ubiquitin-like proteins (UBLs) stands out for its regulatory depth and structural complexity. The small ubiquitin-like modifier (SUMO) family—comprising SUMO1, SUMO2, SUMO3, and the less-characterized SUMO4 and SUMO5—modulates a wide range of processes including genome maintenance, transcriptional regulation, signal transduction, and cellular stress responses [1–3]. SUMOylation influences protein localization, activity, and turnover, and has emerged as a critical node of crosstalk with other PTMs such as ubiquitylation and phosphorylation [4, 5].

Xu Zhang et al. (eds.), *Protein Post-Translational Modifications*, Methods in Molecular Biology, vol. 3018, https://doi.org/10.1007/978-1-0716-5166-7_3,

The SUMOylation pathway follows a highly conserved enzymatic cascade (Fig. 1a) [6, 7]. Mature SUMO proteins, characterized by a C-terminal GG motif (Fig. 1b), are first processed by SUMO-specific proteases (SENPs), which expose the diglycine motif required for conjugation. SUMO activation is mediated by the heterodimeric E1 enzyme SAE1/SAE2, which activates SUMO in an ATP-dependent manner and transfers it to the sole E2 enzyme, UBC9. Substrate specificity is conferred primarily by E3 ligases, such as PIAS and RanBP2, which catalyze the covalent attachment of SUMO to the ε-amino group of lysine residues within consensus motifs (ψKxE) [8, 9]. SENPs also mediate deSUMOylation, thereby dynamically regulating substrate modification [10, 11].

Protein SUMOylation can form a diversity of chains that vary in both modification type and chain architecture (Fig. 1c) [12]. SUMOylation can occur as monomeric, multisite, or poly-SUMOylation on the same substrate. SUMO2 and SUMO3 are particularly prone to forming polySUMO chains via internal lysine residues, whereas SUMO1 typically functions as a chain terminator but can also form nonconsensus linkages [13]. These SUMO chains can be homotypic or heterotypic, branched, hybrid (with other UBLs), or themselves posttranslationally modified (e.g., phosphorylated or acetylated), thus increasing the information-carrying potential of SUMOylation [14]. Such diversity in topology likely encodes distinct signaling outcomes.

The biological significance of SUMOylation extends to critical cellular processes. SUMO-modified substrates play vital roles in DNA damage repair, nuclear transport, chromatin organization, and the regulation of transcription factors [3]. The cellular response to stress often includes robust changes in SUMOylation patterns, particularly involving SUMO2/3-conjugated proteins, which accumulate under proteotoxic or genotoxic conditions [15, 16]. The SUMO pathway is also tightly interwoven with the ubiquitin system; for example, SUMO chains can serve as recruitment signals for SUMO-targeted ubiquitin ligases (STUbLs), such as RNF4, triggering degradation of modified substrates and creating a functional link between SUMOylation and proteostasis [17].

Given the dynamic and low-abundance nature of SUMOylation, unraveling the SUMO-modified proteome requires specialized strategies. Traditional mass spectrometry (MS)-based proteomics is hindered by the large remnants left on peptides after tryptic digestion of native SUMO proteins [18, 19]. Unlike ubiquitin, whose tryptic digestion produces a diagnostic K-εGG motif amenable to immunoaffinity enrichment [20], native SUMO isoforms yield peptides that are difficult to sequence due to their size and branched nature (Fig. 1b). To overcome this, researchers have engineered SUMO constructs with mutated C-termini to yield shorter tryptic remnants, facilitating peptide-level enrichment and identification [21, 22].

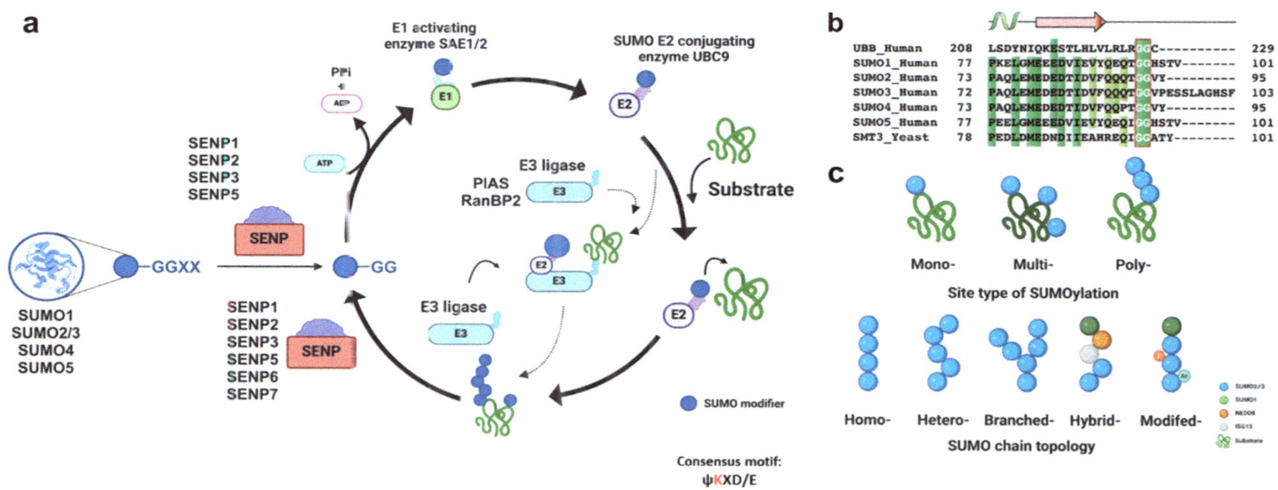

Fig. 1 Overview of protein SUMOylation. (**a**) Schematic representation of the SUMOylation pathway, showing the roles of E1, E2, and E3 enzymes. (**b**) Sequence alignment of five different human SUMO isoforms, and yeast SUMO, highlighting the C-terminal diglycine motif. (**c**) Different forms of SUMO conjugation (mono-, poly-, and multi-SUMOylation) and its modification with ubiquitination and phosphorylation

Proteomic workflows to identify SUMOylated proteins now span multiple levels of resolution. Protein-level enrichment can be achieved through exogenous expression of tagged SUMO variants (e.g., His- or Flag-tagged constructs), or through affinity reagents targeting SUMO-interacting motifs (SIMs) [19, 23]. However, these methods may coenrich binding partners or free SUMO, reducing specificity. Peptide-level strategies, such as the use of mutant SUMO constructs or alternative proteases (e.g., WaLP, Asp-N), generate smaller remnants and support site-specific identification with higher purity [18, 19]. These workflows are often paired with liquid chromatography–tandem MS (LC–MS/MS) using data-dependent (DDA) or data-independent acquisition (DIA), enabling both qualitative and quantitative assessment of SUMOylation dynamics [24].

The functional interplay between SUMO and other PTMs is a growing focus of SUMO proteomics. Phosphorylation can enhance SUMO conjugation at specific motifs (phosphorylation-dependent SUMOylation motifs, PDSMs) [25], while ubiquitylation frequently follows SUMOylation in a hierarchical or cooperative manner [26, 27]. SUMO and ubiquitin can even compete for the same lysine residue or cooccupy distinct sites on a common substrate. These interactions are not merely coincidental but represent an integrated signaling logic regulating protein turnover, cellular stress responses, and transcriptional fidelity [5, 28].

The development of cross-enrichment techniques—such as sequential immunoprecipitation with anti-SUMO and antiubiquitin antibodies—enables the discovery of proteins bearing both modifications [29]. Proximity labeling methods, such as BioID and TurboID, also allow for the capture of SUMO ligase substrates in a native context, albeit with potential for false positives that require orthogonal validation [23].

SUMOylation represents a versatile and essential PTM that orchestrates diverse cellular functions through both direct conjugation and intermodification crosstalk. Figure 1 encapsulates the key principles underlying SUMO biology; from enzymatic machinery and substrate recognition (panel a), to structural signatures of SUMO isoforms (panel b), and the complexity of SUMO chain topologies (panel c). The expanding proteomic toolbox now allows for increasingly precise mapping of SUMO-modified networks, offering unprecedented insights into the regulatory logic of the posttranslational landscape. This chapter presents a mass spectrometry-based method that enables the identification of protein SUMOylation in a site-specific manner, providing a robust platform to interrogate the dynamics and functional relevance of SUMO modification under physiological and stress conditions.

2 Materials

2.1 Cell Culture and Protein Extraction

1. Cell lines and plasmid constructs. HEK293 or MDA-MB-231 cells stably expressing 6×His-SUMO3m (C-terminal Q87R/Q88N mutation) [22] under the control of the cytomegalovirus (CMV) promoter (Fig. 2a) (*see* **Note 1**). These cell lines are generated by transfecting pcDNA3-SUMO3m and selecting with Geneticin (G418).
2. Dulbecco's Modified Eagle Medium (DMEM) supplemented with standard additives, as appropriate for each cell line.
3. Geneticin (G418).
4. Proteasome inhibitor MG132.
5. Phosphate-buffered saline (PBS, ice-cold) 1×.
6. Lysis buffer: 6 M guanidine-HCl, 100 mM NaH_2PO_4, 10 mM Tris-HCl, 10 mM imidazole, 20 mM 2-chloroacetamide, 5 mM 2-mercaptoethanol (pH 8.0).
7. Bradford Protein Assay Kit.
8. 1.5-mL and 2-mL low-retention microfuge tubes (Lobind).

2.2 Ni-NTA Purification, Digestion, and Desalting

1. Ni-NTA agarose beads are used for His-tag capture (Qiagen).
2. Ni-NTA wash buffer: 8 M urea, 100 mM NaH_2PO_4, 10 mM Tris-HCl, 5 mM 2-mercaptoethanol, 20 mM imidazole (pH 6.3) (*see* **Note 2**).
3. 50 mM ammonium bicarbonate (ABC).
4. 1% (vol/vol) trifluoroacetic acid (TFA).
5. Trypsin stock solution: 0.1 μg/μL sequencing-grade modified trypsin in 50 mM ABC.
6. Oasis HLB 30-mg desalting cartridges (for peptide cleanup).
7. Desalting conditioning solution: 0.1% (vol/vol) TFA in HPLC-grade acetonitrile (ACN).
8. Desalting wash solution: 0.1% (vol/vol) TFA in double-distilled water (ddH_2O).
9. Desalting elution solution: 0.1% (vol/vol) TFA and 30% (vol/vol) ddH_2O in ACN.
10. SpeedVac concentrator.

2.3 Affinity Reagents

1. Protein A magnetic beads (MagReSyn).
2. Anti–K(NQTGG) (SUMO3 remnant) antibody (clone UMO177, available upon request to Abcam).
3. Antibody cross-linking wash buffer: 200 mM triethanolamine, pH 8.
4. Cross-linking buffer: 5 mM DMP, 200 mM triethanolamine, pH 8.
5. Quenching buffer: 1 M Tris-HCl, pH 7.5.

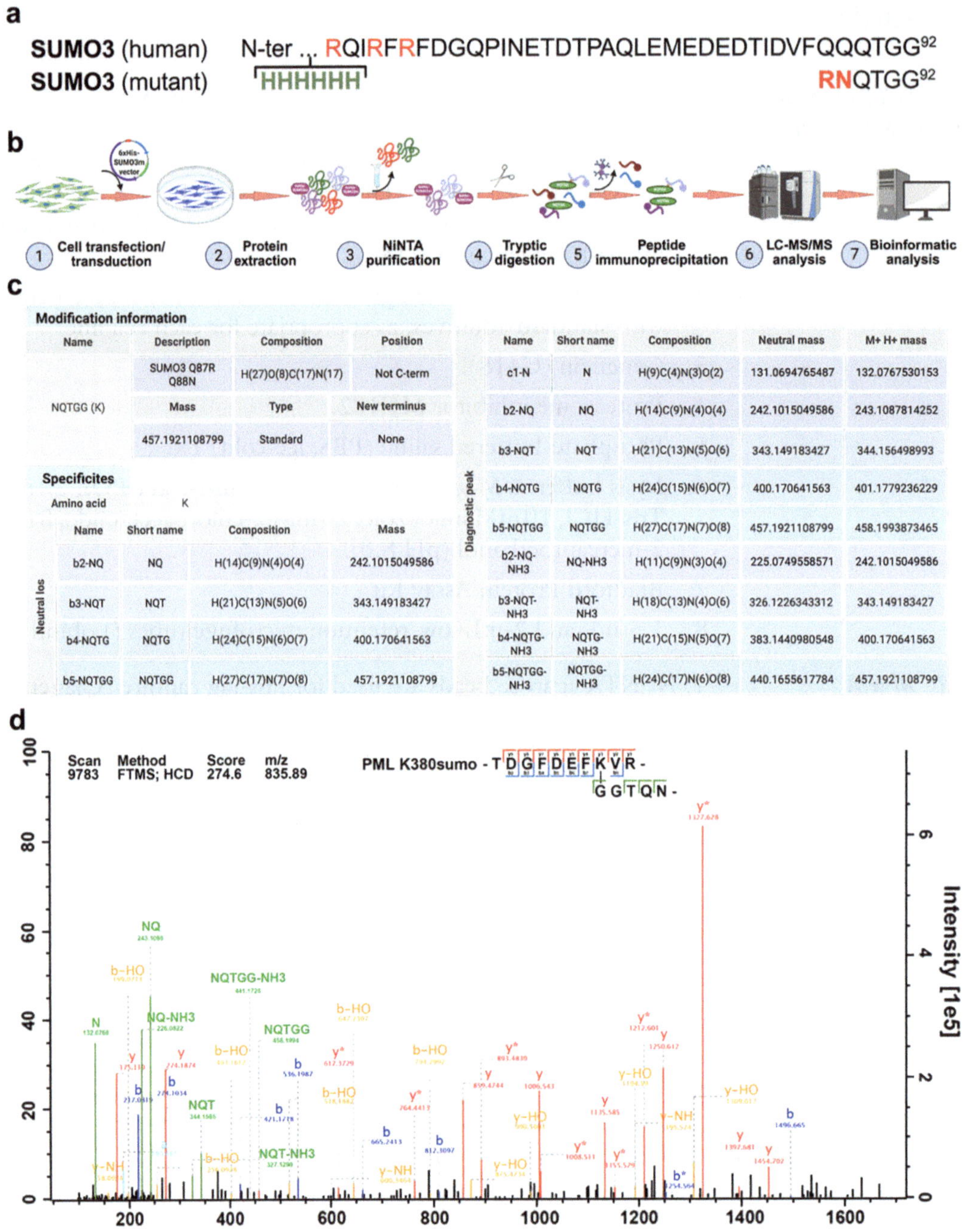

Modification information

Name	Description	Composition	Position
NQTGG (K)	SUMO3 Q87R Q88N	H(27)O(8)C(17)N(17)	Not C-term
	Mass	Type	New terminus
	457.1921108799	Standard	None

Specificites

Amino acid K

Neutral los

Name	Short name	Composition	Mass
b2-NQ	NQ	H(14)C(9)N(4)O(4)	242.1015049586
b3-NQT	NQT	H(21)C(13)N(5)O(6)	343.149183427
b4-NQTG	NQTG	H(24)C(15)N(6)O(7)	400.170641563
b5-NQTGG	NQTGG	H(27)C(17)N(7)O(8)	457.1921108799

Diagnostic peak

Name	Short name	Composition	Neutral mass	M+ H+ mass
c1-N	N	H(9)C(4)N(3)O(2)	131.0694765487	132.0767530153
b2-NQ	NQ	H(14)C(9)N(4)O(4)	242.1015049586	243.1087814252
b3-NQT	NQT	H(21)C(13)N(5)O(6)	343.149183427	344.156498993
b4-NQTG	NQTG	H(24)C(15)N(6)O(7)	400.170641563	401.1779236229
b5-NQTGG	NQTGG	H(27)C(17)N(7)O(8)	457.1921108799	458.1993873465
b2-NQ-NH3	NQ-NH3	H(11)C(9)N(3)O(4)	225.0749558571	242.1015049586
b3-NQT-NH3	NQT-NH3	H(18)C(13)N(4)O(6)	326.1226343312	343.149183427
b4-NQTG-NH3	NQTG-NH3	H(21)C(15)N(5)O(7)	383.1440980548	400.170641563
b5-NQTGG-NH3	NQTGG-NH3	H(24)C(17)N(6)O(8)	440.1655617784	457.1921108799

Fig. 2 Identification of SUMOylation sites by LC–MS/MS. (**a**) Schematic of vector construction for generating 6His-SUMO-NQTGG, showing how the NQTGG remnant is left after tryptic digestion. (**b**) Experimental workflow for SUMO site identification, from sample preparation to bioinformatic analysis. (**c**) MaxQuant settings used to detect NQTGG-modified peptides in SUMOylated proteins. (**d**) MS/MS spectrum of PML tryptic peptide modified at Lysine-380 by SUMO3m. Fragment ions arising from the NQTGG remnant are shown in green. Note that SUMOylation site is a reverse UBC9 consensus motif

6. Immunoprecipitation (IP) elution buffer: 0.2% (vol/vol) formic acid (FA) in ddH_2O.

2.4 LC–MS/MS and Data Analysis

1. LC–MS buffer A: 0.2% (vol/vol) FA in ddH_2O.
2. LC–MS buffer B: 0.2% (vol/vol) FA in ACN.
3. Jupiter C18 bulk (3 μm, 300 Å).
4. MaxQuant (e.g., version 2.2.0).
5. Perseus (e.g., version 2.1.4.0).

3 Methods

The workflow includes cell lysis, Ni-NTA enrichment of SUMOylated proteins, on-bead digestion, immunopurification of SUMOylated peptides, LC–MS/MS analysis (Fig. 2b).

3.1 Cell Culture and Protein Extraction

1. Cells are maintained in Dulbecco's Modified Eagle Medium (DMEM) supplemented with standard additives, as appropriate for each cell line. Stable selection of SUMOylation is maintained using G418 at a concentration of 500 μg/mL. To enhance SUMOylation, cells were treated with the proteasome inhibitor MG132 at a final concentration of 10 μM, as needed.
2. Grow two 15-cm dishes of SUMO3m-expressing cells to ~95% confluence in complete DMEM (*see* **Note 3**).
3. Harvest cells and wash once with 20 mL ice-cold PBS and collect by centrifugation (215 g, 10 min, 4°C). Pellets can be flash-frozen in liquid nitrogen and stored at –80°C.
4. Lyse each pellet in 3 mL ice-cold complete lysis buffer and vortex briefly. Sonicate on ice (15 W microtip, two 10 s pulses) to shear DNA and fully solubilize proteins. Transfer the lysate into 2 mL Lobind tubes, centrifuge the tubes at 17,000 *g* and keep the supernatant (*see* **Note 4**).
5. Determine protein concentration by Bradford assay, according to the manufacturer's instructions (*see* **Note 5**).

3.2 Ni-NTA Purification

1. For each protein extract sample of ~8 mg, prepare 320 μL of Ni-NTA bead slurry. Place beads in a 15-mL tube and wash three times with 10 mL complete lysis buffer (centrifuge at 215 g, 1 min, RT) to equilibrate.
2. Combine the cell lysate with beads, adding additional complete lysis buffer to a final volume of ~5 mL per 8 mg lysate. Incubate the mixture at 4°C for at least 16 h on a tube rotator, allowing His-tagged SUMOylated proteins to bind the resin (*see* **Note 6**).

3. After incubation, pellet the beads (215 g, 1 min, 4°C) and discard the supernatant. Wash the beads once with 10 mL complete lysis buffer, then wash four times with 10 mL Ni-NTA wash buffer, and once with 10 mL of 50 mM ABC (215 g, 1 min, 4°C each).
4. A small aliquot (10 μL) of the bead slurry can be mixed with Bradford reagent to estimate the bound protein amount (*see* **Note 7**).

3.3 On-Bead Digestion and Peptide Desalting

1. Add sequencing-grade trypsin to the Ni-NTA bead slurry at a ratio of 1:20 (w/w, trypsin:protein lysate). Seal the tube and incubate overnight at 37°C on a tube rotator.
2. After digestion, acidify the mixture with 1% TFA (vol/vol) to stop the reaction.
3. Activate a 30-mg Oasis HLB cartridge by passing 1 mL of desalting conditioning solution through the column using a vacuum manifold.
4. Condition the Oasis HLB cartridge by sequentially passing 1 mL of desalting conditioning solution, followed by 1 mL of desalting wash solution, under vacuum.
5. Load the acidified peptide digest onto the cartridge and allow it to flow through completely.
6. Wash the column with 1 mL wash solution to remove residual salts.
7. Elute bound peptides with 1 mL of desalting elution solution.
8. Collect the eluate and dry it to completeness in a SpeedVac (no heating) (*see* **Note 8**).

3.4 SUMO Anti-K(NQTGG) Bead Preparation

1. Prepare SUMO-remnant antibody beads by cross-linking anti–K(NQTGG) antibody to Protein A magnetic beads. Use the following ratio to calculate reagent volumes: for every 4 μg of SUMOylated protein (from Sect. 3.2), use 2 μg of antibody, and 1 μL of Protein A bead slurry.
2. Transfer the required Protein A beads into a 2 mL Lobind tube and wash three times with 1 mL PBS.
3. Add the anti–K(NQTGG) antibody (in μg) to the beads along with 1 mL PBS. Incubate for 1 h at 4°C with gentle rotation to bind the antibody to the beads (*see* **Note 9**).
4. Magnetically separate the beads and remove the supernatant. Wash the beads twice with 1 mL antibody cross-linking wash buffer. Add 10 μL of cross-linking buffer per 1 μL of bead slurry. Incubate for 1 h at room temperature on a rotator to covalently cross-link the antibody.

5. Quench the cross-linking reaction by adding quenching buffer to a final concentration of 5% (v/v), and incubate for 30 min at room temperature.
6. Magnetically separate the beads and remove all supernatant. Wash the beads twice with 1 mL PBS to remove excess buffer.
7. Resuspend the beads in PBS to achieve a final antibody concentration of 1 μg/μL.

3.5 SUMO Anti-K(NQTGG) IP

1. Resuspend the dried peptides from the desalting step in 1 mL PBS. Add the anti–K(NQTGG) antibody–Protein A beads (prepared in Sect. 3.4) to the peptide solution (*see* **Note 10**).
2. Incubate the mixture for 2 h at 4°C with gentle rotation to enrich for SUMOylated peptides.
3. After incubation, place the tube on a magnetic rack and carefully remove the supernatant. Wash the beads twice with 1 mL PBS followed by three washes with 1 mL ice-cold ddH_2O to remove nonspecifically bound peptides.
4. Elute the captured SUMO-peptides by adding 200 μL of IP elution buffer to the beads and incubating 10 min at room temperature
5. Magnetically separate the beads and transfer the eluate to a fresh Lobind tube. Perform this elution step twice more and combine all eluates (*see* **Note 11**).
6. Dry the combined eluates completely in a SpeedVac (no heating). The resulting sample contains enriched SUMO-modified peptides, and is ready for LC–MS/MS analysis.

3.6 MS Sample Injection

1. Reconstitute the dried peptides in 12 μL of LC Buffer A.
2. Pack an 18-cm × 150 μm ID C18 analytical column (Jupiter C18, 3 μm, 300 Å) or use IonOpticks column 25-cm × 75 μm ID C18.
3. Inject ~10 μL (depending on concentration) onto a nanoflow LC–MS system. For example, use an Easy-nLC coupled to an Orbitrap Exploris 480 mass spectrometer (Thermo Fisher) in positive ion mode.
4. Separate peptides with a linear gradient of 5%–30% LC Buffer B over ~105 min at 1 μL/min.
5. Acquire MS survey scans at 120,000 resolution (m/z 200) over m/z 350–1200, using standard mode of automatic gain control (AGC) target and max injection time set as auto mode. Perform data-dependent MS/MS for 3 s on the most intense precursors per cycle with a 1.2 m/z isolation window. Fragment peptides by higher-energy collisional dissociation (HCD) with a normalized collision energy (NCE) of 27%. Collect MS/MS at 45,000 resolution (m/z 200) with a normalized AGC target of 100% and maximum injection 700 ms (*see* **Note 12**).

3.7 Data Processing and Bioinformatic Analyses

1. Analyze raw MS/MS data using MaxQuant (e.g., version 2.2.0), searching against the UniProt/Swiss-Prot human protein database (including isoforms). Set the precursor mass tolerance to ~20 ppm for the first search and ~4.5 ppm for the main search, with a fragment ion mass tolerance of ~7.5 ppm. Use Trypsin/P as the digestion enzyme, allowing up to three missed cleavages. Define carbamidomethylation of cysteine (C) as a fixed modification. Set the following as variable modifications: methionine oxidation, asparagine/glutamine deamidation, serine/threonine phosphorylation, N-terminal acetylation, and lysine SUMO3 remnant (+NQTGG). For instructions on configuring SUMO3m in MaxQuant, refer to Fig. 2c. Specify a minimum peptide length of six amino acids, and control the false discovery rate (FDR) at 1% for both peptide and protein levels. A representative MS/MS spectrum of a promyelocytic leukemia protein (PML)-derived peptide modified at lysine-380 by SUMO3 is shown in Fig. 2d, with fragment ions from the NQTGG remnant highlighted in green.
2. After completing the MaxQuant analysis, export the SUMO3 NQTGG (K)Sites.txt table. Remove all decoy and reversed sequence entries (those beginning with "REV_") to maintain a 1% FDR. Filter the remaining data by site localization probability, retaining only entries with a probability >0.75 to ensure high-confidence site identification.
3. To assess differential SUMOylation between conditions, perform a volcano plot analysis using Perseus (e.g., version 2.1.4.0). Import the filtered SUMO3 NQTGG (K)Sites.txt dataset and log2-transform the LFQ intensity values (*see* **Note 13** for SILAC-specific instructions). Define experimental groups using categorical annotations (e.g., "Control" vs. "MG132"), and perform a two-sample *t*-test to compare conditions. Generate a volcano plot by right-clicking the results table and selecting Add Volcano Plot, setting the *x*-axis to $\log_2$(fold change) and the *y*-axis to $-\log_{10}$(*p*-value). Adjust the FDR and S_0 parameters to define significance thresholds. For example, compare DMSO-treated controls to cells treated with 10 μM MG132 for 16 h ($n=3$ biological replicates). Apply cutoffs of $|\log_2(\text{fold change})| \geq 0.585$ (≈1.5-fold) and $-\log_{10}(p\text{-value}) \geq 1.301$ ($p \leq 0.05$) to identify significantly regulated peptides. In the resulting volcano plot (Fig. 3a), upregulated SUMO peptides are highlighted in green and downregulated SUMO peptides in red.
4. To interpret the biological relevance of MG132-regulated SUMOylated proteins, perform Gene Ontology (GO) enrichment using the DAVID database. Focus on proteins associated with significantly upregulated SUMOylated peptides (green points in Fig. 3a). Extract their UniProt accession numbers

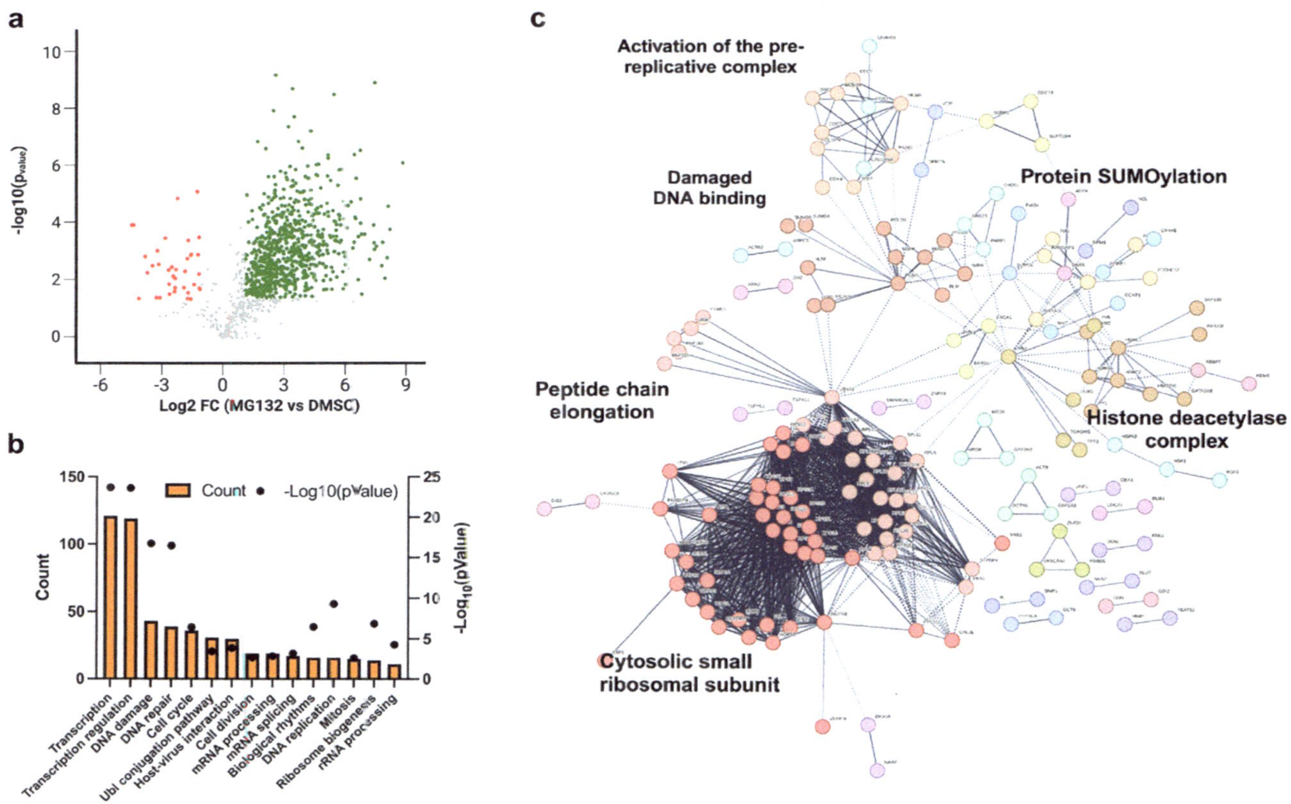

Fig. 3 Differential protein SUMOylation of HEK293SUMO3m following MG132 treatment. (**a**) Volcano plot showing differential SUMOylation levels between control and MG132-treated HEK293SUMO3m cells. (**b**) GO term enrichment analysis: Biological processes of proteins exhibiting increased SUMOylation upon MG132 treatment. (**c**) STRING network analysis of differentially SUMOylated proteins, revealing key interaction hubs and pathways affected by MG132-induced SUMOylation

and input them into DAVID (Step 1: Enter Gene List). Select *Gene List* as the input type and use the appropriate identifier (e.g., "UNIPROT_ACCESSION" or "OFFICIAL_GENE_SYMBOL"). Click *Submit List* and then Continue to Functional Annotation Tool. Under the *Gene_Ontology* section, expand *GOTERM_BP_DIRECT* (biological process) and download the resulting list of enriched GO terms. Rank the GO terms by significance using the Benjamini-adjusted p-value. Terms with Benjamini <0.05 are considered significantly enriched. Visualize enriched terms using GraphPad Prism: plot $-\log_{10}$ (p-value) against gene count (*see* **Note 14**). As shown in Fig. 3b, MG132-induced SUMOylation is associated with biological processes including transcription regulation, DNA damage repair, cell cycle progression, ubiquitin-like conjugation, antiviral response, and RNA processing.

5. To explore interaction networks among MG132-regulated SUMOylated proteins, use the STRING database (https://string-db.org). Submit the list of regulated proteins (by gene symbol, UniProt ID, or Ensembl ID), selecting *Homo sapiens* as the organism. After submission, customize the network by setting the interaction confidence score (e.g., medium or high) and applying clustering via the Markov Cluster Algorithm (MCL). STRING assigns functional annotations to each cluster and color-codes them accordingly. Export cluster details and network images from the Export menu, and annotate each cluster with its primary biological function (*see* **Note 15**). The interaction map shown in Fig. 3c highlights key modules, including the cytosolic small ribosomal subunit, prereplication complex, histone deacetylase complex, DNA damage recognition, SUMOylation machinery, and peptide chain elongation.

4 Notes

1. Other cell types can be engineered to stably express the 6× His-SUMO3m construct. For cell lines that are refractory to conventional transfection methods, lentiviral transduction can be used as an alternative. All SUMO3m-expressing cell lines should be regularly authenticated and screened for mycoplasma to ensure identity and purity.
2. Prepare chaotropic lysis buffers (e.g., guanidine-HCl with 2-chloroacetamide, imidazole, and 2-mercaptoethanol) immediately before use. Urea-containing buffers degrade over time, forming cyanate, which can modify proteins; therefore, prepare fresh 8 M urea solutions on the day of the experiment. Add imidazole and 2-mercaptoethanol freshly to

the Ni-NTA wash buffer. Note that these buffers are stable for only ~1 day at room temperature, so plan accordingly.

3. As a reference, two 150-mm dishes of HEK293 cells expressing SUMO3m at ~95% confluence typically yield ~40 million cells, corresponding to ~8–10 mg of total protein. To enhance SUMOylation levels, cells may be treated with 10 μM MG132 for 12–16 h or subjected to a 1-h heat shock at 42°C. Perform a minimum of three independent biological replicates for statistical robustness.
4. After sonication or mechanical shearing, the lysate should be fluid and easily pipettable using a P1000 tip. A viscous lysate indicates incomplete DNA shearing. If this occurs, add 1 mL of complete lysis buffer and repeat the sonication or shearing step until the solution becomes uniformly fluid. Proper DNA disruption is critical for accurate pipetting and downstream handling.
5. Since strong denaturants such as guanidine or urea can interfere with absorbance-based protein quantification, dilute lysates at least ten-fold in ddH_2O (not lysis buffer) before performing the Bradford assay. If protein quantification or downstream processing will be delayed, lysates may be stored at –80°C for up to 6 months.
6. Proteolysis directly on the bead slurry can be inefficient. To ensure complete digestion, add sequencing-grade trypsin and incubate the mixture overnight (~16 h) at 37°C with gentle rotation. For improved efficiency and reduced missed cleavages, a second aliquot of trypsin can be added and incubated for an additional 4 h.
7. Ni-NTA agarose does not absorb at 595 nm, and thus, does not interfere with Bradford measurements. However, be aware that Ni-NTA enrichment of SUMOylated proteins, typically, yields only 1%–2% of the total input protein. Adjust input amounts accordingly to ensure sufficient material for analysis.
8. After desalting, snap-freeze peptide eluates in liquid nitrogen and lyophilize them. The resulting dried peptides should appear as a fine white powder. Store lyophilized peptides at –20°C for up to 6 months; they are stable under these conditions.
9. Steps involving antibody–bead coupling (protocol Sects. 3.4.1–3.4.3) must be performed at cold temperatures. Carry out all washes and incubations on ice or at 4°C using prechilled PBS. For example, after washing Protein A beads three times with 1 mL PBS, add the anti-NQTGG antibody and incubate for 1 h at 4°C with gentle rotation. Maintaining cold conditions preserves antibody integrity and minimizes protease activity.

10. Similarly, SUMO immunoprecipitation (Sects. 3.5.1–3.5.3) should be performed using ice-cold buffers. Dissolve the dried peptides in 1 mL of ice-cold PBS and add to the antibody–bead complex. Incubate for 2 h at 4°C with rotation. After binding, wash the beads with ice-cold PBS and ddH_2O to remove non-specific binders. Check the pH of the peptide solution using pH paper and adjust to ~7.4 prior to IP, if necessary, as extreme pH can impair binding efficiency.
11. Prior to LC–MS/MS analysis, remove any residual beads or antibody fragments that may clog the column. For this, an optional cleanup step can be performed using a homemade C18 StageTip: insert one C18 disk (~8 mg lysate equivalent) into a 200 μL pipette tip. Equilibrate with ~100 μL ACN, load the peptide sample, wash with 50 μL of 0.1% FA, and elute with 50 μL of 30% ACN/0.1% FA into a Lobind tube. This additional desalting step helps remove particulates and improves chromatographic performance.
12. The LC–MS/MS parameters provided in the protocol (e.g., MS resolution, gradient length, AGC targets) serve as general starting points. Optimize these settings based on the specific mass spectrometer, column, and sample complexity. Regular instrument tuning and calibration are essential for ensuring high-quality data acquisition.
13. For SILAC-based datasets, $\log_2$-transform the H/L and M/L ratios and perform a one-sample t-test to determine $\log_2$ fold changes and corresponding $-\log_{10}$(p-values).
14. Alternative platforms for Gene Ontology (GO) enrichment analysis include g:Profiler (https://biit.cs.ut.ee/gprofiler), Metascape (https://metascape.org), and PANTHER (http://pantherdb.org). GO analysis can also be conducted programmatically in R using the clusterProfiler package from Bioconductor.
15. Protein–protein interaction (PPI) networks can also be visualized using Cytoscape (https://cytoscape.org), which allows flexible formatting, annotation, and integration of functional data.

Acknowledgments

This work was carried out with financial support from the Natural Sciences and Engineering Research Council (NSERC 311598). The Institute for Research in Immunology and Cancer (IRIC) receives infrastructure support from IRICoR, the Canadian Foundation for Innovation, and the Fonds de Recherche du Québec–Santé (FRQS).

References

1. Ryu HY, Ahn SH, Hochstrasser M (2020) SUMO and cellular adaptive mechanisms. Exp Mol Med 52(6):931–939. https://doi.org/10.1038/s12276-020-0457-2
2. Vertegaal ACO (2022) Signalling mechanisms and cellular functions of SUMO. Nat Rev Mol Cell Biol 23(11):715–731. https://doi.org/10.1038/s41580-022-00500-y
3. Zhao X (2018) SUMO-mediated regulation of nuclear functions and signaling processes. Mol Cell 71(3):409–418. https://doi.org/10.1016/j.molcel.2018.07.027
4. Rinfret Robert C, McManus FP, Lamoliatte F, Thibault P (2020) Interplay of ubiquitin-like modifiers following arsenic trioxide treatment. J Proteome Res 19(5):1999–2010. https://doi.org/10.1021/acs.jproteome.9b00807
5. Zhang T et al (2022) Crosstalk between SUMOylation and ubiquitylation controls DNA end resection by maintaining MRE11 homeostasis on chromatin. Nat Commun 13(1):5133. https://doi.org/10.1038/s41467-022-32920-x
6. Gareau JR, Lima CD (2010) The SUMO pathway: emerging mechanisms that shape specificity, conjugation and recognition. Nat Rev Mol Cell Biol 11(12):861–871. https://doi.org/10.1038/nrm3011
7. Geiss-Friedlander R, Melchior F (2007) Concepts in sumoylation: a decade on. Nat Rev Mol Cell Biol 8(12):947–956. https://doi.org/10.1038/nrm2293
8. Sampson DA, Wang M, Matunis MJ (2001) The small ubiquitin-like modifier-1 (SUMO-1) consensus sequence mediates Ubc9 binding and is essential for SUMO-1 modification. J Biol Chem 276(24):21664–21669. https://doi.org/10.1074/jbc.M100006200
9. Yang SH, Galanis A, Witty J, Sharrocks AD (2006) An extended consensus motif enhances the specificity of substrate modification by SUMO. EMBO J 25(21):5083–5093. https://doi.org/10.1038/sj.emboj.7601383
10. Claessens LA, Vertegaal ACO (2024) SUMO proteases: from cellular functions to disease. Trends Cell Biol 34(11):901–912. https://doi.org/10.1016/j.tcb.2024.01.002
11. Nayak A, Muller S (2014) SUMO-specific proteases/isopeptidases: SENPs and beyond. GenomeBiol15(7):422.https://doi.org/10.1186/s13059-014-0422-2
12. Jansen NS, Vertegaal ACO (2021) A chain of events: regulating target proteins by SUMO polymers. Trends Biochem Sci 46(2):113–123. https://doi.org/10.1016/j.tibs.2020.09.002
13. Wang W, Matunis MJ (2023) Paralogue-specific roles of SUMO1 and SUMO2/3 in protein quality control and associated diseases. Cells 13(1):8. https://doi.org/10.3390/cells13010008
14. Hay RT (2005) SUMO: a history of modification. Mol Cell 18(1):1–12. https://doi.org/10.1016/j.molcel.2005.03.012
15. Karhausen J, Ulloa L, Yang W (2021) SUMOylation connects cell stress responses and inflammatory control: lessons from the gut as a model organ. Front Immunol 12:646633. https://doi.org/10.3389/fimmu.2021.646633
16. Paasch F et al (2018) Failed mitochondrial import and impaired proteostasis trigger SUMOylation of mitochondrial proteins. J Biol Chem 293(2):599–609. https://doi.org/10.1074/jbc.M117.817833
17. Chang YC, Oram MK, Bielinsky AK (2021) SUMO-targeted ubiquitin ligases and their functions in maintaining genome stability. Int J Mol Sci 22(10):5391. https://doi.org/10.3390/ijms22105391
18. Hendriks IA et al (2018) Site-specific characterization of endogenous SUMOylation across species and organs. Nat Commun 9(1):2456. https://doi.org/10.1038/s41467-018-04957-4
19. Lumpkin RJ et al (2017) Site-specific identification and quantitation of endogenous SUMO modifications under native conditions. Nat Commun 8(1):1171. https://doi.org/10.1038/s41467-017-01271-3
20. Fulzele A, Bennett EJ (2018) Ubiquitin diGLY proteomics as an approach to identify and quantify the ubiquitin-modified proteome. Methods Mol Biol 1844:363–384. https://doi.org/10.1007/978-1-4939-8706-1_23
21. Hendriks IA et al (2015) System-wide identification of wild-type SUMO-2 conjugation sites. Nat Commun 6: 7289. https://doi.org/10.1038/ncomms8289
22. Lamoliatte F et al (2014) Large-scale analysis of lysine SUMOylation by SUMO remnant immunoaffinity profiling. Nat Commun 5: 5409. https://doi.org/10.1038/ncomms6409
23. Barroso-Gomila O et al (2021) Identification of proximal SUMO-dependent interactors using SUMO-ID. Nat Commun 12(1):6671. https://doi.org/10.1038/s41467-021-26807-6
24. Bekes M et al (2011) The dynamics and mechanism of SUMO chain deconjugation by SUMO-specific proteases. J Biol Chem 286(12):10238–10247. https://doi.org/10.1074/jbc.M110.205153
25. Hietakangas V et al (2006) PDSM, a motif for phosphorylation-dependent SUMO modification. Proc Natl Acad Sci U S A 103(1):45–50. https://doi.org/10.1073/pnas.0503698102

26. Jin J (2019) Interplay between ubiquitylation and SUMOylation: empowered by phase separation. J Biol Chem 294(42):15235–15236. https://doi.org/10.1074/jbc.H119.011037
27. Nie M, Boddy MN (2016) Cooperativity of the SUMO and ubiquitin pathways in genome stability. Biomolecules 6(1):14. https://doi.org/10.3390/biom6010014
28. Gill G (2004) SUMO and ubiquitin in the nucleus: different functions, similar mechanisms. Genes Dev 18(17):2046–2059. https://doi.org/10.1101/gad.1214604
29. Lamoliatte F et al (2017) Uncovering the SUMOylation and ubiquitylation crosstalk in human cells using sequential peptide immunopurification. Nat Commun 8: 14109. https://doi.org/10.1038/ncomms14109

Chapter 4

Computational Pipelines for Protein Ubiquitylation Analysis and Prediction

Anastasiya Potapenko, Lyndal Henden, Ghasem Azemi, Antonio Di Ieva, Angela S. Laird, Roger Chung, Albert Lee and Jennilee M. Davidson

Abstract

Ubiquitylation is a crucial posttranslational modification that regulates cellular homeostasis and has been linked to a range of diseases. Identification and analysis of ubiquitylated proteins and ubiquitylation sites have historically been successfully performed by mass spectrometry from in vitro and in vivo samples. However, to address some of the challenges and improve the efficiency of ubiquitylation analysis by mass spectrometry, computational tools have been developed and are becoming increasingly popular. In this chapter, we summarize the available computational tools for predicting ubiquitylation in silico and the common approaches used to enrich ubiquitylation in samples for mass spectrometry and computational analysis. We subsequently provide simple steps for molecular biologists to follow for predicting ubiquitylation sites on proteins in silico, and we provide a computational approach to identify average ubiquitin branch sites across samples from ubiquitin-enriched mass spectrometry data.

Keywords Ubiquitylation, Ubiquitylation sites, Mass spectrometry, Ubiquitylation prediction, Computational analysis, In silico, In vitro

1 Introduction

Ubiquitylation is a posttranslational modification (PTM) that alters the lifespan, subcellular localization, and protein structure and function. Most biological processes, including gene transcription, cell growth, immunity, DNA repair, and metabolism are regulated by ubiquitylation [1, 2], making it a crucial PTM for cellular homeostasis. Given the diversity of biological processes controlled by ubiquitylation, it is unsurprising that dysregulated ubiquitylation has been linked to a range of diseases, including cancer [3–6], inflammatory diseases [7], and neurodegeneration [8, 9]. For these reasons, ubiquitylation is one of the most studied PTMs [10].

The addition of ubiquitin to target proteins is reversible and occurs via isopeptide linkage of the C-terminal carboxyl group of

Xu Zhang et al. (eds.), *Protein Post-Translational Modifications*, Methods in Molecular Biology, vol. 3018,
https://doi.org/10.1007/978-1-0716-5166-7_4,

ubiquitin typically to lysine (K) residues within the target protein [11]. This process is regulated by a cascade of ubiquitin-activating (E1), ubiquitin-conjugating (E2), and ubiquitin-ligase (E3) enzymes [12], as well as deubiquitinase (DUB) enzymes that effectively reverse ubiquitylation and ubiquitin-like modification of proteins [13, 14]. Briefly, ubiquitin is activated by ATP and conjugated to an E1, then transferred to an E2, and subsequently transferred to the lysine residue of the target protein by an E3. Polyubiquitin chains sequentially assemble on proteins via formation of covalent linkages between the C-terminus of one ubiquitin and the ε-amino group of any of the seven lysine residues (K6, K11, K27, K33, K48, K63), or the amino group of the methionine (M1) residue of another ubiquitin [15]. Ubiquitylation can occur at a single lysine residue (monoubiquitylation) or at multiple lysine residues (polyubiquitylation), and chain topology can be complex as chains can comprise multiple ubiquitin linkage types (heterotypic chains) [16, 17], be branched [18], and contain ubiquitin-like proteins [19]. While polyubiquitylation via other lysine residues within ubiquitin is less well characterized, this form of polyubiquitylation is involved in a variety of important biological processes, as reviewed by Tracz and Bialek [20]. Ubiquitylation of cysteine, serine, and threonine residues, and the N-terminal group of proteins is also less well characterized, though biologically important, as reviewed by McClellan et al. [21].

1.1 Computational Prediction Tools for Ubiquitylation Analysis

Identification of ubiquitylated proteins and ubiquitylation sites has been successfully performed in vitro and in vivo experimentally in wet-lab settings using mass spectrometry (MS). However, such wet-lab-based approaches are costly, laborious, and time intensive. To improve the efficiency and practicality of identifying ubiquitylated proteins and ubiquitylation sites, computational predictive methodologies have been developed and more are under investigation. Computational predictive methods offer advantages including, among others, the capacity to integrate diverse data, such as physicochemical properties of proteins and evolutionary information to improve prediction accuracy, as well as the ability to handle large datasets simultaneously, reducing dimensionality, and facilitating high-throughput studies.

Relying on artificial feature selection makes many of the ubiquitylation-focused machine learning-based methods susceptible to bias. The Evolutionary Screening Algorithm (ESA)-UbiSite [22] uses machine learning methodologies such as support vector machine (SVM) as an optimization approach for selecting features for physiochemical properties, and thus improving prediction accuracy. In addition, while many predictive models of ubiquitylation sites are trained to assume that previously unvalidated ubiquitylation sites are nonubiquitylated (negative sample), the ESA [22], based on evolutionary selection [23], selects a set of effective negatives among nonvalidated sites. These effective negatives have the least likelihood of being a PTM. The ESA also produces informative physiochemical property features and the optimized SVM prediction model, such that ESA-UbiSite applies an optimization approach for feature selection, integrating physiochemical properties from the AAindex [24] with SVM together with ubiquitin site prediction. The ESA-UbiSite [22] method can, therefore, be used to predict ubiquitylation sites in silico for further validation. More details about this method are presented in the next section.

The pattern of ubiquitylation sites is not conserved across species [25, 26]. Hence, recent ubiquitylation-focused predictors have been developed to be either species-focused or multispecies prediction models. Transfer deep learning techniques, namely DeepTL-Ubi [27], enhance species-specific ubiquitylation site prediction by leveraging extensive amounts of human data and transferring common knowledge to other species. This effectively mitigates the problem of insufficient nonhuman species data for training purposes in deep learning models. Another deep learning model made significant improvements for protein ubiquitylation site prediction by using a capsule network, Caps-Ubi [28]. In Caps-Ubi, protein fragments are first encoded using both one-hot encoding (a method of representing categorical data as binary vectors) and amino acid continuous encoding (which considers the physicochemical properties and evolutionary information of proteins). Three convolutional layers and a capsule network layer subsequently extract protein features and make predictions. We have summarized the available methods and tools for computational analysis of ubiquitylation in Table 1.

1.2 Mass Spectrometry Methods to Characterize Ubiquitylome

Identification of putative ubiquitylation sites helps reveal the mechanisms driving biological processes. In silico prediction methods train models on experimentally validated ubiquitylation sites and offer a basis for testing hypotheses and exploring potential candidates experimentally. Mass spectrometry is a powerful experimental approach to identify and characterize PTMs including ubiquitylation [43] from in vitro and in vivo samples. The C-terminus of ubiquitin consists of an arginine (R) followed by two glycine (G) residues, and the C-terminal glycine is conjugated to the lysine of the target protein. Upon protein digestion with trypsin, ubiquitin is cleaved from the target protein at the arginine residue, leaving behind a diglycine (diGLY) remnant on the formerly ubiquitylated lysine. This diGLY remnant is a signature of ubiquitylation that can be detected by MS as a mass shift of 114.1 kDa [44–48]. As trypsin also cleaves proteins modified by the ubiquitin-like proteins NEDD8, ISG15, and SUMO, producing a diGLY remnant that is indistinguishable from that of ubiquitylation, sequential digestion with LysC followed by trypsin cleavage is the preferred method for identifying ubiquitylation sites [49–51]. This is because LysC cleaves peptides after lysines to produce a ubiquitin remnant that is unique to ubiquitin, enabling distinction between

Table 1
A summary of the available computational prediction tools for ubiquitylation analysis

Computational tool	Application
Caps-Ubi [28]	Multispecies ubiquitylation site prediction.
Plant PTMviewer [29]	Compiles various PTMs for tens of thousands of plant proteins; can be used to search for PTMs in user-defined amino acid sequence motifs of interest.
RUBI [30]	Sequence-based ubiquitylation predictor.
DeepTL-Ubi [27]	Deep learning method for predicting ubiquitylation sites of multiple species.
ESA-UbiSite [22, 31]	UbiSite identifies ubiquitin-conjugation sites on lysines based on large-scale proteome datasets. ESA-UbiSite has an additional evolutionary screening algorithm (ESA) that retrieves physicochemical properties of proteins from protein sequences to improve identification of human ubiquitylation sites.
UbSite [32]	Identifies protein ubiquitylation sites using a neural network.
UbiPred [33]	First tool for ubiquitylation site prediction; an SVM-based predictor of ubiquitylation sites that employs an informative physicochemical property mining algorithm (IPMA) to mine physicochemical properties from protein sequences.
hUbiPred [34]	Ubiquitylation predictor that was built by the integration of two types of neural networks and uses binary encoding and physicochemical properties of amino acids as training input.
UbPred [35]	A random forest predictor of ubiquitylation sites in yeast that uses protein sequence, physicochemical properties, and evolutionary features.
CKSAAP_UbSite [36]	Predicts ubiquitylation sites based on protein sequence using the same dataset of yeast ubiquitylation sites as UbPred; an SVM-based predictor that considers composition of k-spaced amino acid pairs surrounding potential ubiquitylation sites.
hCKSAAP_UbSite [37]	Updated form of CKSAAP_UbSite for prediction of human ubiquitylation sites.
[38]	Ubiquitylation site predictor based on a nearest-neighbor algorithm.
UbiProber [39]	SVM-based predictor of general and species-specific ubiquitylation sites; uses information from key amino acids as input.
iUbiq-Lys [40]	SVM-based predictor of ubiquitylation sites; uses sequence features extracted from sequence evolution information by a gray system model.
GPS-Uber [41]	Prediction of general and E3-specific lysine ubiquitylation sites based on protein sequence and structural features; uses three algorithm types: penalized logistic regression (PLR), deep neural network (DNN), and convolutional neural network (CNN).
Musite Deep [42]	Rapid prediction and visualization of PTMs including ubiquitylation from protein sequences; uses a deep-learning framework and only uses protein sequence as input.

ubiquitylation and modification by ubiquitin-like proteins. Nonetheless, as only 6% of diGLY remnants are estimated to be produced by ubiquitin-like modifications [46], trypsin digestion alone can be sufficient; however, the possibility of ubiquitin-like modifications must be considered when analyzing the MS data.

Characterization of the ubiquitylome by MS, however, faces several challenges including that ubiquitylation is a transient and rapidly changing modification, is sensitive to sample preparation steps, the proportion of ubiquitylated proteins in the proteome is low, and chain topology can be complex [52]. To address these challenges and enhance studies of the ubiquitylome, an array of in vitro and in vivo approaches to enrich ubiquitylated proteins has been developed. Broadly, ubiquitin enrichment approaches can be categorized as antibody-based, tag-based, and ubiquitin binding domain-based. A common antibody-based enrichment approach is to immunoprecipitate ubiquitylated proteins using a pan-ubiquitin (e.g., α-FK2) or linkage-specific antibody (e.g., α-K48). Tag-based enrichment strategies involve expression and subsequent purification of epitope-tagged ubiquitin. A commonly used example of this is the expression of hexahistidine-tagged ubiquitin (HIS_6-Ub), followed by Ni-NTA purification under denaturing conditions to isolate proteins ubiquitylated by HIS_6-Ub [53]. An example of a high-throughput tag-based enrichment approach is StUbEx PLUS [54], which involves knockdown of endogenous ubiquitin in cells and replacement with stably expressed HIS_6-FLAG-Ub [55]. Ubiquitin binding domain-based enrichment strategies include the use of tandem ubiquitin binding entities (TUBEs), which selectively bind polyubiquitin chains and protect them from degradation and deubiquitylation [56]. Table 2 provides examples of studies utilizing the various ubiquitylation enrichment approaches outlined. Moreover, ubiquitylated substrates (or polyubiquitin chains) can be treated with linkage-specific deubiquitylating enzymes prior to MS analysis, a technique termed UbiCrest [57].

Following in vitro or in vivo ubiquitin enrichment, the ubiquitylated proteins are preferably digested with LysC and trypsin

Table 2
Examples of studies that have utilized the various ubiquitylation enrichment approaches

Enrichment strategy	
Antibody-based enrichment	Immunoprecipitation of endogenous ubiquitylated proteins [63–65]
	DiGLY antibody [45, 66–69]
	UbiSite antibody [49, 70]
Tag-based enrichment	Ni-NTA purification of HIS_6-Ub ubiquitylated proteins [53, 71, 72]
	Two-step purification with Histidine-biotin tag [73]
Ubiquitin-binding domain-based enrichment	Tandem ubiquitin-binding entities (TUBEs) [51]
	Tandem hybrid ubiquitin-binding domains (ThUBDs) [74]

[49–51], and the diGLY-modified peptides are further enriched with anti-diGLY antibody for MS analysis [46]. To enrich ubiquitin-modified sites specifically, the UbiSite antibody can be used [49]. An additional advantage of the UbiSite antibody is that it can also enrich N-terminally ubiquitylated proteins, which diGLY antibodies cannot. The most common method for ubiquitylation analysis is data-dependent acquisition (DDA)-MS that involves an initial MS scan to select precursors based on peptide intensity, followed by a second MS run to acquire fragment ion spectra for these individual precursors. Although a relatively new approach, data-independent acquisition (DIA)-MS, which involves simultaneous fragmentation of all peptide ions within a narrow mass-to-charge window, is an alternative approach and has been found to identify a greater number of ubiquitylation sites [51, 58]. MS-based techniques enabling quantification of the forms of ubiquitin on proteins include Stable Isotope Labelling with Amino Acids in Cell Culture (SILAC), and Ubiquitin Absolute QUAntification (Ub-AQUA) [55, 59]. Additionally, isobaric chemical tags such as the Tandem Mass Tag (TMT) system can be used for ubiquitylation studies with the UbiFAST method [60], which overcomes the protein quantity limitations when using TMT after diGLY enrichment [61]. Another approach to identify in vivo PTM sites, including ubiquitylation, without the need for either transgene expression or antibody enrichment of ubiquitylated proteins or peptides is Ub COFRADIC (COmbined FRAction Diagonal Chromatography) [62]. Here, free amines are blocked by acetylation, the ubiquitin is enzymatically removed such that the newly free amines (where the lysine was ubiquitylated) can be chemically modified with a handle that allows specific isolation of the ubiquitylated peptides during the COFRADIC chromatographic runs.

This chapter provides simple steps for predicting ubiquitylation sites on proteins in silico using ESA-UbiSite as an example, as well as a computational method to identify average ubiquitin branch sites from MS data of in vitro and in vivo samples with complementary wet-lab methods to enrich site-specific and linkage-specific ubiquitylation.

2 Materials

2.1 ESA-UbiSite Method

1. https://github.com/NctuICLab/ESA-UbiSite/tree/master
2. A C++ compiler, e.g., g++.
3. Perl interpreter.

2.2 Endogenous Site-Specific Enrichment Using UbiSite

1. Lysis buffer: 8 M guanidine-HCl (pH 8.5), 25 mM Ammonium Bicarbonate (AmBic), protease inhibitor, phosphatase inhibitor.

2. 2 mM dithiothreitol (DTT).
3. 11 mM chloroacetamide (CAA).
4. 25 mM AmBic.
5. Protein quantification assay kit.
6. PVDF membrane (0.45 μM pore).
7. Endoproteinase LysC.
8. C18 cartridges.
9. 150 mM sodium chloride (NaCl).
10. Immunoprecipitation buffer: 50 mM MOPS (pH 7.5), 10 mM sodium phosphate, 50 mM NaCl, 0.1% Triton-X100.
11. UbiSite antibody conjugated and crosslinked to Protein G beads.
12. 0.1% trifluoroacetic acid (TFA).
13. 1 M AmBic.
14. Trypsin.

2.3 Immuno-precipitation of Endogenous Linkage-Specific Ubiquitylated Proteins

1. Proteasome inhibitor.
2. Tris-buffered saline (TBS).
3. Lysis buffer: 1% (v/v) NP-40 in TBS, 2 mM EDTA, 10 mM *N*-ethylmaleimide, protease inhibitor, phosphatase inhibitor.
4. Protein quantification assay kit.
5. Ubiquitin linkage-specific antibody (i.e., α-K48) and α-IgG control antibody of the corresponding species.
6. Protein A/G magnetic bead slurry.
7. 1× Laemmli sample buffer.

2.4 In-Gel Tryptic Digestion

1. Coomassie brilliant blue stain.
2. 50 mM AmBic.
3. 50% acetonitrile (ACN) in 50 mM AmBic.
4. 100% ACN.
5. 10 mM DTT in 50 mM AmBic.
6. 20 mM iodoacetamide (IAA) in 50 mM AmBic.
7. Trypsin.
8. 50% ACN in 2% formic acid.
9. 0.1% formic acid.
10. C18 OMIX tips.
11. 50% ACN in 0.1% formic acid.
12. Glass vials for LC–MS/MS analysis.

2.5 LC–MS/MS

1. High-performance liquid chromatography (e.g., Ultimate 3000 NanoLC).
2. Tandem mass spectrometer (e.g., Thermo Q-Exactive Plus Orbitrap).
3. Precolumn: Acclaim™ PepMap™ C18 Reversed Phase HPLC Column, 5 μm, 0.3 mm × 5 mm.
4. Analytical column: Acclaim PepMap RSLC column particle size of 2 μm, diameter of 0.075 mm and length of 150 mm.
5. Buffer A: 0.1% (v/v) formic acid in water.
6. Buffer B: 80% (v/v) ACN, 0.1% (v/v) formic acid in water.

2.6 R Computational Analysis of Ubiquitin Branch Sites Between Samples

1. https://gitlab.com/mq-mnd/grp_williams/ubiquitin_analysis.git
2. https://github.com/ruma1974/UPSRM/ [75].
3. The Comprehensive R Archive Network (CRAN).
4. RStudio.
5. R packages: ComplexHeatmap, tidyverse, plyr.
6. MaxQuant data: "GlyGly (K)Sites.txt" and "proteinGroups.txt," or download example published dataset (PXD010161) [76] from ProteomeXchange. Only "insoluble fraction" "proteinGroups.txt" from the insol_global folder and "GlyGly (K)Sites.txt" from the insol_digly folder are required for this example.

3 Methods

3.1 ESA-UbiSite to Predict Ubiquitylation Sites from Human Protein Sequences In Silico

3.1.1 ESA-UbiSite Method [22]

Here, we present a detailed guide to running our preferred, simple method of predicting ubiquitylation sites on a protein of interest in silico, ESA-UbiSite.

The workflow of ESA-UbiSite involves first encoding protein sequences using amino acid indices (from aaindex1.tab file) to generate numerical feature vectors representing various physicochemical properties of the amino acids. These feature vectors are then scaled and used as input for an SVM model, implemented with libsvm, to predict ubiquitylation sites. The prediction results are processed and visualized, typically outputting a final HTML report of the predicted ubiquitylation sites.

3.1.2 Operating System

While ESA-UbiSite can run on any operating system, Unix-like systems such as Linux and macOS are preferred.

3.1.3 Requirements

Before proceeding, ensure you have the following installed:

1. A C++ compiler, e.g., g++.
2. Perl interpreter.
3. PHP interpreter.

3.1.4 Steps to Setup ESA-UbiSite

1. Clone the Repository: Start by cloning the ESA-UbiSite repository from GitHub.

```
git clone https://github.com/NctuICLab/ESA-UbiSite.git
```

2. Navigate to the ESA-UbiSite Directory: Change to the newly cloned directory.

```
cd ESA-UbiSite
```

3. Compile the aaindex Program: Navigate to the src/aaindex directory and compile the aaindex program.

```
cd src/aaindex
make
```

4. Compile the libsvm Tools: Move to the libsvm_320 directory and compile the svm-train, svm-predict, and svm-scale tools.

```
cd ../libsvm_320
make
```

5. Return to the Main Directory: Go back to the main ESA-UbiSite directory.

```
cd ../..
```

3.1.5 Preparing Your Input and Output

1. Save your .fa file: Ensure that your input.fa file is saved in the main ESA-UbiSite directory. This file, e.g., example.fa, should be in FASTA format and contains the protein sequences you wish to analyze.
2. Create an output directory: Create a directory where the script will save the results in. For example:

```
mkdir example_output
```

3.1.6 Running ESA-UbiSite

Once the setup is complete, you can run the ESA-UbiSite pipeline with the following command:

```
perl ESAUbiSite_main.pl example.fa example_output
```

3.1.7 Viewing the Results

After running the script, the results will be saved in the specified example_output directory.

To view the results, open the HTML file generated by the script.

3.1.8 Sample Runs

Example 1: Running ESA-UbiSite for the example.fa data already available. Run:

```
perl ESAUbiSite_main.pl example.fa example_output
```

The results are displayed in Fig. 1.

Prediction results of ESA-UbiSite:

>sp|A6ZKI3|F127A_HUMAN Protein FAM127A OS=Homo sapiens GN=FAM127A PE=1 SV=1

MAVESRVTQEEIKKEPEKPIDREKTCPLLLRVFTTNNGRHHRMDEFSRGNVPSSELQIYTWMDATLKELT
SLVKEVYPEARKKGTHFNFAIVFTDVKRPGYRVKEIGSTMSGRKGTDDSMTLQSQKFQIGDYLDIAITPP
NRAPPPSGRMRPY

Position	Sequence	Ubiquitination	Score
13	VESRVTQEEIKKEPEKPIDRE	N	0.23
14	ESRVTQEEIKKEPEKPIDREK	N	0.11
18	TQEEIKKEPEKPIDREKTCPL	N	0.25
24	KEPEKPIDREKTCPLLLRVFT	N	0.38
67	QIYTWMDATLKELTSLVKEVY	N	0.34
74	ATLKELTSLVKEVYPEARKKG	N	0.47
82	**LVKEVYPEARKKGTHFNFAIV**	**Y**	**0.60**
83	**VKEVYPEARKKGTHFNFAIVF**	**Y**	**0.61**
97	**FNFAIVFTDVKRPGYRVKEIG**	**Y**	**0.74**
104	TDVKRPGYRVKEIGSTMSGRK	N	0.28
114	KEIGSTMSGRKGTDDSMTLQS	N	0.37
126	**TDDSMTLQSQKFQIGDYLDIA**	**Y**	**0.63**

>sp|O43164|PJA2_HUMAN E3 ubiquitin-protein ligase Praja-2 OS=Homo sapiens GN=PJA2 PE=1 SV=4

MSQYTEKEPAAMDQESGKAVWPKPAGGYQTITGRRYGRRHAYVSFKPCMTRHERSLGRAGDDYEVLELDD
VPKENSSGSSPLDQVDSSLPSEPIFEKSETEIPTCGSALNQTTESSQSFVAVHHSEEGRDTLGSSTNLHN
HSEGEYIPGACSASSVQNGIALVHTDSYDPDGKHGEDNDHLQLSAEVVEGSRYQESLGNTVFELENREAE
AYTGLSPPVPSFNCEVRDEFEELDSVPLVKSSAGDTEFVHQNSQEIQRSSQDEMVSTKQQNNTSQERQTE
HSPEDAACGPGHICSEQNTNDREKNHGSSPEQVVRPKVRKLISSSQVDQETGFNRHEAKQRSVQRWREAL
EVEESGSDDLLIKCEEYDGEHDCMFLDPPYSRVITQRETENNQMTSESGATAGRQEVDNTFWNGCGDYYQ
LYDKDEDSSECSDGEWSASLPHRFSGTEKDQSSSDESWETLPGKDENEPELQSDSSGPEEENQELSLQEG
EQTSLEEGEIPWLQYNEVNESSSDEGNEPANEFAQPAFMLDGNNNLEDDSSVSEDLDVDWSLFDGFADGL
GVAEAISYVDPQFLTYMALEERLAQAMETALAHLESLAVDVEVANPPASKESIDGLPETLVLEDHTAIGQ
EQCCPICCSEYIKDDIATELPCHHFFHKPCVSIWLQKSGTCPVCRRHFPPAVIEASAAPSSEPDPDAPPS
NDSIAEAPi

Position	Sequence	Ubiquitination	Score
7	MSQYTEKEPAAMDQESG	N	0.39
18	EPAAMDQESGKAVWPKPAGGY	N	0.29
23	**DQESGKAVWPKPAGGYQTITG**	**Y**	**0.77**
46	YGRRHAYVSFKPCMTRHERSL	N	0.37
73	YEVLELDDVPKENSSGSSPLD	N	0.35
97	SSLPSEPIFEKSETEIPTCGS	N	0.30
173	**VHTDSYDPDGKHGEDNDHLQL**	**Y**	**0.69**
240	FEELDSVPLVKSSAGDTEFVH	N	0.34
268	RSSQDEMVSTKQQNNTSQERQ	N	0.31
304	**CSEQNTNDREKNHGSSPEQVV**	**Y**	**0.52**
317	**GSSPEQVVRPKVRKLISSSQV**	**Y**	**0.68**
320	PEQVVRPKVRKLISSSQVDQE	N	0.46
339	QETGFNRHEAKQRSVQRWREA	N	0.32
363	EESGSDDLLIKCEEYDGEHDC	N	0.41
424	GCGDYYQLYDKDEDSSECSDG	N	0.25
449	**SLPHRFSGTEKDQSSSDESWE**	**Y**	**0.77**
464	**SDESWETLPGKDENEPELQSD**	**Y**	**0.83**
610	**DVEVANPPASKESIDGLPETL**	**Y**	**0.59**
643	CCPICCSEYIKDDIATELPCH	N	0.04
658	TELPCHHFFHKPCVSIWLQKS	N	0.15
667	HKPCVSIWLQKSGTCPVCRRH	N	0.35

Fig. 1 Example output from ESA-UbiSite of the provided example data in Github package

Example 2: Running ESA-UbiSite for a protein of interest (i.e., SQSTM1)

1. Obtain sequence for SQSTM1
   a. Navigate to https://www.uniprot.org/uniprotkb?query=SQSTM1

   b. Select Q13501
   c. Navigate to Sequence and Isoforms and copy sequence
2. Create a new file for SQSTM1 named SQSTM1.fa in the ESA-UbiSite folder
3. Paste sequence copied from Sequence and Isoforms page into SQSTM1.fa file
4. Create a directory for the analysis results to be saved in, e.g., results_directory
5. Run

```
perl ESAUbiSite_main.pl SQSTM1.fa results_directory
```

The results are displayed in Fig. 2, with comparison to known ubiquitylation sites of Sequestosome-1/p62 in Fig. 3.

3.2 Examples of Various Ubiquitin Enrichment Strategies for Computational Analysis of Ubiquitylation from MS Data of In Vitro Samples

With the most common approach to analyzing protein ubiquitylation being from DDA-MS data, we here present the most common and simple methods required to enrich ubiquitin from samples for MS and subsequent computational analysis.

3.2.1 Endogenous Site-Specific Ubiquitylation Enrichment Using UbiSite

Protocol adapted from [70].

1. Lyse cells in denaturing lysis buffer (8 M guanidine-HCl, 25 mM AmBic, containing protease and phosphatase inhibitors, pH 8.5).
2. Probe sonicate samples, centrifuge at 15,000 × *g* for 30 min to pellet cellular debris, then transfer supernatant to new tubes.
3. Quantify protein concentration.
4. Reduce proteins with 2 mM dithiothreitol (DTT) for 30 min at room temperature.
5. Alkylate with 11 mM chloroacetamide (CAA) for 30 min at room temperature.
6. Dilute samples with 25 mM AmBic to 2 M guanidine-HCl, then clear through a 0.45 μM PVDF membrane.
7. Digest with LysC (LysC:protein ratio 1:100) overnight at room temperature.
8. Purify peptides by C18 cartridges, then lyophilize.
9. Dissolve peptides in immunoprecipitation buffer (50 mM MOPS pH 7.5, 10 mM sodium phosphate, 50 mM NaCl) containing 0.1% Triton X-100.

Prediction results of ESA-UbiSite:

>sp|Q13501|SQSTM_HUMAN Sequestosome-1 OS=Homo sapiens OX=9606 GN=SQSTM1 PE=1 SV=1

```
MASLTVKAYLLGKEDAAREIRRFSFCCSPEPEAEAEAAAGPGPCERLLSRVAALFPALRPGGFQAHYRDE
DGDLVAFSSDEELTMAMSYVKDDIFRIYIKEKKECRRDHRPPCAQEAPRNMVHPNVICDGCNGPVVGTRY
KCSVCPDYDLCSVCEGKGLHRGHTKLAFPSPFGHLSEGFSHSRWLRKVKHGHFGWPGWEMGPPGNWSPRP
PRAGEARPGPTAESASGPSEDPSVNFLKNVGESVAAALSPLGIEVDIDVEHGGKRSRLTPVSPESSSTEE
KSSSQPSSCCSDPSKPGGNVEGATQSLAEQMRKIALESEGRPEEQMESDNCSGGDDDWTHLSSKEVDPST
GELQSLQMPESEGPSSLDPSQEGPTGLKEAALYPHLPPEADPRLIESLSQMLSMGFSDEGGWLTRLLQTK
NYDIGAALDTIQYSKHPPPL
```

Position	Sequence	Ubiquitination	Score
7	MASLTVKAYLLGKEDAA	N	0.42
13	**SLTVKAYLLGKEDAAREIRRF**	**Y**	**0.70**
91	EELTMAMSYVKDDIFRIYIKE	N	0.41
100	VKDDIFRIYIKEKKECRRDHR	N	0.27
102	DDIFRIYIKEKKECRRDHRPP	N	0.25
103	DIFRIYIKEKKECRRDHRPPC	N	0.22
141	CNGPVVGTRYKCSVCPDYDLC	N	0.09
157	DYDLCSVCEGKGLHRGHTKLA	N	0.17
165	**EGKGLHRGHTKLAFPSPFGHL**	**Y**	**0.55**
187	**EGFSHSRWLRKVKHGHFGWPG**	**Y**	**0.54**
189	FSHSRWLRKVKHGHFGWPGWE	N	0.36
238	**PSEDPSVNFLKNVGESVAAAL**	**Y**	**0.55**
264	**EVDIDVEHGGKRSRLTPVSPE**	**Y**	**0.61**
281	VSPESSSTEEKSSSQPSSCCS	N	0.09
295	QPSSCCSDPSKPGGNVEGATQ	N	0.08
313	ATQSLAEQMRKIALESEGRPE	N	0.50
344	**GDDDWTHLSSKEVDPSTGELQ**	**Y**	**0.56**
378	DPSQEGPTGLKEAALYPHLPP	N	0.45
420	**GGWLTRLLQTKNYDIGAALDT**	**Y**	**0.65**
435	**GAALDTIQYSKHPPPL**	**Y**	**0.63**

Fig. 2 ESA-UbiSite HTML output of SQSTM1.fa

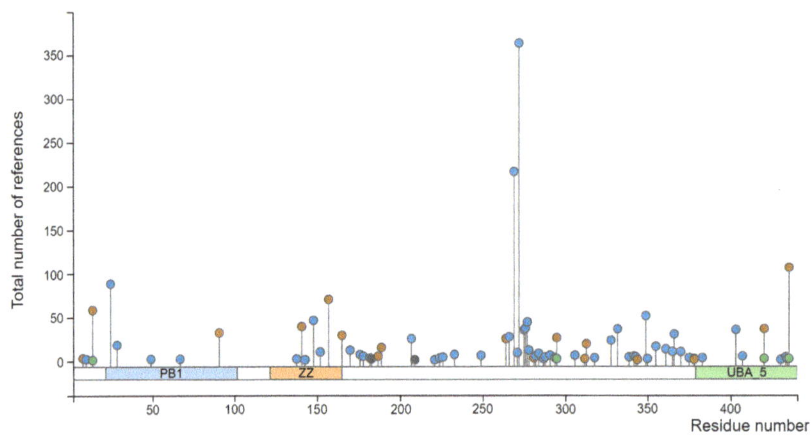

Fig. 3 Phosphosite.org identified PTMs on protein Sequestosome-1/p62 (encoded by SQSTM1) as reported in the literature, including ubiquitylation (brown). Of the 17 reported ubiquitylation sites on Sequestosome-1/p62 (as of 08/2024), ESA-UbiSite predicted eight of the reported ubiquitylation sites and one unreported site (see Fig. 2 for site prediction). PTM sites: Blue, Phosphorylation; Green, Acetylation; Brown, Ubiquitylation; Gray, Unknown

10. Add UbiSite antibody conjugated and crosslinked to Protein G beads to peptides, incubate for 5 h at 4°C (approximately 100 μg UbiSite antibody for every 10 mg starting material).

11. Wash beads thrice with immunoprecipitation buffer without Triton X-100 added, then thrice more with 150 mM NaCl.
12. Elute purified peptides with 0.1% TFA thrice, each for 5 min.
13. Neutralize peptides with 1 M AmBic, then dilute to 25 mM AmBic.
14. Digest peptides with trypsin overnight at 37°C.
15. *See* **Note 1** for subsequent steps.

3.2.2 Immunoprecipitation of Endogenous Linkage-Specific Ubiquitylated Proteins

Protocol adapted from Lee et al. [63].

1. Prior to lysis, treat cells with proteasome inhibitor (*see* **Note 2**).
2. Lyse cells in NP-40 lysis buffer (1% [v/v] Nonidet P-40 in TBS containing 2 mM EDTA), also containing 10 mM *N*-ethylmaleimide, protease inhibitor, phosphatase inhibitor (*see* **Note 3**).
3. Probe sonicate samples, centrifuge at 15,000 × *g* for 30 min at 4°C to pellet cellular debris, then transfer supernatant to new tubes.
4. Boil supernatant for 10 min at 95°C to denature proteins (*see* **Note 4**).
5. Quantify protein concentration.
6. Immunoprecipitate endogenous ubiquitylated proteins by incubating equal amounts of lysate with linkage-specific antibody (i.e., α-K48) or equivalent amount α-IgG control antibody, rotating end-over-end overnight at 4°C.
7. Capture antibody:protein complexes by adding equivalent amount of protein A/G magnetic bead slurry to each sample, rotate end-over-end at 4°C for 3 h.
8. Use magnet to capture bead:antibody:protein complexes.
9. Wash bead:antibody:protein complexes three times in NP-40 lysis buffer, using a magnet to separate and discard supernatant each time.
10. Elute proteins from beads in 1× Laemmli sample buffer for 10 min at 95°C for in-gel digestion. If reducing agent is required, add to sample prior to loading.
11. Refer to in-gel tryptic digestion section.

3.2.3 In-Gel Tryptic Digestion

1. Separate proteins by gel electrophoresis at 180 V for 10 min (*see* **Note 5**).
2. Perform Coomassie brilliant blue staining, destain in 25% methanol overnight.

3. Cut gel lanes into approximately 1-mm pieces, destain in 50% methanol in 50 mM AmBic until colorless (*see* **Note 6**).
4. Slowly dry gel pieces with 50% ACN in 50 mM AmBic, followed by 100% ACN.
5. Reduce with 10 mM DTT in 50 mM AmBic.
6. Alkylate with 20 mM IAA in 50 mM AmBic in the dark (*see* **Note 7**).
7. Wash gel pieces in 50 mM AmBic, followed by 50% ACN, and 100% ACN.
8. Digest with trypsin overnight at 37°C.
9. Collect any remaining liquid.
10. Extract remaining peptides with 50% ACN in 2% formic acid, then 100% ACN, add to the remaining liquid in Step 9.
11. Dry samples under vacuum centrifugation at 25°C.
12. Resuspend dried peptides in 0.1% formic acid, bath sonicate.
13. Desalt peptides with C18 OMIX tips.
14. Wash peptides thrice with 0.1% formic acid, then elute with 50% ACN in 0.1% formic acid.
15. Dry under vacuum centrifugation at 25°C.
16. Resuspend dried peptides in 12 μL 0.1% formic acid, bath sonicate, centrifuge at 14,000 × *g* to pellet debris.
17. Load supernatant into glass vials for LC-MS/MS analysis as recently described [8].

3.2.4 Liquid Chromatography–Tandem Mass Spectrometry (LC–MS/MS)

LC–MS/MS of purified ubiquitylated peptides can be carried out on most tandem mass spectrometers with fragmentation capabilities. The settings used for peptide identification are also easily applied to identify enriched ubiquitylated peptides. We have used the following protocol and mass spectrometry settings on a Thermo Q-Exactive Plus orbitrap mass spectrometer coupled to an Ultimate 3000 NanoLC (Tables 3 and 4). Each investigator should evaluate their LC-MS settings depending on the instrument used and their suitable application.

3.3 Computational Analysis to Identify Average Ubiquitin Branch Sites Across Samples from MS Data of In Vitro and In Vivo Samples

This method uses the programming language R to analyze MaxQuant database search outputs from MS data of ubiquitin enriched samples. The code below has been adapted from the R package UPSRM (https://github.com/ruma1974/UPSRM/) [75] for human data analysis. We have provided an example method of how to adapt this analysis in R with published or new datasets. The example provided is from the insoluble fraction (PXD010161) mouse dataset to demonstrate how to modify the code to suit your purpose for both human and mouse ubiquitin protein sequence analysis (*see* **Note 8**). This

Table 3
Example setting parameters for nanoflow liquid chromatography

Nanoflow liquid chromatography	Parameters
Column	**Precolumn:** Acclaim™ PepMap™ C18 Reversed Phase HPLC Column, 5 μm, 0.3 mm × 5 mm
	Analytical column: Acclaim PepMap RSLC column particle size of 2 μm, diameter of 0.075 mm and length of 150 mm (Thermo Fisher Scientific)
Flow rate	300 nL/min
Buffer and gradient	Buffer A: 0.1% (v/v) formic acid in water
	Buffer B: 80% (v/v) acetonitrile, 0.1% (v/v) formic acid in water
	Gradient (0–60 min)
	0–40 min 2%–33% Buffer B
	40–50 min 33%–63% Buffer B
	50–52 min 63%–95% Buffer B
	52–57 min 95% Buffer B
	57–60 min 2% Buffer B

Table 4
Example setting parameters for tandem mass spectrometry

Thermo Q-Exactive Plus	Parameters
Full MS range	400–2000 m/z
Polarity	Positive
Emitter tip and voltage	10 μm (New Objectives), 1.6 kV
Capillary temperature	250°C
Loop count	10
Full MS resolution	70,000
AGC target	1×10^6
Maximum IT	30 ms
dd-MS2 resolution	17,500
AGC target	2×10^4
Maximum IT	50 ms
Isolation width	2.0 m/z
Normalized collision energy	27
Dynamic exclusion	20 s

can be adapted to MS datasets derived from essentially any ubiquitin enrichment strategy. You can download an R script with the code below from the GitLab repository (https://gitlab.com/mq-mnd/grp_williams/ubiquitin_analysis.git). The script will be periodically updated to include minor changes.

Requirements

To begin, download the UBL dataset from this link: (https://github.com/ruma1974/UPSRM/blob/main/data/UBL.rda).

Before proceeding, ensure you have the following R packages installed; each package only needs to be installed once. Once the packages have been installed, load them into R.

```
# Install packages
install.packages("ComplexHeatmap")
install.packages("tidyverse")
install.packages("plyr")

# Load packages
library("ComplexHeatmap")
library("tidyverse")
library("plyr")
```

Set your working directory to the location of your data files on your computer using the setwd() function in R or navigate to Session > Set Working Directory > Choose Directory in the toolbar

```
setwd("/path/to/files/")
```

Load the MaxQuant database search output files into R using the read.delim() function (or a similar function). The files loaded here are named GlyGly (K)Sites.txt and proteinGroups.txt. If a file name contains a space, ensure you include a backslash (\) before the space in the file name so that R reads the file correctly.

```
DFub <- read.delim("GlyGly\ (K)Sites.txt", header
= TRUE)
DFpro <- read.delim("proteinGroups.txt", header =
TRUE)
```

The UBL data are stored in an RData format (with extensions like .rdata or .rda). Use the load() function to load this dataset into R.

```
load("UBL.rda")
```

To choose a group for analysis, first view the available options using the names() function. For this analysis, the E3 group is selected.

```
names(UBL$UPSL)
## [1] "Proteasome" "SUMO"  "DUBs"
## [4] "immunoproteasome" "ProteasomeRegulators"
"ProteasomeActivators"
## [7] "Ub"  "E1"  "E1sumo"
## [10] "E2" "E3"  "Ubl"
ups <- "E3"
```

When exporting MaxQuant data, ensure that "Gene names" are selected and exported in the protein dataset. If "Gene names" are missing from your protein dataset but are included in the Fasta headers, you can add them to the dataset using the following codes.

We have provided the published dataset PXD010161 missing the "Gene names" column as an example to demonstrate this step. Use the following code chunk to extract and add "Gene names" from the Fasta.headers column to your dataset.

```
if (!"Gene.names" %in% colnames(DFpro)) {
  DFpro$Gene.names <- NA
    for (i in seq_len(nrow(DFpro))) {
        fasta_header <- DFpro$Fasta.headers[i]
        # Check for 'GN=' in the header
        if (grepl("GN=", fasta_header)) {
            # Split the header at 'GN=' and extract
gene names
            split_1 <- unlist(strsplit(fasta_header,
"GN="))
            # Extract and combine gene names
            gene_i <- sapply(split_1[-1], function(x)
strsplit(x, " ")[[1]][1])
            gene_combined <- paste(gene_i, collapse
= ";")
            # Assign combined gene names to the Gene.
names column
            DFpro$Gene.names[i] <- gene_combined
        }
    }
}
```

Next, clean your MaxQuant data by removing any unwanted entries, such as reverse decoy database entries and potential contaminants.

```
DFub_sub <- DFub[-c(grep("REV_", DFub$Leading.pro-
teins), grep("CON_", DFub$Leading.proteins)),]
```

Subset the quantitative LFQ data to focus on genes in the UPS group.

```
# Identify rows to keep based on gene names matching
the UPS group
keep_i <- which(sapply(DFpro$Gene.names, func-
tion(genes) {
   any(toupper(unlist(strsplit(genes, ";"))) %in%
toupper(UBL$UPSL[[ups]]))
}))
# Subset the data based on the identified rows and
relevant columns
DFpro_sub <- DFpro[keep_i, grep("LFQ.|Gene.names",
names(DFpro))]
# Calculate the mean for each gene across columns
DFpro_sub <- ddply(DFpro_sub, "Gene.names",
numcolwise(mean))
```

Before visualizing the quantitative LFQ data, rename the columns of DFpro_sub to remove any unnecessary text. For example, columns named LFQ.intensity.in_dg_M1, LFQ.intensity.in_dg_M2, etc., can be renamed as M1, M2, and so on.

```
colnames(DFpro_sub) <- gsub("LFQ.intensity.in_dg_",
"", colnames(DFpro_sub))
```

Create a vector that represents the groups of interest, matching the order of columns in DFpro_sub. For instance, if the columns represent four mutant groups (M1, M2, etc.) and four wild-type groups (W1, W2, etc.), the vector would look like this:

```
Fac <- c("M", "M", "M", "M", "W", "W", "W", "W")
```

Finally, to plot a heatmap with clustering, define a color for each group using the HeatmapAnnotation() function. Ensure the group names match those in your dataset. Then, use the Heatmap() function to create the visualization. See Fig. 4.

```
# Define heatmap parameters
ha <- HeatmapAnnotation(
    Sample = Fac,
    col = list(Sample = c(M = "pink", W =
"dodgerblue")),
    na_col = "black")
# Generate heatmap
Heatmap(as.matrix(log2(DFpro_sub[,2:ncol(DFpro_
sub)] + 1)),
            col = c("turquoise", "orange"),
            name = "LFQ",
            top_annotation = ha,
            show_row_names = FALSE)
```

To compare the average branch topology of ubiquitin between groups, start by defining the group comparisons. For example, to compare the wild-type group with the mutant group, create a list like this:

```
comp <- list(c("M", "W"))
```

If you need to include an additional group for comparison (e.g., group B), simply add it to the list:

```
comp <- list(c("M", "W"), c("M", "B"))
```

Next, define the reference sequences and their corresponding positions (*see* **Notes 8 and 9**):

```
ref_seq <- data.frame(
    position = c(6, 11, 27, 29, 33, 48, 63),
    seq = c(
      "MQIFVK()TLTGK()TITLEVEPSDTIENVK",
      "TLTGK()TITLEVEPSDTIENVK",
      "TITLEVEPSDTIENVK()AK",
      "AK()IQDK",
      "IQDK()EGIPPDQQR",
      "LIFAGK()QLEDGR",
      "TLSDYNIQK()ESTLHLVLR"
    )
)
```

The protein data needs to be reformatted to select GG sequences that match the reference sequences. The following code does this by

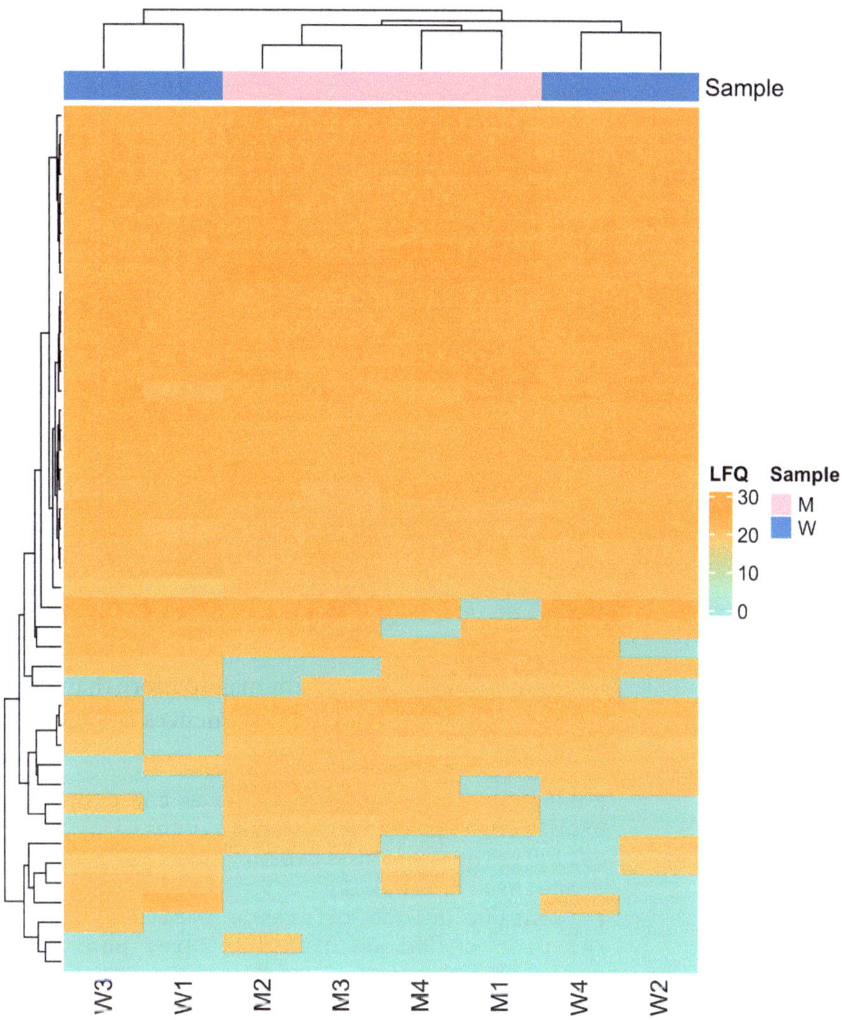

Fig. 4 Heatmap with sample clustering of E3 ligase expression between mutant (M) and wildtype (W) samples in the insoluble fraction

subsetting the reference sequences and removing sequences with a C-terminal K-GG:

```
# Remove values inside parentheses within GG sequences
DFub_sub$seq   <-   gsub("\\s*\\([^\\)]+\\)",   "_",
as.character(DFub_sub$GlyGly..K..Probabilities))
# Select relevant columns
intensity_index <- c(
   which(grepl("Intensity.",       names(DFub_sub))
&!grepl("___", names(DFub_sub))),
    grep("Positions.within.proteins",
names(DFub_sub)),
    grep("GlyGly..K..Probabilities",
names(DFub_sub)),
```

```
    grep("seq", names(DFub_sub))
)
# Select GG sequences that match reference sequences
DFub_sub_2 <- data.frame()
for (i in 1:nrow(ref_seq)) {
    ref_ <- gsub("\\(\\)", "_", ref_seq[i, "seq"])
    matching_rows <- DFub_sub[grep(ref_, DFub_sub-
$seq), intensity_index]
    DFub_sub_2 <- rbind.data.frame(
      DFub_sub_2,
      cbind.data.frame(
        matching_rows,
        data.frame(ref_position = ref_seq[i, "posi-
tion"], ref_seq = ref_seq[i, "seq"])
    )
  )
}
# Remove C-terminal K-GG
DFub_sub_2 <- DFub_sub_2[!endsWith(DFub_sub_2$seq,
"_"),]
# Remove temporary column 'seq'
DFub_sub_2 <- DFub_sub_2[,colnames(DFub_sub_2)!=
"seq"]
```

Some protein sequences might map to more than one reference sequence. To handle this, combine sequences that map to multiple reference sequences:

```
# Extract the first position from the protein sequence
DFub_sub_2$Positions.within.proteins_2    <-   do.
call('rbind',strsplit(DFub_sub_2$Positions.within.
proteins, ";"))[, 1]
# Identify unique reference positions
ref_plot <- unique(DFub_sub_2$ref_position)
# Iterate through each sequence and combine those
mapping to multiple reference positions
for (i in 1:nrow(DFub_sub_2)) {
    pos_i  <-  unique(unlist(DFub_sub_2[i,  c("ref_
position", "Positions.within.proteins_2")]))
    pos_i <- pos_i[order(as.numeric(pos_i))]
    # If multiple positions are found, combine them
and update the reference plot
    if (length(pos_i) > 1) {
        ref_plot <- ref_plot[!(ref_plot %in% pos_i)]
        pos_i <- paste(pos_i, collapse = ";")
        ref_plot <- c(ref_plot, pos_i)
    }
}
# Output the combined positions for joint analysis
ref_plot
## (1) "6;11" "29;33" "48;63"
```

The average ubiquitin linkage comparisons will be conducted jointly for positions K6 & K11, K29 & K33, and K48 & K63. To perform the comparisons and generate boxplots of the results (Fig. 5),

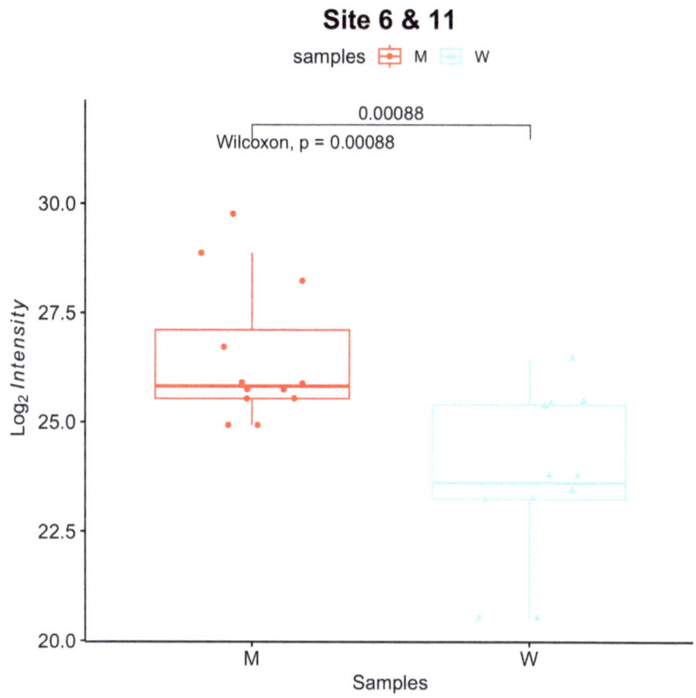

Fig. 5 Example of average ubiquitin branch sites (K6 & K11) output across mutant (M) and wild-type (W) conditions

run the following code. Keep in mind that this code will save the results as a PDF to your computer.

```
# Set the filename for output
filename <- "analysis_2/Q_Ubsites.pdf"
# Plot positions
if (!is.null(Fac)) {
    # Open a PDF device for plotting
    pdf(filename, width = 10, height = 10)
    # Loop through each reference plot position
    for (i in seq_along(ref_plot)) {
        # Extract the current positions
        position_i <- strsplit(ref_plot[i], ";")[[1]]
        # Calculate log2 intensities for the positions
        log2_intensity <- log2(1 + DFub_sub_2[DFub_sub
_2$ref_position %in% position_i, grep("Intensity",
names(DFub_sub_2))])
        names(log2_intensity) <- Fac
        # Transform data for plotting
        ggDF <- log2_intensity %>% pivot_
longer(everything(), names_to = "samples", values_to
= "y")
        # Create a boxplot with jittered points
        p <- ggpubr::ggboxplot(
```

```
            ggDF, x = "samples", y = "y",
            color = "samples", palette = rain-
bow(length(levels(as.factor(Fac)))),
            add = "jitter", shape = "samples",
            main = paste("Site", paste(position_i,
collapse = " & "))
        )
        # Adjust plot aesthetics
        p <- p +
            ggplot2::theme(plot.title = ggplot2::ele-
ment_text(hjust = 0.5)) +
            ggplot2::theme(plot.title = ggplot2::ele-
ment_text(size = 20, face = "bold")) +
            ggplot2::labs(x = "Samples", y =
bquote(~Log(2) ~ italic(Intensity)))
        # Add statistical comparisons if provided
        if (!is.null(comp)) {
          p <- p +
            ggpubr::stat_compare_means(comparisons =
comp) +
            ggpubr::stat_compare_means(label.y =
max(ggDF$y))
        }
        # Print the plot to the PDF
        print(p)
    }
    # Close the PDF device
    dev.off()
}
```

4 Notes

1. Subject samples to high pH fractionation, load onto C18 stage-tips, then fractionate by stepwise elution of 10 mM ammonium hydroxide with increasing concentrations of ACN.
2. Treatment of cells with proteasome inhibitors prior to harvest preserves ubiquitylation modifications [77].
3. Best practice to preserve ubiquitylation on proteins is to add EDTA or EGTA to the lysis buffer to remove traces of heavy metal ions, and NEM or IAA to block the activity of deubiquitinases [77]. The addition of NEM is particularly important in immunoprecipitation experiments where cell lysates are kept under nondenaturing conditions for several hours.
4. Immunoprecipitation is performed under denaturing conditions to dissociate noncovalent cointeractors. An alternative to boiling the lysate is to use a denaturing lysis buffer.
5. As an alternative to in-gel digestion, on-bead digestion can also be performed as in [70].

6. As an alternative to cutting out the full gel lane, the gel can be cut above the heavy chain of the immunoprecipitation antibody to exclude the antibody fragments from MS analysis, as performed in [64].
7. Alkylation of cysteines with IAA produces a 2-acetamidoacetamide covalent adduct with a molecular mass identical to that of the diGLY peptide remaining after digestion with trypsin [78] and can result in false-positive identification of ubiquitylated proteins by mass spectrometry. Alternatively, a less reactive chloroacetamide can be used, such as CAA (chloroacetamine), or alkylation with IAA can be performed at a lower reaction temperature or concentration [79].
8. The ubiquitin peptide sequences, here, are consistent between mouse and human, so the linkage types will not be affected in the provided example.
9. These peptide sequences can be manually modified to account for the number of missed cleavages according to your selected MaxQuant settings.

Acknowledgments

This work was supported by the Macquarie University Research Fellowship, ALS Scholars in Therapeutics Award from the Sean M. Healey and AMG Center for ALS at Massachusetts General Hospital and FightMND (in partnership with ALS Finding a Cure), the Macquarie University Research Excellence Scholarship, Motor Neuron Disease Research Australia (PhD Top-Up Scholarship), National Health and Medical Research Council, Machado-Joseph Disease Foundation, and donations to the Macquarie University MND Research Centre

References

1. Pickart CM (2001) Mechanisms underlying ubiquitination. Annu Rev Biochem 70:503–533. https://doi.org/10.1146/annurev.biochem.70.1.503
2. Weissman AM (2001) Themes and variations on ubiquitylation. Nat Rev Mol Cell Biol 2(3):169–178. https://doi.org/10.1038/35056563
3. Cockram PE, Kist M, Prakash S, Chen S-H, Wertz IE, Vucic D (2021) Ubiquitination in the regulation of inflammatory cell death and cancer. Cell Death Differ 28(2):591–605. https://doi.org/10.1038/s41418-020-00708-5
4. Dang F, Nie L, Wei W (2021) Ubiquitin signaling in cell cycle control and tumorigenesis. Cell Death Differ 28(2):427–438. https://doi.org/10.1038/s41418-020-00648-0
5. Humphreys LM, Smith P, Chen Z, Fouad S, D'Angiolella V (2021) The role of E3 ubiquitin ligases in the development and progression of glioblastoma. Cell Death Differ 28(2):522–537. https://doi.org/10.1038/s41418-020-00696-6
6. Masclef L, Ahmed O, Estavoyer B, Larrivée B, Labrecque N, Nijnik A et al (2021) Roles and mechanisms of BAP1 deubiquitinase in tumor suppression. Cell Death Differ 28(2):606–625. https://doi.org/10.1038/s41418-020-00709-4
7. Beck DB, Werner A, Kastner DL, Aksentijevich I (2022) Disorders of ubiquitylation: unchained inflammation. Nat Rev Rheumatol 18(8):435–447. https://doi.org/10.1038/s41584-022-00778-4

8. Davidson JM, Wu SSL, Rayner SL, Cheng F, Duncan K, Russo C et al (2023) The E3 ubiquitin ligase SCF cyclin F promotes sequestosome-1/p62 insolubility and foci formation and is dysregulated in ALS and FTD pathogenesis. Mol Neurobiol 60:5034–5054. https://doi.org/10.1007/s12035-023-03355-2
9. Rayner SL, Yang S, Farrawell NE, Jagaraj CJ, Cheng F, Davidson JM et al (2022) TDP-43 is a ubiquitylation substrate of the SCFcyclin F complex. Neurobiol Dis 167:105673. https://doi.org/10.1016/j.nbd.2022.105673
10. Hershko A, Ciechanover A (1998) The ubiquitin system. Annu Rev Biochem 67:425–479. https://doi.org/10.1146/annurev.biochem.67.1.425
11. Hochstrasser M (1998) There's the rub: a novel ubiquitin-like modification linked to cell cycle regulation. Genes Dev 12(7):901–907
12. Ciechanover A (2015) The unravelling of the ubiquitin system. Nat Rev Mol Cell Biol 16(5):322–324. https://doi.org/10.1038/nrm3982
13. Wilkinson KD (1997) Regulation of ubiquitin-dependent processes by deubiquitinating enzymes. FASEB J 11(14):1245–1256. https://doi.org/10.1096/fasebj.11.14.9409543
14. Nijman SM, Luna-Vargas MP, Velds A, Brummelkamp TR, Dirac AM, Sixma TK et al (2005) A genomic and functional inventory of deubiquitinating enzymes. Cell 123(5):773–786. https://doi.org/10.1016/j.cell.2005.11.007
15. Hochstrasser M (2006) Lingering mysteries of ubiquitin-chain assembly. Cell 124(1):27–34. https://doi.org/10.1016/j.cell.2005.12.025
16. Emmerich CH, Ordureau A, Strickson S, Arthur JSC, Pedrioli PG, Komander D et al (2013) Activation of the canonical IKK complex by K63/M1-linked hybrid ubiquitin chains. Proc Natl Acad Sci U S A 110(38):15247–15252
17. Kristariyanto YA, Rehman SAA, Campbell DG, Morrice NA, Johnson C, Toth R et al (2015) K29-selective ubiquitin binding domain reveals structural basis of specificity and heterotypic nature of k29 polyubiquitin. Mol Cell 58(1):83–94
18. Meyer H-J, Rape M (2014) Enhanced protein degradation by branched ubiquitin chains. Cell 157(4):910–921
19. Dikic I, Schulman BA (2022) An expanded lexicon for the ubiquitin code. Nat Rev Mol Cell Biol 24:273–287. https://doi.org/10.1038/s41580-022-00543-1
20. Tracz M, Bialek W (2021) Beyond K48 and K63: non-canonical protein ubiquitination. Cell Mol Biol Lett 26(1):1. https://doi.org/10.1186/s11658-020-00245-6
21. McClellan AJ, Laugesen SH, Ellgaard L (2019) Cellular functions and molecular mechanisms of non-lysine ubiquitination. Open Biol 9(9):190147. https://doi.org/10.1098/rsob.190147
22. Wang JR, Huang WL, Tsai MJ, Hsu KT, Huang HL, Ho SY (2017) ESA-UbiSite: accurate prediction of human ubiquitination sites by identifying a set of effective negatives. Bioinformatics 33(5):661–668. https://doi.org/10.1093/bioinformatics/btw701
23. Bürger R (1999) Evolution of genetic variability and the advantage of sex and recombination in changing environments. Genetics 153(2):1055–1069. https://doi.org/10.1093/genetics/153.2.1055
24. Kawashima S, Kanehisa M (2000) AAindex: amino acid index database. Nucleic Acids Res 28(1):374. https://doi.org/10.1093/nar/28.1.374
25. Chen Z, Zhou Y, Zhang Z, Song J (2015) Towards more accurate prediction of ubiquitination sites: a comprehensive review of current methods, tools and features. Brief Bioinform 16(4):640–657. https://doi.org/10.1093/bib/bbu031
26. Kumar VS, Vellaichamy A (2019) Sequence and structure-based characterization of ubiquitination sites in human and yeast proteins using Chou's sample formulation. Proteins 87(8):646–657. https://doi.org/10.1002/prot.25689
27. Liu Y, Li A, Zhao X-M, Wang M (2021) DeepTL-Ubi: a novel deep transfer learning method for effectively predicting ubiquitination sites of multiple species. Methods 192:103–111. https://doi.org/10.1016/j.ymeth.2020.08.003
28. Luo Y, Jiang J, Zhu J, Huang Q, Li W, Wang Y et al (2022) A Caps-Ubi model for protein ubiquitination site prediction. Front Plant Sci 13:884903. https://doi.org/10.3389/fpls.2022.884903
29. Willems P (2022) Exploring posttranslational modifications with the plant PTM viewer. In: Klemenčič M, Stael S, Huesgen PF (eds) Plant proteases and plant cell death: methods and protocols. Springer US, New York, NY, pp 285–296. https://doi.org/10.1007/978-1-0716-2079-3_23
30. Walsh I, Di Domenico T, Tosatto SCE (2014) RUBI: rapid proteomic-scale prediction of lysine ubiquitination and factors influencing predictor performance. Amino Acids 46(4):853–862. https://doi.org/10.1007/s00726-013-1645-3
31. Huang C-H, Su M-G, Kao H-J, Jhong J-H, Weng S-L, Lee T-Y (2016) UbiSite: incorporating two-layered machine learning method

with substrate motifs to predict ubiquitin-conjugation site on lysines. BMC Syst Biol 10(1):S6. https://doi.org/10.1186/s12918-015-0246-z
32. Lee TY, Chen SA, Hung HY, Ou YY (2011) Incorporating distant sequence features and radial basis function networks to identify ubiquitin conjugation sites. PLoS One 6(3):e17331. https://doi.org/10.1371/journal.pone.0017331
33. Tung C-W, Ho S-Y (2008) Computational identification of ubiquitylation sites from protein sequences. BMC Bioinformatics 9(1):310. https://doi.org/10.1186/1471-2105-9-310
34. Wang X, Yan R, Wang Y (2021) Computational identification of human ubiquitination sites using convolutional and recurrent neural networks. Mol Omics 17(6):948–955. https://doi.org/10.1039/d0mo00183j
35. Radivojac P, Vacic V, Haynes C, Cocklin RR, Mohan A, Heyen JW et al (2010) Identification, analysis, and prediction of protein ubiquitination sites. Proteins 78(2):365–380. https://doi.org/10.1002/prot.22555
36. Chen Z, Chen YZ, Wang XF, Wang C, Yan RX, Zhang Z (2011) Prediction of ubiquitination sites by using the composition of k-spaced amino acid pairs. PLoS One 6(7):e22930. https://doi.org/10.1371/journal.pone.0022930
37. Chen Z, Zhou Y, Song J, Zhang Z (2013) hCKSAAP_UbSite: improved prediction of human ubiquitination sites by exploiting amino acid pattern and properties. Biochim Biophys Acta 1834(8):1461–1467. https://doi.org/10.1016/j.bbapap.2013.04.006
38. Cai Y, Huang T, Hu L, Shi X, Xie L, Li Y (2012) Prediction of lysine ubiquitination with mRMR feature selection and analysis. Amino Acids 42(4):1387–1395. https://doi.org/10.1007/s00726-011-0835-0
39. Chen X, Qiu JD, Shi SP, Suo SB, Huang SY, Liang RP (2013) Incorporating key position and amino acid residue features to identify general and species-specific ubiquitin conjugation sites. Bioinformatics 29(13):1614–1622. https://doi.org/10.1093/bioinformatics/btt196
40. Qiu W-R, Xiao X, Lin W-Z, Chou K-C (2015) iUbiq-Lys: prediction of lysine ubiquitination sites in proteins by extracting sequence evolution information via a gray system model. J Biomol Struct Dyn 33(8):1731–1742. https://doi.org/10.1080/07391102.2014.968875
41. Wang C, Tan X, Tang D, Gou Y, Han C, Ning W et al (2022) GPS-Uber: a hybrid-learning framework for prediction of general and E3-specific lysine ubiquitination sites. Brief Bioinform 23(2):bbab574. https://doi.org/10.1093/bib/bbab574
42. Wang D, Liu D, Yuchi J, He F, Jiang Y, Cai S et al (2020) MusiteDeep: a deep-learning based webserver for protein post-translational modification site prediction and visualization. Nucleic Acids Res 48(W1):W140–W146. https://doi.org/10.1093/nar/gkaa275
43. Larance M, Lamond AI (2015) Multidimensional proteomics for cell biology. Nat Rev Mol Cell Biol 16(5):269–280. https://doi.org/10.1038/nrm3970
44. Peng J, Schwartz D, Elias JE, Thoreen CC, Cheng D, Marsischky G et al (2003) A proteomics approach to understanding protein ubiquitination. Nat Biotechnol 21(8):921–926. https://doi.org/10.1038/nbt849
45. Xu G, Paige JS, Jaffrey SR (2010) Global analysis of lysine ubiquitination by ubiquitin remnant immunoaffinity profiling. Nat Biotechnol 28(8):868–873. https://doi.org/10.1038/nbt.1654
46. Kim W, Bennett EJ, Huttlin EL, Guo A, Li J, Possemato A et al (2011) Systematic and quantitative assessment of the ubiquitin-modified proteome. Mol Cell 44(2):325–340. https://doi.org/10.1016/j.molcel.2011.08.025
47. Lee KA, Hammerle LP, Andrews PS, Stokes MP, Mustelin T, Silva JC et al (2011) Ubiquitin ligase substrate identification through quantitative proteomics at both the protein and peptide levels. J Biol Chem 286(48):41530–41538. https://doi.org/10.1074/jbc.M111.248856
48. Fulzele A, Bennett EJ (2018) Ubiquitin diGLY proteomics as an approach to identify and quantify the ubiquitin-modified proteome. Methods Mol Biol 1844:363–384. https://doi.org/10.1007/978-1-4939-8706-1_23
49. Akimov V, Barrio-Hernandez I, Hansen SVF, Hallenborg P, Pedersen A-K, Bekker-Jensen DB et al (2018) UbiSite approach for comprehensive mapping of lysine and N-terminal ubiquitination sites. Nat Struct Mol Biol 25(7):631–640. https://doi.org/10.1038/s41594-018-0084-y
50. McDonald WH, Ohi R, Miyamoto DT, Mitchison TJ, Yates JR (2002) Comparison of three directly coupled HPLC MS/MS strategies for identification of proteins from complex mixtures: single-dimension LC-MS/MS, 2-phase MudPIT, and 3-phase MudPIT. Int J Mass Spectrom 219(1):245–251. https://doi.org/10.1016/S1387-3806(02)00563-8
51. Steger M, Demichev V, Backman M, Ohmayer U, Ihmor P, Müller S et al (2021) Time-resolved in vivo ubiquitinome profiling by DIA-MS reveals USP7 targets on a proteome-wide scale. Nat Commun 12(1):5399. https://doi.org/10.1038/s41467-021-25454-1
52. Olsen JV, Mann M (2013) Status of large-scale analysis of post-translational modifications

by mass spectrometry. Mol Cell Proteomics 12(12):3444–3452. https://doi.org/10.1074/mcp.O113.034181
53. Kirkpatrick DS, Weldon SF, Tsaprailis G, Liebler DC, Gandolfi AJ (2005) Proteomic identification of ubiquitinated proteins from human cells expressing His-tagged ubiquitin. Proteomics 5(8):2104–2111. https://doi.org/10.1002/pmic.200401089
54. Akimov V, Olsen LCB, Hansen SVF, Barrio-Hernandez I, Puglia M, Jensen SS et al (2018) StUbEx PLUS—a modified stable tagged ubiquitin exchange system for peptide level purification and in-depth mapping of ubiquitination sites. J Proteome Res 17(1):296–304. https://doi.org/10.1021/acs.jproteome.7b00566
55. Kirkpatrick DS, Hathaway NA, Hanna J, Elsasser S, Rush J, Finley D et al (2006) Quantitative analysis of in vitro ubiquitinated cyclin B1 reveals complex chain topology. Nature Cell Biol 8(7):700–710. https://doi.org/10.1038/ncb1436
56. Hjerpe R, Aillet F, Lopitz-Otsoa F, Lang V, England P, Rodriguez MS (2009) Efficient protection and isolation of ubiquitylated proteins using tandem ubiquitin-binding entities. EMBO Rep 10(11):1250–1258. https://doi.org/10.1038/embor.2009.192
57. Hospenthal MK, Mevissen TET, Komander D (2015) Deubiquitinase-based analysis of ubiquitin chain architecture using ubiquitin chain restriction (UbiCRest). Nat Protoc 10(2):349–361. https://doi.org/10.1038/nprot.2015.018
58. Hansen FM, Tanzer MC, Brüning F, Bludau I, Stafford C, Schulman BA et al (2021) Data-independent acquisition method for ubiquitinome analysis reveals regulation of circadian biology. Nat Commun 12(1):254. https://doi.org/10.1038/s41467-020-20509-1
59. Ohtake F, Tsuchiya H, Tanaka K, Saeki Y (2019) Chapter five–methods to measure ubiquitin chain length and linkage. In: Hochstrasser M (ed) Methods in enzymology. vol 618. Academic Press, Cambridge, MA, pp 105–133. https://doi.org/10.1016/bs.mie.2018.12.019
60. Udeshi ND, Mani DC, Satpathy S, Fereshetian S, Gasser JA, Svinkina T et al (2020) Rapid and deep-scale ubiquitylation profiling for biology and translational research. Nat Commun 11(1):359. https://doi.org/10.1038/s41467-019-14175-1
61. Rose CM, Isasa M, Ordureau A, Prado MA, Beausoleil SA, Jedrychowski MP et al (2016) Highly multiplexed quantitative mass spectrometry analysis of ubiquitylomes. Cell Syst 3(4):395–403.e4. https://doi.org/10.1016/j.cels.2016.08.009
62. Stes E, Laga M, Walton A, Samyn N, Timmerman E, De Smet I et al (2014) A COFRADIC protocol to study protein ubiquitination. J Proteome Res 13(6):3107–3113. https://doi.org/10.1021/pr4012443
63. Lee A, Rayner SL, Gwee SSL, De Luca A, Shahheydari H, Sundaramoorthy V et al (2018) Pathogenic mutation in the ALS/FTD gene, CCNF, causes elevated Lys48-linked ubiquitylation and defective autophagy. Cell Mol Life Sci 75(2):335–354. https://doi.org/10.1007/s00018-017-2632-8
64. Chadchankar J, Korboukh V, Conway LC, Wobst HJ, Walker CA, Doig P et al (2019) Inactive USP14 and inactive UCHL5 cause accumulation of distinct ubiquitinated proteins in mammalian cells. PLoS One 14(11):e0225145. https://doi.org/10.1371/journal.pone.0225145
65. Schwertman P, Bezstarosti K, Laffeber C, Vermeulen W, Demmers JAA, Marteijn JA (2013) An immunoaffinity purification method for the proteomic analysis of ubiquitinated protein complexes. Anal Biochem 440(2):227–236. https://doi.org/10.1016/j.ab.2013.05.020
66. Yang H, Li N, Chen L, Zhou L, Zhou Y, Liu J et al (2023) Ubiquitinomics revealed disease- and stage-specific patterns relevant for the 3PM approach in human sigmoid colon cancers. EPMA J 14(3):503–525. https://doi.org/10.1007/s13167-023-00328-2
67. Zhan X, Lu M, Yang L, Yang J, Zhan X, Zheng S et al (2022) Ubiquitination-mediated molecular pathway alterations in human lung squamous cell carcinomas identified by quantitative ubiquitinomics. Front Endocrinol (Lausanne) 13:970843. https://doi.org/10.3389/fendo.2022.970843
68. Lu M, Chen W, Zhuang W, Zhan X (2020) Label-free quantitative identification of abnormally ubiquitinated proteins as useful biomarkers for human lung squamous cell carcinomas. EPMA J 11(1):73–94. https://doi.org/10.1007/s13167-019-00197-8
69. Udeshi ND, Mertins P, Svinkina T, Carr SA (2013) Large-scale identification of ubiquitination sites by mass spectrometry. Nat Protoc 8(10):1950–1960. https://doi.org/10.1038/nprot.2013.120
70. Trulsson F, Akimov V, Robu M, van Overbeek N, Berrocal DAP, Shah RG et al (2022) Deubiquitinating enzymes and the proteasome regulate preferential sets of ubiquitin substrates. Nat Commun 13(1):2736. https://doi.org/10.1038/s41467-022-30376-7
71. Jeon HB, Choi ES, Yoon JH, Hwang JH, Chang JW, Lee EK et al (2007) A proteomics approach to identify the ubiquitinated proteins

in mouse heart. Biochem Biophys Res Commun 357(3):731–736. https://doi.org/10.1016/j.bbrc.2007.04.015
72. Oshikawa K, Matsumoto M, Oyamada K, Nakayama K I. (2012) Proteome-wide identification of ubiquitylation sites by conjugation of engineered lysine-less ubiquitin. J Proteome Res 11(2):796–807. https://doi.org/10.1021/pr200668y
73. Tagwerker C, Flick K, Cui M, Guerrero C, Dou Y, Auer B et al (2006) A tandem affinity tag for two-step purification under fully denaturing conditions: application in ubiquitin profiling and protein complex identification combined with in vivocross-linking. Mol Cell Proteomics 5(4):737–748. https://doi.org/10.1074/mcp.M500368-MCP200
74. Gao Y, Li Y, Zhang C, Zhao M, Deng C, Lan Q et al (2016) Enhanced purification of ubiquitinated proteins by engineered tandem hybrid ubiquitin-binding domains (ThUBDs). Mol Cell Proteomics 15(4):1381–1396. https://doi.org/10.1074/mcp.o115.051839
75. Matthiesen R, Rodriguez MS, Carvalho AS (2023) A computational tool for analysis of mass spectrometry data of ubiquitin-enriched samples. In: Manuel SR, Rosa B (Eds.) The ubiquitin code. Methods in molecular biology. 2602. Springer US, New York, NY, pp 205–214. https://doi.org/10.1007/978-1-0716-2859-1_15
76. Sap KA, Guler AT, Bezstarosti K, Bury AE, Juenemann K, Demmers JA et al (2019) Global proteome and ubiquitinome changes in the soluble and insoluble fractions of Q175 huntington mice brains. Mol Cell Proteomics 18(9):1705–1720. https://doi.org/10.1074/mcp.RA119.001486
77. Emmerich CH, Cohen P (2015) Optimising methods for the preservation, capture and identification of ubiquitin chains and ubiquitylated proteins by immunoblotting. Biochem Biophys Res Commun 466(1):1–14. https://doi.org/10.1016/j.bbrc.2015.08.109
78. Nielsen ML, Vermeulen M, Bonaldi T, Cox J, Moroder L, Mann M (2008) Iodoacetamide-induced artifact mimics ubiquitination in mass spectrometry. Nat Methods 5(6):459–460. https://doi.org/10.1038/nmeth0608-459
79. Sylvestersen KB, Young C, Nielsen ML (2013) Advances in characterizing ubiquitylation sites by mass spectrometry. Curr Opin Chem Biol 17(1):49–58. https://doi.org/10.1016/j.cbpa.2012.12.009

Check for updates

Chapter 5

Extensive Characterization of N-Glycosylation from Major Histocompatibility Complex with Hydrophilic Interaction Chromatography and Mass Spectrometry

Rui Chen

Abstract

The major histocompatibility complex (MHC) is an essential part of the adaptive immune systems that present antigenic peptides at the cell surface for recognition and activation of circulating T lymphocytes, which is vital for our body to recognize and respond to foreign pathogens, including viruses, bacteria, and cancerous cells. Human leukocyte antigen (HLA) genes encode proteins located on the surface of most cells in the body, and studies have shown that protein glycosylation plays a significant role in the folding of human leukocyte antigen (HLA) proteins, loading of peptides, and forming of HLA-peptide complex. In the meantime, some of those peptides presented by the MHC system are also glycosylated. Thus, glycosylation has a huge impact on cellular uptake, proteolytic processing, presentation by MHC, and subsequent T-cell priming. However, glycosylation characterization of the whole MHC system has often been neglected during conventional immunopeptidomics analysis due to the lower abundance of glycosylated peptides and increased complexity of the mass spectrum. To tackle this problem, we utilized hydrophilic interaction chromatography to enrich glycopeptides from both peptides bound to HLA and digest of HLA proteins that obtained from immunoprecipitation of MHC complex from cell lysate. Enriched glycopeptides are analyzed by mass spectrometry that enables the characterization of glycosylation of MHC by database search.

Keywords Human leucocyte antigen, Major histocompatibility complex, Glycosylation, LC-MS/MS

1 Introduction

The cell-mediated adaptive immune response is regulated by the major histocompatibility complex (MHC) or human leukocyte antigen (HLA) in humans [1]. HLA molecules are cell surface glycoproteins whose primary function is to present endogenous and exogenous antigens to T lymphocytes for recognition and response [2]. The HLA molecules that present antigen to T lymphocytes are divided into two main classes: (1) HLA class I and (2) HLA class II molecules. HLA class I molecules play an essential role in the immune defense against intracellular pathogens, whereas HLA class

Xu Zhang et al. (eds.), *Protein Post-Translational Modifications*, Methods in Molecular Biology, vol. 3018,
https://doi.org/10.1007/978-1-0716-5166-7_5,

II molecules are predominantly involved in displaying peptides from extracellular pathogens [3]. All HLA genes carry *N*-glycosylation sites but the nature of the glycans attached have not been well characterized. Cellular protein *N*-glycosylation is a complex multistep process in which many enzymes in the endoplasmic reticulum (ER) and Golgi are involved [4]. Therefore, HLA complexes will undergo their glycosylation while traversing through the ER and Golgi compartments, which happens during the protein folding and antigen loading process [5, 6]. Thus, characterization of the glycosylation pattern can help understand the role of glycans on the antigen presentation. However, there was only limited research on the glycans attached to HLA proteins at proteomics scale [7].

On the other hand, glycosylation has also been identified from the peptides bound to HLA proteins, termed as MHC associated peptides (MAP) or immunopeptides. Despite the advances in mass spectrometry techniques that enables the sensitive detection of low abundant glycosylated immunopeptides and new bioinformatic tools that correlate MS/MS spectra with peptide sequence and glycan composition [8], the knowledge about the structure and immunogenicity of glycosylated MAPs is still limited. Aberrant glycosylation has been recognized as a diagnostic biomarker in cancer [9] and many pathogens also have distinct glycosylation patterns [10, 11]. Identification of glycosylated peptides from MHC would be an efficient and straightforward approach in finding candidates for vaccine development for both immunotherapy of cancer and infectious disease caused by pathogen infection since glycosylation could be more specific in triggering immune response from cytotoxic T cells [12]. Although some glycosylated MAPs have been detected directly from the liquid chromatography tandem mass spectrometry (LC-MS/MS) analysis of whole immunopeptidome that was isolated

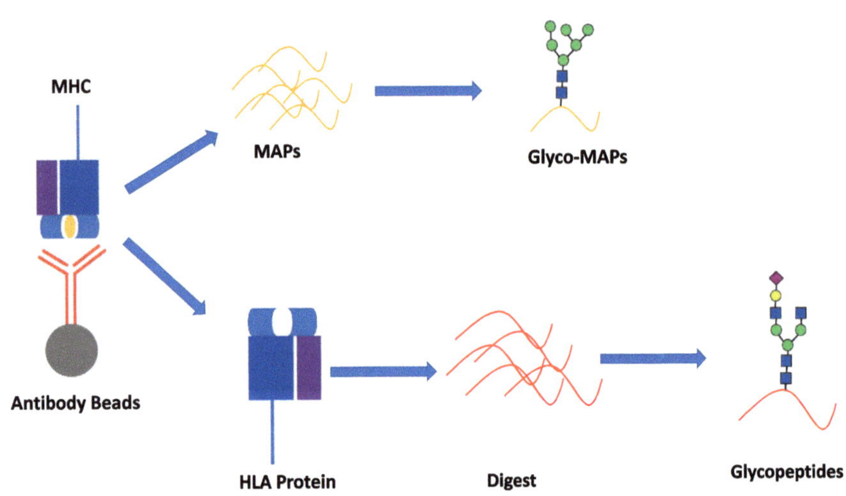

Fig. 1 Scheme for the workflow of extensive characterization of glycosylation from MHC with HILIC SPE

Table 1
List of glycosylated class II MAPs identified from one experiment

Gene	Peptide	Probability	Glycan composition (MW)
ARSA	LLPTLAALAGAPLPN#VT	1	Hex[5]HexNAc[2](1216.422800)
ST8SIA4	N#DSVLWIPAFMVKGGEK	1	Hex[5]HexNAc[2](1216.422800)
LGMN	DHGSTGILVFPNEDLHVKDLNET	1	Hex[4]HexNAc[2](1054.370000)
			Hex[6]HexNAc[2](1378.475600)
	DHGSTGILVFPNEDLHVKDLN#	1	Hex[5]HexNAc[2](1216.422800)
	DHGSTGILVFPNEDLHVKDLN#E	1	Hex[5]HexNAc[2](1216.422800)
LIPA	N#VSEIISYWGFPSEE	1	Hex[2]HexNAc[2](730.264300)
			Hex[3]HexNAc[2](892.317200)
			Hex[4]HexNAc[2](1054.370000)
			Hex[5]HexNAc[2](1216.422800)
	N#VSEIISYWGFPSE	1	Hex[5]HexNAc[2](1216.422800)
	N#VSEIISYWGFPSEEY	1	Hex[5]HexNAc[2](1216.422800)
LAMP1	DPAFKAAN#GSLRALQ	1	Hex[4]HexNAc[2](1054.370000)
			Hex[5]HexNAc[2](1216.422800)
	AN#GSLRALQATVGNSYK	1	Hex[5]HexNAc[2](1216.422800)
	DPAFKAAN#GSLRA	0.99	Hex[5]HexNAc[2](1216.422800)
	LPDARDPAFKAAN#GSLRAL	0.87	Hex[5]HexNAc[2](1216.422800)
	IAN#FTKAASTYSIDS	1	Hex[2]HexNAc[2](730.264300)
			Hex[3]HexNAc[2](892.317200)
NAPSA	AN#GTKFAIQYGTGRVDG	1	dHex[1]Hex[3]HexNAc[2](1038.375100)
	QAN#GTKFAIQYGTGRVDG	0.82	dHex[1]Hex[3]HexNAc[2](1038.375100)
	N#GTKFAIQYGTGRVDG	1	Hex[2]HexNAc[2](730.264300)
			Hex[2]HexNAc[2]dHex[1](876.322300)
	N#GTKFAIQYGTGRVD	1	Hex[2]HexNAc[2](730.264300)
			Hex[3]HexNAc[2](892.317200)
PSAP	N#STFVQALVEHVK	1	Hex[2]HexNAc[2]dHex[1](876.322300)
	RTN#STFVQALVEHVKEE	0.92	Hex[2]HexNAc[2]dHex[1](876.322300)
	TN#STFVQALVEHVKEE	1	Hex[2]HexNAc[2]dHex[1](876.322300)
			dHex[1]Hex[3]HexNAc[2](1038.375100)
			HexNAc[1]dHex[1](349.137200)
			HexNAc(203.079300)
	N#STFVQALVEHVKE	1	Hex[2]HexNAc[2]dHex[1](876.322300)
			HexNAc(203.079300)

Table 2
List of glycosylated peptides from HLA proteins identified

Gene	Peptide	Probability	Glycan composition (MW)
HLA-DRA	FTPPVVN#VTWLR	1	Hex[6]HexNAc[2](1378.475600)
			Hex[7]HexNAc[2](1540.528500)
			Hex[8]HexNAc[2](1702.581300)
			Hex[9]HexNAc[2](1864.634100)
HLA-DRB1	FLEQVKHECHFFNGTER	1	dHex[2]Hex[2]HexNAc[6]Sulf[1](1914.654500)
			Hex[8]HexNAc[2](1702.581300)
			Hex[6]HexNAc[2](1378.475600)
			dHex[1]Hex[4]HexNAc[4](1606.586600)
			dHex[1]Hex[5]HexNAc[5](1971.718800)
			dHex[1]Hex[5]HexNAc[4]NeuAc[1](2059.734900)
			dHex[1]Hex[5]HexNAc[4]NeuAc[2](2350.830300)
HLA-DRB1	FLEYSTSECHFFNGTER	1	Hex[6]HexNAc[2](1378.475600)
			Hex[7]HexNAc[2](1540.528500)
			Hex[8]HexNAc[2](1702.581300)
			dHex[1]Hex[4]HexNAc[3]NeuAc[1](1694.602700)
			dHex[1]Hex[5]HexNAc[4](1768.639500)
			dHex[2]Hex[5]HexNAc[4](1914.697400)
			dHex[1]Hex[5]HexNAc[4]NeuAc[1](2059.734900)
			dHex[1]Hex[5]HexNAc[5](1971.718800)
			dHex[2]Hex[4]HexNAc[3](1549.565200)
HLA-B:51	IALRYYN#QSEAGSHTWQTMYGCDVGPDGR	1	Hex[5]HexNAc[3](1419.502200)

Table 2
continued

Gene	Peptide	Probability	Glycan composition (MW)
	YYN#QSEAGSHTWQTMYGCDVGPDGR	1	Hex[6]HexNAc[3](1581.555000)
HLA-C:07	NLRGYYNQSEDGSHTLQR	1	Hex[6]HexNAc[2](1378.475600)
			Hex[7]HexNAc[2](1540.528500)
			Hex[8]HexNAc[2](1702.581300)
			Hex[9]HexNAc[2](1864.634100)
			dHex[1]Hex[2]HexNAc[4]NeuAc[2](1864.671800)
HLA-C:15	GYYN#QSEAGSHIIQR	1	Hex[5]HexNAc[2](1216.422800)
			Hex[6]HexNAc[2](1378.475600)
			Hex[7]HexNAc[2](1540.528500)
			Hex[8]HexNAc[2](1702.581300)
			Hex[9]HexNAc[2](1864.634100)
			dHex[1]Hex[2]HexNAc[4]NeuAc[2](1864.671800)
			dHex[1]Hex[3]HexNAc[4](1444.533800)
			dHex[1]Hex[3]HexNAc[5](1647.613200)
			dHex[1]Hex[3]HexNAc[5](1647.613200)
			dHex[1]Hex[5]HexNAc[4]NeuAc[1](2059.734900)
			dHex[1]Hex[5]HexNAc[4]NeuAc[2](2350.830300)
			dHex[1]Hex[5]HexNAc[5](1971.718800)
			dHex[2]Hex[5]HexNAc[4](1914.697400)

from cells or tissues with immunoprecipitation [13, 14], the analysis of glycopeptides are still challenging since peptides with both of these posttranslational modifications usually have lower abundance and ionization efficiency than those without modifications [15]. Furthermore, database searching for the whole immunopeptidome might increase the false discovery rate (FDR) of identifications. Thus, enrichment of glycosylated MAPs prior to LC-MS/MS analysis is crucial to increase the specificity and accuracy of their identification.

Herein, we present a mass spectrometric approach to comprehensively characterize the glycosylation of the MHC system, including HLA proteins and their associated peptides (Fig. 1). HLA complexes are isolated by immunoprecipitation and captured by a C18 cartridge as described in most conventional immunopeptidomics workflows. Following the elution of presented antigenic peptides from the HLA complex for immunopeptidomic analysis, the more hydrophobic HLA proteins can be recovered from the cartridge by a second elution step with higher percentage of organic solvent. These HLA proteins can then be digested and intact glycopeptides of HLAs are further enriched by hydrophilic interaction chromatography solid phase extraction (HILIC SPE) [16] and analyzed by LC-MS/MS. By database searching with MSFraggerGlyco [17], both glycosylated MAPs (Table 1) and HLA peptides (Table 2) can be identified which provides with more information to the study the impact of expression and post-translational modification (PTM) from each allele on the antigen presentation.

2 Materials

Prepare all solutions using Millipore water and reagents of analytical grade for cell lysis and immunoprecipitation. For MAPs and HLA proteins purification, HLA protein digestion and glycopeptides enrichment, use liquid chromatography mass spectrometry (LC-MS) grade water and solvents to prepare all solutions. Prepare and store all reagents at room temperature (unless indicated otherwise).

2.1 Cell Culture

1. Cells: human lung cancer cell line Calu-3 obtained from the American Type Culture Collection (ATCC) or any other cell line.
2. Cell culture media: Dulbecco's modified eagle medium (DMEM) media with 10% fetal bovine serum and penicillin–streptomycin solution (final concentration of 1000 U/ml and 1 mg/ml, respectively), store at 4°C.

2.2 Immunoprecipitation

1. Cell lysis buffer: 1.5% Triton X-100, 0.5% CHAPS, and 200 mM sodium chloride in 20 mM Tris-HCl (pH 7.4), 1×protease inhibitor cocktail (add before each cell lysis), store at 4°C.

2. Antibodies: pan anti-HLA monoclonal antibody (W6/32) for immunoprecipitation of MHC class I complex and anti-HLA-DR monoclonal antibody (L243) for immunoprecipitation of MHC class II complex, store at 4°C (*see* **Note 1**).
3. Aldehyde-activated crosslinked 4% beaded agarose (amino-link plus, Thermo Scientific), store at 4°C.
4. Phosphate buffered saline (PBS): 8.1 mM Na_2HPO_4, 1.5 mM KH_2PO_4, 137 mM NaCl, and 2.7 mM KCl, pH = 7.4, store at 4°C.
5. Immunoprecipitation washing buffer: 200 mM sodium chloride, 20 mM Tris-HCl (pH 7.4), store at 4°C
6. MHC eluting solution: 10% acetic acid.

2.3 Purification and Desalting of MHC Peptides and HLA Protein

1. Acidifying solution: 10% trifluoroacetic acid (TFA).
2. MAP elution solution: 30% acetonitrile (ACN) with 0.1% TFA.
3. HLA protein elution solution: 80% ACN with 0.1% TFA.
4. Washing solution: 0.1% TFA.
5. C18 solid phase extraction (SPE) cartridge (100 mg of resin).

2.4 Digestion of HLA Protein

1. Protein denaturing solution: 8 M urea in 100 mM Tris-HCl, pH 7.4.
2. Reduction agent: 1 M dithiothreitol (DTT) in water, store at −20°C.
3. Alkylation agent: 1 M iodoacetamide (IAA) in water, store at −20°C.
4. Diluting solution: 50 mM Tris-HCl, pH 7.4.
5. Trypsin: Sequence or proteomics grade trypsin from Promega, store at −20°C.

2.5 Desalting of HLA Protein Digest

1. Acidifying solution: 10% TFA.
2. Digest elution solution: 50% ACN with 0.1% TFA.
3. Washing solution: 0.1% TFA.
4. C18 solid phase extraction (SPE) cartridge.

2.6 Enrichment of Glycosylated Peptides

1. Polyhydroxylethyl A beads (5 μm).
2. Toptip EMPTY tips with frits.
3. HILIC loading and washing solution: 80% ACN, 1% TFA.
4. HILIC elution solution: 30% ACN with 0.1% TFA.

2.7 LC-MS/MS Analysis

1. Mass spectrometer: mass spectrometer that is capable of acquiring MS/MS spectrum at high resolution, such as Orbitrap Astral, Orbitrap Exploris 480, Bruker TIMS TOF, AB SCIEX 6600 TripleTOF, etc.

2. Nano-LC system: ultra-high performance liquid chromatography (UHPLC) systems such as Thermo Ultimate 3000, Waters M class, Evosep One, etc.
3. Reversed phase columns: analytical nano-LC column with 75 μ m i.d. that is packed with 1.9 μm (or under) C18 beads. It can be purchased (such as Waters nanoEase M/Z Peptide BEH C18 Column, Thermo Acclaim™ PepMap™ 100 C18 HPLC columns) or packed in-house with C18 beads of similar grade. Trapping column with 300 μm i.d. that is packed with 5 μm (or under) C18 beads. It can be purchased (such as Waters nanoEase M/Z Symmetry C18 trap column, Thermo Acclaim™ PepMap™ 100 C18 trap columns) or packed in-house.
4. Mobile phases: solvent A (0.1% formic acid in water) and solvent B (0.1% formic acid in ACN).

2.8 Database Search

1. Protein database: latest version of Uniprot human protein database that can be downloaded from https://www.uniprot.org/.
2. Database search engine: latest version of MSFragger [18] that can be downloaded from https://fragpipe.nesvilab.org/.

3 Methods

3.1 Cell Culture

Grow 10 T175 flasks of Calu-3 cells in cell culturing media at 37°C with 5% CO_2 until 80% confluent (*see* **Note 2**).

3.2 Immunoprecipitation

1. Conjugate 4 mg of W6/32 and L243 antibodies separately 1 ml of AminoLink™ Plus coupling resin following the instruction from the vendor (*see* **Note 3**).
2. Wash the cells in each flask with 10 ml of cold PBS for three times. Add 5 ml of cell lysis buffer with fresh protease inhibitor and detach the cells by gently shaking the flask until cells are detached. Transfer the detached cells in the lysis buffer to a Dounce homogenizer. Break the cells with a loose piston and transfer the homogenate to a 50 ml tube. Combine all the homogenates from 10 flaks.
3. Centrifuge at 20,000×*g* for 15 min to remove cell debris. Transfer the supernatant to a clean tube. Store 100 μl of lysate and use the rest for immunoprecipitation.
4. Incubate the lysate with W6/32 antibody conjugated beads overnight at 4°C with gentle rotation to capture MHC class I complex. After incubation, centrifuge at 3,000×*g* for 3 min to pellet the beads. Collect the supernatant as flow-through. Wash the beads with 50 ml of washing buffer by rotating for

30 min and repeat it for three times. Keep the supernatant from each wash.

5. Incubate the flow-through of MHC class I immunoprecipitation with 1 ml of L243 antibody conjugated beads (*see* **Note 4**) overnight at 4°C with gentle rotation. Centrifuge at 3,000×*g* for 3 min to pellet the beads. Collect the supernatant as flow-through. Wash the beads with 50 ml of washing buffer by rotating for 30 min and repeat it for three times. Keep the supernatant from each wash.
6. Elute MHC class I and class II complex, respectively, by adding 10 ml of eluting solution to the agarose beads and incubate at room temperature for 15 min. Centrifuge at 3,000×*g* for 3 min, collect the supernatant as elution. Use 100 μl of the elution for validation of immunoprecipitation results (*see* **Note 5**).

3.3 Purification of MAPs and HLA Proteins

1. Adjust the pH of eluents from W6/32 antibody conjugated beads which contains the class I MAPs and HLA-ABC proteins by adding 10% TFA to achieve a final concentration of 0.1% (i.e., 50 μl of 10% TFA to 5 ml of elution).
2. Activate the C18 cartridge by passing 1 ml of ACN though the C18 beads and discard flow-through.
3. Clean the C18 cartridge by passing 1 ml of 0.1% TFA through the C18 beads and discard flow-through. Repeat this step for two more times.
4. Load acidified sample solution to the C18 cartridge by passing the sample solution through the C18 beads. Discard flow-through.
5. Wash the C18 cartridge by passing 1 ml of 0.1% TFA through the C18 beads for three times. Discard the flow-through.
6. Elute the bound MAPs by passing 1 ml of MAP elution solution and collect the eluents with clean Eppendorf tubes.
7. Elute the remaining HLA proteins by passing 1 ml of HLA-ABC protein elution solution and collect the eluents with clean Eppendorf tubes.
8. Desalt and purify of MHC class II associated peptides and HLA proteins with the same steps described above.
9. Dry all the eluents down with Speed Vac.

3.4 Enrichment of Glycosylated MAPs

1. Pack 5 mg of polyhydroxyethyl A into each of two (*see* **Note 6**) Toptip EMPTY tips by centrifugation at 2,000×*g* for 3 min. Discard the flow-through and the beads stay on top of the frit to form the HILIC SPE column. Wash the spin columns by adding 50 μl of H_2O, centrifugating at 2,000×*g* for 3 min, and discard the flow-through. Equilibrate the column by adding 50 μl of HILIC loading and washing solution and centrifuging at 2,000×*g* for

3 min, discard the flow-through. Repeat the equilibration steps for additional two times.

2. Dissolve the protein digests in 50 μl of HILIC loading and washing solution and then pipette onto the prepared HILIC SPE columns.
3. Centrifuge at 2,000×*g* for 3 min, collect the flow-through with a clean Eppendorf tube.
4. Add 50 μl of loading buffer and centrifugate at 2,000×*g* for 3 min. Repeat this step for an additional two times, combining the flow-through from sample loading to washing steps (*see* **Note 7**).
5. Add 50 μl of elution solution and centrifugate at 2,000×*g* for 3 min. Repeat once. Combine the eluates (*see* **Note 8**).
6. Speed Vac the flow-through and elution to dryness.

3.5 Digestion of HLA Proteins

1. Solubilize dried HLA proteins (MHC-I and MHC-II, respectively) with 100 μl of protein denaturing solution.
2. Add 5 μl of reduction agent to the sample and vortex to mix well. Incubate the sample at 56°C for 45 min.
3. Add 10 μl of alkylation agent to sample and vortex to mix well. Incubate the sample at room temperature for 30 min in dark.
4. Add 400 μl of diluting solution to the sample and add 1 μg of trypsin to each sample. Vortex to mix well and incubate the sample at 37°C for 12–16 h.

3.6 Desalting of HLA Protein Digest and Enrichment of Glycopeptides

1. Add 10% TFA to the sample solution to adjust TFA concentration to a final concentration of 0.1% (i.e., 5 μl of 10% TFA to 500 μl of sample).
2. Activate the C18 cartridge by passing 1 ml of ACN though the C18 beads and discard flow through.
3. Clean the C18 cartridge by passing 1 ml of 0.1% TFA through the C18 beads and discard flow through. Repeat this step for two more times.
4. Load acidified sample solution to the C18 cartridge by passing the sample solution through the C18 beads. Discard flow through.
5. Wash the C18 cartridge by passing 1 ml of 0.1% TFA through the C18 beads for three times. Discard the flow through.
6. Elute the bound peptides by passing 1 ml of 50% ACN in 0.1% TFA through the C18 beads. Collect the elution with Eppendorf tube.
7. Dry the elution with Speed Vac to dryness and store the dried peptides at −20°C.

8. Enrichment of glycopeptides from HLA protein digest with the same procedure described above for MAPs.

3.7 LC-MS/MS Analysis

1. Solubilize dried peptides with 10 µl of solvent A by brief vortex. Centrifugate at 17,000×*g* for 1 min and transfer the sample to a glass sample vial.
2. Load the sample to LC-MS/MS system and run a 1 h gradient from 5%–35% solvent B at a flow rate between 200 nl/min and 500 nl/min.
3. Acquire MS/MS data in data-dependent mode with the following parameters that are typical for Orbitrap Exploris 480 mass spectrometer (Table 3). All data were recorded with the Xcalibur software and exported as .raw files for further analysis.

3.8 Database Search

1. Import MS data into FragPipe to start database search (*see* **Note 9**)
2. Load workflow. Select the N-glyco-HCD workflow.
3. Import protein database, for both searches, select uniprot human protein database file, add common Repository of Adventitious Proteins (cRAP), and generate decoy database.
4. Import Nglyco309_Byonic glycan masses as variable modifications on N-X-S/T (*see* **Note 10**). Use the default parameters

Table 3
Typical parameters for Orbitrap Exploris 480 mass spectrometer

Name	Parameter
Spray voltage	1700–2200 V
Capillary temperature	275°C
Full-scan MS range	300–1800 m/z
MS1 resolution	60,000
Ion target	5.00×10^5
Maximum injection time	120 ms
Cycle time	3 s
Precursor charge state	2–4+
Higher-energy collisional dissociation (HCD)	28%
MS2 accumulation time	120 ms
MS/MS resolution (Orbitrap)	30,000
MS2 target	1.00×10^5
Isolation width	1.2 Da

set by MSFragger for analysis of HLA protein. For search of MAPs, change the cleavage setting to nonspecific.

5. Launch the search. Once the search is finished, the result file can be found in the folder where the raw files were placed.

4 Notes

1. W6/32 antibody is pan-HLA antibody that recognizes most of class I HLAs, including HLA-A, HLA-B, HLA-C, and non-canonical HLAs like HLA-E, HLA-F, and HLA-G. This antibody can be produced in-house with HB-95 hybridoma. It is also available from various vendors. L243 antibodies react with human HLA-DR but not HLA-DQ or HLA-DP. If there is specific need for HLA-DQ or HLA-DP, additional antibodies are needed. L243 antibodies can be produced in-house with HB-55 hybridoma.
2. Ten T175 flasks of Calu-3 cells contain around 2×10^8 of cells, depending on the confluency and viability of cells. The number of cells can be increased up to 1×10^9, if the number of glycopeptides identification is low.
3. One milliliter of AminoLink™ Plus coupling resin can conjugate around 4 mg of antibody. The conjugating efficiency can be measured by comparing the antibody concentration before and after conjugation. Low conjugating efficiency will result in poor capture of MHC complexes and very few numbers of both MAPs and HLA peptides.
4. The antibody conjugated beads can be packed into a cartridge and the immunoprecipitation can be operated as SPE. The beads can also be packed in a spin column or HPLC column. The choice of methods for immunoprecipitation depends on the availability of equipment and users' preference.
5. The success of immunoprecipitation needs to be checked before glycopeptide enrichment either and it can be done either by western blotting or LC-MS/MS analysis. For western blotting, the intensity of HLA protein should be much stronger from the elution than from the cell lysate while the intensity of actin should be opposite. It is also recommended to collect the flow-though from the washing step and include this fraction for western blotting. For LC-MS/MS validation, the sequence coverage of HLA protein should be higher than 50% and HLA proteins should range among the top 10 proteins by intensity. The identification of glycopeptides will be compromised, if the efficiency of immunoprecipitation is low.

6. One of the SPE spin column is for class I MAPs and the other is for class II MAPs. The enrichment can be done in parallel and together with the enrichment of glycopeptides from HLA proteins digest, which increases the numbers of samples to four at the same time.
7. The majority of peptides from the flow-through and wash fractions are not glycosylated. The sequence and abundance can be used for the identification and quantification of specific HLA alleles.
8. The majority of peptides from elution should be glycosylated. If non-glycosylated peptides are detected from the elution, more washing steps are needed.
9. Searching parameters for MAPs and HLA protein are different. When loading .raw data to the search engine, load the .raw files from each sample separately and search with different workflows. There will be four separate searches: class I MAPs, class I HLAs, class II MAPs, and class II HLAs.
10. Several glycan databases are available from the search engine and the largest one was chosen to include as many glycan structures as possible. However, increasing the number of potential targets also increases the chance of false positive identifications which possibly leads to few identifications. It is suggested to test with all glycan databases to achieve optimal identifications. It is also possible to build a customized glycan database for MAPs to remove some of glycans that are not reported among MAPs.

References

1. Chi H, Pepper M, Thomas PG (2024) Principles and therapeutic applications of adaptive immunity. Cell 187:2052–2078
2. Neefjes J, Jongsma ML, Paul P, Bakke O (2011) Towards a systems understanding of MHC class I and MHC class II antigen presentation. Nat Rev Immunol 1:823–836
3. Williams TM (2001) Human leukocyte antigen gene polymorphism and the histocompatibility laboratory. J Mol Diagn 3:98–104
4. Ryan SO, Cobb BA (2012) Roles for major histocompatibility complex glycosylation in immune function. Semin Immunopathol 34:425–441
5. Ilca FT, Louise HB (2021) The glycosylation status of MHC class I molecules impacts their interactions with TAPBPR. Mol Immunol 139:168–176
6. Martayan A, Sibilio L, Setini A et al (2008) N-linked glycosylation selectively regulates the generic folding of HLA-Cw1. J Biol Chem 283:16469–16476
7. Hoek M, Demmers LC, Wu W et al (2021) Allotype-specific glycosylation and cellular localization of human leukocyte antigen class I proteins. J Proteome Res 20:4518–4528
8. Bedran G, Polasky DA, Hsiao Y et al (2023) Unraveling the glycosylated immunopeptidome with HLA-Glyco. Nat Commun 14:3461
9. Block TM, Comunale MA, Lowman M et al (2005) Use of targeted glycoproteomics to identify serum glycoproteins that correlate with liver cancer in woodchucks and humans. Proc Natl Acad Sci USA 102:779–784
10. Nothaft H, Szymanski C (2010) Protein glycosylation in bacteria: sweeter than ever. Nat Rev Microbiol 8:765–778
11. Vigerust DJ, Shepherd VL (2007) Virus glycosylation: role in virulence and immune interactions. Trends Microbiol 15:211–218
12. Malaker SA, Penny SA, Steadman LG et al (2017) Identification of glycopeptides as post-translationally modified neoantigens in leukemia. Cancer Immunol Res 5:376–384

13. Bassani-Sternberg M, Bräunlein E, Klar R et al (2016) Direct identification of clinically relevant neoepitopes presented on native human melanoma tissue by mass spectrometry. Nat Commun 7:13404
14. Laumont CM, Daouda T, Laverdure J-P et al (2016) Global proteogenomic analysis of human MHC class I-associated peptides derived from non-canonical reading frames. Nat Commun 7:10238
15. Olsen JV, Mann M (2013) Status of large-scale analysis of post-translational modifications by mass spectrometry. Mol Cell Proteomics 12:3444–3452
16. Chen R, Fulton KM, Tran A et al (2023) Integrated immunopeptidomics and proteomics study of SARS-CoV-2–infected Calu-3 cells reveals dynamic changes in allele-specific HLA abundance and antigen presentation. Mol Cell Proteomics 22:100645
17. Polasky DA, Yu F, Teo GC et al (2020) Fast and comprehensive N-and O-glycoproteomics analysis with MSFragger-Glyco. Nat Methods 17:1125–1132
18. Kong AT, Leprevost FV, Avtonomov DM et al (2017) MSFragger: ultrafast and comprehensive peptide identification in mass spectrometry–based proteomics. Nat Methods 14:513–520

Check for updates

Chapter 6

Endogenous Phosphopeptidomics Analysis by Using Mesoporous IMAC Materials

Ci Wu, Bing Liu and Junfeng Ma

Abstract

The analysis of endogenous phosphopeptides is crucial for identifying biomarkers of cellular signaling alterations and monitoring disease-related pathophysiological changes in vivo. While phosphoproteomics has seen remarkable progress in recent years, the field of endogenous phosphopeptidomics has lagged behind. A major challenge in endogenous phosphopeptidomics analysis lies in the presence of highly abundant nonmodified peptides and proteins within complex biological matrices, such as biofluids, which can obscure the detection of low-abundance phosphopeptides. In this chapter, we describe a protocol of a recently developed approach by utilizing Zirconium (IV)-grafted mesoporous beads for selective phosphopeptide enrichment, followed by high-resolution nanoRPLC–MS/MS analysis, enabling enhanced sensitivity and specificity in the detection of endogenous phosphopeptides from serum samples.

Keywords Endogenous phosphopeptides, Phosphopeptidomics, Serum, IMAC, Mesoporous material, LC–MS/MS

1 Introduction

Endogenous phosphopeptides present in biofluids, such as serum, either derived from proteolytic degradation of larger proteins or actively secreted by cells and tissues, serve as important indicators of physiological and pathological processes [1–3]. Growing evidence highlights the potential of profiling endogenous phosphopeptides in human serum to gain insights into an individual's health status and to uncover promising biomarkers for disease diagnosis and monitoring [4]. However, unlike phosphoproteomics, the comprehensive analysis of endogenous phosphopeptides in biofluids remains technically challenging [5]. Despite the activity of atypical kinases, such as those from the Fam20 family, phosphorylation levels in serum proteins are generally very low due to the high activity of endogenous phosphatases [6]. This, coupled with the complexity of biological matrices, makes the detection of endogenous phosphopeptides particularly difficult. To facilitate their identification, abundant

Xu Zhang et al. (eds.), *Protein Post-Translational Modifications*, Methods in Molecular Biology, vol. 3018,
https://doi.org/10.1007/978-1-0716-5166-7_6,

serum proteins must often be removed through techniques such as organic solvent-assisted precipitation, ultrafiltration, or the use of mesoporous beads [7]. For the sensitive detection of low-abundance endogenous phosphopeptides, efficient and selective enrichment strategies are essential.

While substantial progress has been made for deep phosphoproteomics, largely through methods such as metal oxide affinity chromatography (e.g., TiO_2) [8, 9] and immobilized metal ion affinity chromatography (e.g., Zr^{4+}/Ti^{4+}-IMAC, Fe-NTA) [10–12], these approaches are not directly transferable to endogenous phosphopeptidomics due to the unique challenges posed by biofluid samples. To enhance their enrichment efficiency, various sample preprocessing techniques have been explored to remove proteins and nonmodified peptides in serum, including organic solvent assisting protein precipitation, filter-mediated ultracentrifugation, and mesoporous beads [13, 14]. Among enrichment strategies, mesoporous materials with phosphate-specific affinity have shown significant promise. These materials can effectively exclude high-molecular-weight protein species, enabling selective enrichment of endogenous phosphopeptides. For instance, porous metal–organic frameworks (MOFs) modified with metal oxide or metal ions (e.g., Ti^{4+}) have been employed for this purpose [15–17]. While these approaches have demonstrated success in enriching a panel of high-abundance phosphopeptides in serum, their performance toward lower abundant endogenous phosphopeptides remains largely unexplored. Zou et al. [18] conducted an in-depth study by combining selective phosphopeptide enrichment, offline HPLC fractionation into nine fractions, and nano-flow RPLC–MS/MS with complementary fragmentation techniques. This comprehensive workflow enabled the identification of 143 endogenous phosphopeptides from 500 μL of control serum. Very recently, we developed an effective strategy for the analysis of endogenous phosphopeptides in serum samples based on a novel mesoporous material [19]. In brief, the enrichment material was synthesized by immobilization of Zr^{4+} onto mesoporous silica MCM-48 beads (with a pore diameter of ~3 nm), as shown in Fig. 1. The MCM-48–based Zr^{4+}-IMAC was then applied for direct analysis of the endogenous phosphopeptidome of serum samples. By combining the substantially simplified one-pot enrichment procedure with high-resolution mass spectrometric identification and quantification, our method enabled deep analysis of endogenous phosphopeptides (Fig. 2). Of note, the method also allowed simultaneous phosphoproteomics for serum samples.

In this chapter, we provide a step-by-step description of our protocol to facilely synthesize Zr^{4+}-grafted mesoporous silica MCM-48 beads and apply them for endogenous phosphopeptidomics analysis. We hope the simplified and sensitive workflow can help facilitate endogenous phosphopeptidomics studies and accelerate endogenous phosphopeptides-based biomarker discovery from biofluids, such as serum.

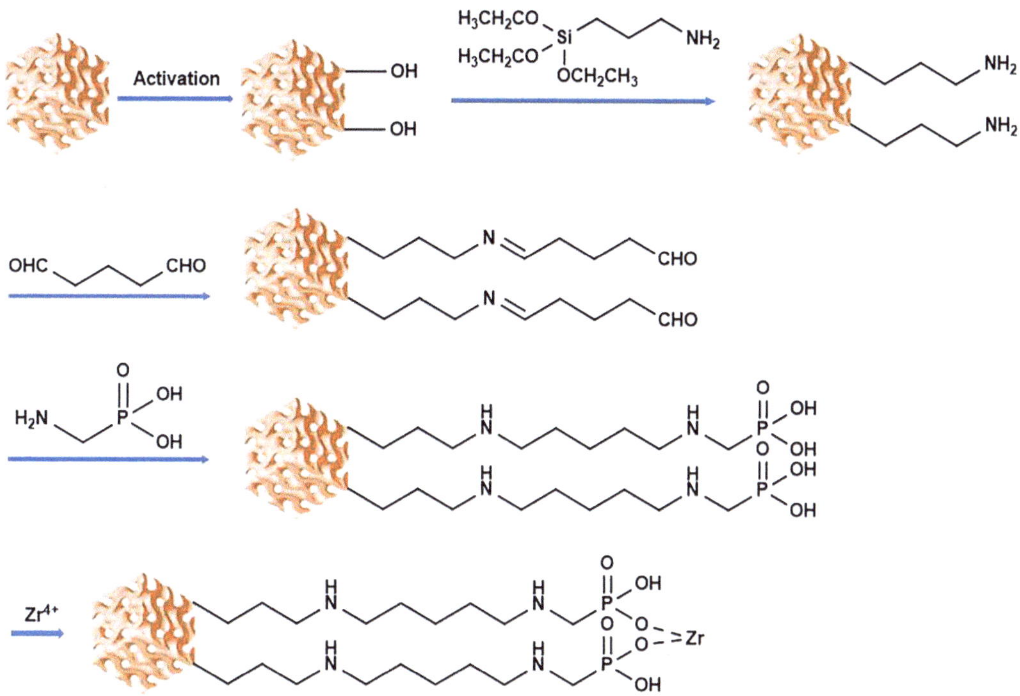

Fig. 1 Scheme for the preparation of MCM-48–based Zr-IMAC particles

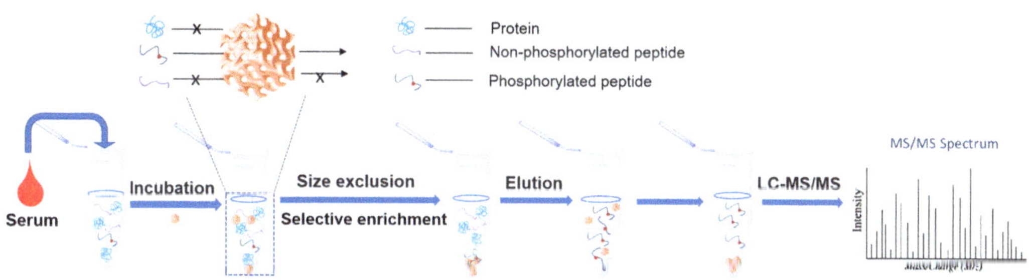

Fig. 2 Procedure for the analysis of endogenous phosphopeptides from serum (adapted from [19], an open access article distributed under the terms of the Creative Commons CC-BY 4.0 license, which permits unrestricted use, distribution, and reproduction)

2 Materials

2.1 Synthesis of MCM-48-Based Zr4+-IMAC

1. Mesoporous silica MCM-48 (15 μm of particle size, 3 nm of pore size, Sigma [St. Louis, MO]).
2. 3-Aminopropyltriethoxysilane (APTES, 97%).
3. (Aminomethyl) phosphonic acid (AMPA, 99%).

4. 100 mM phosphate buffer saline, PBS (pH 8).
5. Zirconium (IV) sulfate $Zr(SO_4)_2$.
6. Sodium cyanoborohydride ($NaCNBH_3$).
7. Methane sulfonic acid (98%).
8. Glutaraldehyde (GA, 50% solution).
9. Reaction solution: 2 mg/mL AMPA and 10 mg/mL $NaCNBH_3$ in 100 mM PBS (pH 8).

2.2 Enrichment and Analysis of Endogenous Phosphopeptides

1. Pyrrolidine.
2. Ammonium hydroxide solution (v/v, ~28%).
3. Acetonitrile (ACN, LC/MS grade).
4. Glycolic acid.
5. Protease inhibitor cocktail I.
6. Phosphotase inhibitor cocktail.
7. A mixture of inhibitors: 2 μM pepstatin A, 10 μM bestatin hydrochloride, and 1 μM marimastat.
8. Loading buffer: 80% ACN, 5% TFA, and 0.1 M glycolic acid.
9. Washing buffer1: 80% acetonitrile, 1% TFA in water.
10. Washing buffer2: 0.1% TFA in water.
11. Pooled human serum (Innovative Research, Inc [Novi, MI]).
12. C18 Trap column (Waters Acquity UPLC M-Class Trap, Symmetry C18, 100 Å, 5 μm, 180 μm × 20 mm).
13. Analytical column (Waters Acquity UPLC M-Class, peptide BEH C18 column, 300 Å, 1.7 μm, 75 μm × 150 mm).
14. Buffer A: 2% ACN, 0.1% formic acid in water.
15. Buffer B: 0.1% formic acid in ACN.
16. Orbitrap Lumos Fusion mass spectrometer (Thermo Fisher Scientific).
17. Proteome Discoverer 2.5 (Thermo Fisher Scientific).

3 Methods

3.1 Synthetic Procedure for Zirconium (IV)-Immobilized MCM-48

1. Weigh out the appropriate amount of MCM-48 (about 1 g), and activate it by refluxing in 33% (v/v) methanesulfonic acid for 24 h (*see* **Note 1**).
2. Wash the activated MCM-48 material thoroughly with deionized H_2O and dried at 70°C for 12 h.
3. Reflux the activated MCM-48 material with 2 mL of APTES and 40 mL of dry toluene for 16 h under continuous

stirring and heating (110°C), resulting APTES@MCM-48 (*see* **Note 2**).

4. Wash the APTES@MCM-48 material with ethanol followed by drying at 70°C for 12 h (*see* **Note 3**).
5. Stir the resulting APTES@MCM-48 with 10 mL of 10% glutaraldehyde in 100 mM PBS (pH 8.0) for 6 h at 25°C.
6. Wash the glutaraldehyde-activated material with deionized water until the pH reaches neutral.
7. Stir the glutaraldehyde-activated material with 10 mL of a reaction solution for 12 h (*see* **Note 4**).
8. Wash the AMPA-modified material with deionized water until the pH reaches a neutral condition.
9. Immobilize the zirconium (IV) ions by stirring the AMPA modified material with 10 mL 100 mM $Zr(SO_4)_2$ in water at 25°C for 6 h (*see* **Note 5**).
10. Wash the zirconium (IV)–immobilized MCM-48 with deionized water for five times.
11. Wash the zirconium (IV)–immobilized MCM-48 with ethanol followed by drying at 70°C for 12 h.

3.2 Enrichment and Analysis of Endogenous Phosphopeptides

3.2.1 Serum Sample Preparation

1. Add a panel of mixture solution of inhibitors into 500 μL of the human serum sample (*see* **Note 6**).
2. Dilute the serum with 5 mL of the loading buffer.
3. Centrifuge the solution at 12,000×*g* for 10 min at 4°C, with the supernatant collected for further processing.

3.2.2 Enrichment of Endogenous Phosphopeptides

1. Weight 2 mg of Zr-IMAC@MCM-48 particles, then equilibrate them with 500 μL loading buffer by vibrating at room temperature for 5 min, centrifuge at 5000 rpm for 2 min, and remove the supernatant (*see* **Note 7**).
2. Repeat the equalization step twice.
3. Incubate the prepared serum sample with the equalized Zr-IMAC@MCM-48 beads for 1 h at room temperature.
4. Centrifugate at 5000 rpm for 2 min, and remove the supernatant.
5. Wash the beads with 400 μL of washing bufferl by vibrating at 1000 rpm for 5 min.

6. Centrifugate at 5000 rpm for 2 min, and remove the supernatant.
7. Wash the beads with 400 μL of washing buffer2 by vibrating at 1000 rpm for 5 min.
8. Centrifugate at 5000 rpm for 2 min, and remove the supernatant.
9. Elute the peptides from the beads with 50 μL of 1% ammonium hydroxide by vibrating at 1000 rpm for 5 min.
10. Centrifugate at 5000 rpm for 2 min, and collect the supernatant as eluent 1.
11. Elute the peptides from the beads with 50 μL of 5% pyrrolidine by vibrating at 1000 rpm for 5 min.
12. Centrifugate at 5000 rpm for 2 min, and collect the supernatant as eluent 2.
13. Combine eluents 1 and 2, and desalt with a C18 micro spin column. Dry peptides to completion with a SpeedVac.

3.3 nanoRPLC–MS/MS and Data Analysis

1. Resuspend peptides in 0.1% formic acid, and load onto a C18 Trap column at 10 μL/min for 4 min (*see* **Note 8**).
2. Separate the peptides with an analytical column, which is temperature-controlled at 40°C. Set the flow rate as 400 nL/min, and use 150-min gradient of buffer A and buffer B for separation: 1% buffer B at 0 min, 5% buffer B at 1 min, 30% buffer B at 80 min, 50% buffer B at 120 min, 98% buffer B at 130 min, 98% buffer B at 133 min, 1% buffer B at 133.1 min, and 1% buffer B at 150 min (*see* **Note 9**).
3. Acquire MS data by Orbitrap Fusion Lumos mass spectrometer using an ion spray voltage of 2.4 kV and an ion transfer temperature of 275°C. Record mass spectra with Xcalibur 4.0. Turn on advanced peak determination for MS analyses. Set DDA (data-dependent acquisition) parameters as below: detector type: Orbitrap; mass range: 375–1500 *m/z*; Orbitrap resolution: 120,000; scan range: 375–1500 *m/z*; RF lens: 30%; AGC target: standard; maximum injection time mode: auto; microscans: 1. Charge state: 2–7; exclusion duration: 40 s; cycle time: 3 s. MS/MS parameters were set as below: isolation mode: quadrupole; isolation window: 1.6 *m/z*; HCD normalized collision energy: 30%; detector type: Orbitrap; resolution: 30,000; normalized AGC target: 200%. DIA (data-independent acquisition) data were acquired with the following parameters: scan range = 400–1250 *m/z*, MS resolution of 120,000 at *m/z* 200, an AGC target = 4e5, and maximum injection time = 50 ms. The DIA-MS/MS scan was performed in the HCD mode with the following parameters: isolation window of 20 Da with 1 Da overlap, precursor range = 400–1000 *m/z*, fragment scan range

Table 1
Phosphopeptides identified from the tryptic digest of α-casein and β-casein and its mixture spiked in human serum enriched by Zr-IMAC@MCM-48 beads

No.	Amino acid sequence	Peptide *m/z*	No. of phosphosites	Standard protein digest	Standard digests spiked in serum
α1	EKVNELSpKDIGSpESpTEDQAMEDIK	2935.16	3	Y	Y
α2	VNELSpKDIGSpESpTEDQAMEDIK	2678.023	3	Y	Y
α3	EQLSpTpSEENSKKTVDMESpTEVFTK	2987.192	3	Y	
α4	NAVPITPTLNREQLSTpSpEENSK	2588.164	2	Y	Y
α5	DIGSpESpTEDQAMEDIK	1927.692	2	Y	Y
α6	NAVPITPTpLNREQLSpTSEENSKK	2716.259	2	Y	Y
α7	EKVNELSKDIGSpESpTEDQAMEDIK	2855.194	2	Y	Y
α8	VNELSKDIGSpESpTEDQAMEDIK	2598.057	2	Y	
α9	NAVPITPTLNREQLSpTSEENSKK	2636.293	1	Y	Y
α10	KTVDMESpTEVFTKK	1722.802	1	Y	Y
α11	TVDMESpTEVFTK	1466.612	1	Y	Y
α12	TVDMESpTEVFTKK	1594.707	1	Y	Y
α13	DIGSESpTEDQAMEDIK	1847.725	1	Y	Y
α14	KYKVPQLEIVPNSpAEER	2080.047	1	Y	Y
α15	YKVPQLEIVPNSpAEER	1951.952	1	Y	Y
α16	YKVPQLEIVPNSpAEERLHSMK	2548.263	1	Y	Y
α17	VPQLEIVPNSpAEER	1660.794	1	Y	Y
α18	VPQLEIVPNSpAEERLHSMK	2257.105	1	Y	
α19	KTVDMoESpTEVFTKK	1738.797	1	Y	Y
α20	TVDMoESpTEVFTKK	1610.702	1	Y	Y
α21	TVDMoESpTEVFTK	1482.607	1	Y	Y
α22	EKVNELSpKDIGSpESpTEDQAMoEDIK	2951.155	3	Y	
α23	YKVPQLEIVPNSpAEERLHSMoK	2564.258	1	Y	Y
α24	VNELSKDIGSpESpTEDQAMoEDIK	2614.051	2	Y	Y
α25	VPQLEIVPNSpAEERLHSMoK	2273.1	1	Y	
α26	DIGSpESpTEDQAMoEDIK	1943.686	2	Y	Y
α27	DIGSpESTEDQAMoEDIK	1863.72	1	Y	
β1	KIEKFQSpEEQQQTEDELQDK	2560.145	1	Y	Y
β2	IEKFQSpEEQQQTEDELQDK	2432.05	1	Y	Y
β3	FQSpEEQQQTEDELQDK	2061.828	1	Y	Y

S or T with p, phosphorylated site; *Mo*, oxidation on methionine

110–1600 *m/z*; resolution = 30,000 with maximum injection time of 54 ms, AGC target = 5e4; normalized collision energy = 30%. All data were acquired in profile mode using positive polarity (*see* **Note 10**).

4. Perform database searching of the raw files in Proteome Discoverer 2.5 with the Sequest HT search engine by using the database of *Homo sapiens* (TaxID 9606, Uniprot). Set the database-searching parameters as below: nonspecific enzyme, the precursor mass tolerance was set at 10 ppm, whereas the fragment-mass tolerance was set at 0.02 Da. Variable modifications of methionine oxidation (+15.9949 Da), acetyl (N-terminus, +42.011 Da), and phosphorylation (T, S, or Y, + 79.966 Da) are allowed. The false-discovery rate (FDR) is determined by using a target-decoy search strategy. On the peptide level, the corresponding FDR is less than 1%. Phosphosites with >0.75 localization probabilities are regarded as confident identification. Label-free quantification (LFQ) method is used for quantifying the phosphopeptides (*see* **Note 11**).

A comparison test for the phosphopeptides enrichment of Zr-IMAC@MCM-48 beads by using α-casein and β-casein digest with or without serum as samples is shown in Table 1. As seen, 30 phosphopeptides (including 27 phosphopeptides from α-casein and three from β-casein) could be identified from the sample without serum, and 24 phosphopeptides (including 21 phosphopeptides from α-casein and three phosphopeptides from β-casein) when the digest of α-casein and β-casein was spiked into the serum sample.

4 Notes

1. Use methanesulfonic acid only in a well-ventilated area, preferably in a fume hood. If large amounts of fumes are produced, wear an appropriate respirator.
2. Use appropriate personal protective equipment: wear safety goggles, gloves, and a lab coat to prevent skin and eye contact.
3. Sodium cyanoborohydride ($NaCNBH_3$) should be freshly prepared. $NaCNBH_3$ is a toxic and potentially hazardous substance due to the presence of cyanide. Handle it with extreme care to avoid exposure to skin, eyes, or respiratory system. Wear a mask when weighing $NaCNBH_3$ and use protective gloves (preferably nitrile or neoprene) when handling $NaCNBH_3$ to avoid direct skin contact. To avoid exposing $NaCNBH_3$ to coworkers, cover the weigh boat containing $NaCNBH_3$ with another weigh boat of similar size during transport to the fume hood to prevent accidental exposure to coworkers or the environment.

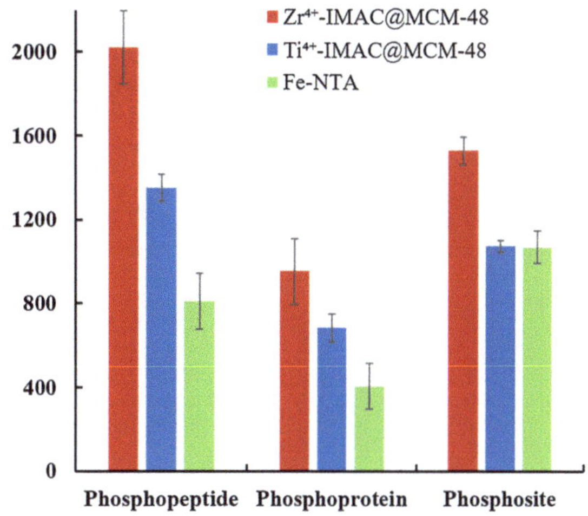

Fig. 3 Comparison of Zr-IMAC@MCM-48 with Ti-IMAC@MCM-48 and the commercial Fe-NTA beads for the enrichment of phosphopeptides from the digest of HeLa cell lysates ($n = 3$)

4. Perform all washing steps during the material synthesis by using 50 mL of solvent each time. Vibrate at 1200 rpm, followed by centrifugation at 5000 rpm for 2 min. Then, remove the supernatant.
5. Other metals besides zirconium, such as titanium and iron, may be applicable, but different performance might be achieved. For example, we prepared Zr-IMAC@MCM-48, Ti-IMAC@MCM-48, and Fe-IMAC@MCM-48 with the same approach. Their performance, together with that of the commercial Fe-NTA beads, was assessed by analyzing enriched phosphopeptides with a nanoHPLC-MS/MS system (Fig. 3). The average numbers of identified phosphopeptides, phosphoproteins, and phosphosites by the three IMAC materials were: Zr-IMAC@MCM-48 (2024, 953, and 1529), Ti-IMAC@MCM-48 (1354, 684, and 1074), and Fe-NTA (810, 407, and 1070), respectively. Unfortunately, no phoshopeptides were captured by Fe-IMAC@MCM-48, likely due to the low valence state (+3) of iron ions, and thus insufficient complexation with both the backbone of the AMPA-based material and the phosphate group on peptides. Thus, caution should be taken when other metals are being used, with appropriate assessment recommended.
6. The inclusion of various kinds of protease inhibitors (pepstatin A, bestatin hydrochloride, and marimastat) into the loading buffer is recommended to eliminate potential breakdown of proteins into smaller peptides during sample preparation. Among them, pepstatin A is a highly specific inhibitor of acid

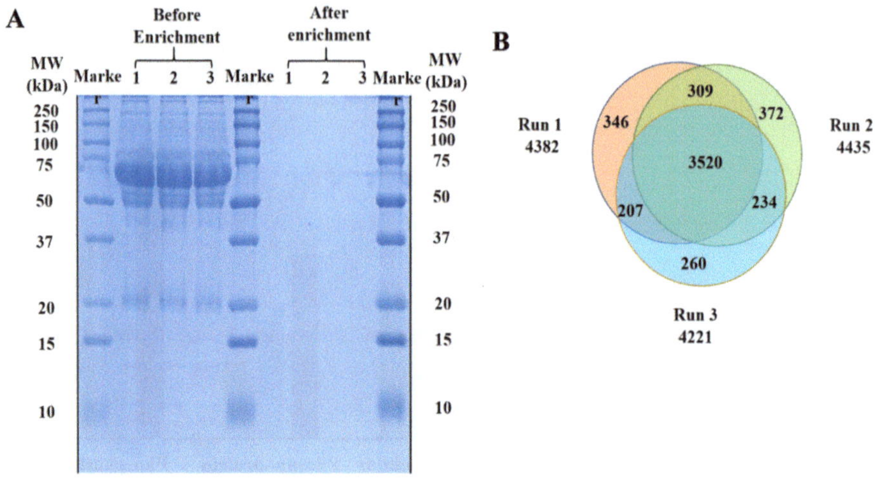

Fig. 4 (A) SDS-PAGE analysis of serum samples before and after enrichment of endogenous phosphopeptides; (B) The performance evaluation of the HPLC column before, during, and after analysis of 10 serum endogenous phosphopeptide samples (QC sample: 200 ng standard HeLa cell digests obtained from Fisher Scientific)

proteases, in particular aspartate proteases; bestatin hydrochloride is an inhibitor of aminopeptidases and some metalloproteases; and marimastat is a broad-spectrum inhibitor of matrix metalloproteinases (MMPs).

7. To validate the protein removal performance of the mesoporous material, SDS-PAGE can be used to evaluate the presence of proteins before and after enrichment of endogenous phosphopeptides. In comparison to nonenriched samples, no protein bands could be visualized in the enriched samples, indicating a high efficiency of protein removal (Fig. 4A). In addition, after the running of multiple endogenous phosphopeptide samples extracted from serum, the C18 separation column still exhibited high identification efficiency for QC samples (RSD: 2.6%, n = 3) (Fig. 4B), further demonstrating the bi-functional material's ability to remove protein interference from the samples.
8. Before loading, samples can be further filtered or centrifuged with high speed to remove the remaining beads that may clog the separation column.
9. HPLC conditions can be adjusted according to the instruments/columns available.
10. Data acquisition parameters, such as dynamic exclusion, MS/MS fragmentation, collision energy, should be optimized according to the specific instruments used.
11. The material and method can also be used for phosphoproteomics in biofluids and other types of samples.

Acknowledgments

This project was partially supported by NIH/NCI P30-CA051008. The instrument Orbitrap Lumos Tribrid mass spectrometer was partially supported by Dekelbaum Foundation. This project was grateful for support from the National Natural Science Foundation of China (Grant no. 21705065), the Scientific Research Foundation of the Higher Education Institutions of Liaoning Province China (No. LJ212410165049), and Liaoning Normal University Doctoral Research Start-up Fund (Grant No. 2024BSL002).

References

1. Rubinsztein DC (2006) The roles of intracellular protein-degradation pathways in neurodegeneration. Nature 443:780–786
2. Peng J, Zhang H, Niu H et al (2020) Peptidomic analyses: the progress in enrichment and identification of endogenous peptides. TrAC Trends Anal Chem 125:115835
3. Li XS, Yuan BF, Feng YQ (2016) Recent advances in phosphopeptide enrichment: strategies and techniques. TrAC Trends Anal Chem 78:70–83
4. Li Y, Sun N, Hu X et al (2019) Recent advances in nanoporous materials as sample preparation techniques for peptidome research. TrAC Trends Anal Chem 120:115658
5. Yan Y, Deng C (2019) Recent advances in nanomaterials for sample pre-treatment in phosphoproteomics research. TrAC Trends Anal Chem 120:115655
6. Worby CA, Mayfield JE, Pollak AJ et al (2021) The ABCs of the atypical Fam20 secretory pathway kinases. J Biol Chem 296:100267
7. Hu L, Zhou H, Li Y et al (2019) Profiling of endogenous serum phosphorylated peptides by titanium (IV) immobilized mesoporous silica particles enrichment and MALDI-TOFMS detection. Anal Chem 81:94–104
8. Deng J, Lazar IM (2016) Proteolytic digestion and TiO_2 phosphopeptide enrichment microreactor for fast MS identification of proteins. J Am Soc Mass Spectrom 27:686–698
9. Ino Y, Kinoshita E, Kinoshita-Kikuta E et al (2022) Evaluation of four phosphopeptide enrichment strategies for mass spectrometry-based proteomic analysis. Proteomics 22:2100216
10. Wang F, Veth T, Kuipers M et al (2023) Optimized suspension trapping method for phosphoproteomics sample preparation. Anal Chem 95:9471–9479
11. Martínez-Val A, Fort K, Koenig C et al (2023) Hybrid-DIA: intelligent data acquisition integrates targeted and discovery proteomics to analyze phospho-signaling in single spheroids. Nat Commun 14:3599
12. Tsai CF, Wang YT, Yen HY et al (2015) Large-scale determination of absolute phosphorylation stoichiometries in human cells by motif-targeting quantitative proteomics. Nat Commun 6:6622
13. La Barbera G, Capriotti AL, Cavaliere C et al (2018) Development of an enrichment method for endogenous phosphopeptide characterization in human serum. Anal Bioanal Chem 410:1177–1185
14. Qin H, Gao P, Wang F et al (2011) Highly efficient extraction of serum peptides by ordered mesoporous carbon. Angew Chem 123:12426–12429
15. Chen Y, Xiong Z, Peng L et al (2015) Facile preparation of core–shell magnetic metal–organic framework nanoparticles for the selective capture of phosphopeptides. ACS Appl Mater Interfaces 7:16338–16347
16. Su J, He X, Chen L et al (2018) Adenosine phosphate functionalized magnetic mesoporous graphene oxide nanocomposite for highly selective enrichment of phosphopeptides. ACS Sustainable Chem Eng 6:2188–2196
17. Yao J, Sun N, Wang J et al (2017) Rapid synthesis of titanium(IV)-immobilized magnetic mesoporous silica nanoparticles for endogenous phosphopeptides enrichment. Proteomics 17:1600320
18. Zhu J, Wang F, Cheng K et al (2013) Analysis of human serum phosphopeptidome by a focused database searching strategy. J Proteomics 78:389–397
19. Wu C, Zhang S, Hou C et al (2024) In-depth endogenous phosphopeptidomics of serum with zirconium(IV)-grafted mesoporous silica enrichment. Anal Chem 96:8254–8262

Chapter 7

O- and N-Phosphoproteomic Analysis Using Online Alkaline-pH Reversed-Phase Nanoelectrospray-Tandem Mass Spectrometry

Yuqiu Wang, Minchu Tang and Wenfan Xie

Abstract

Protein phosphorylation (O-linked and N-linked) is associated with a wide range of biological processes and cell signaling pathways. Aberrant phosphorylation is often observed as a hallmark of diseases, including cancer, neurodegenerative disorders, and metabolic syndromes. Mass spectrometry (MS) is an indispensable tool for studying protein phosphorylation. This is primarily due to its high sensitivity and throughput, which allows for comprehensive identification and quantification of phosphorylation events in complex biological samples. In this chapter, we describe a detailed protocol for phosphoproteomic analysis using one-dimensional online alkaline-pH reversed-phase nanoelectrospray-tandem MS (alkaline-pH-MS/MS). It complements traditional online low-pH reversed-phase nanoelectrospray-tandem MS (low-pH-MS/MS) by modulating the charge state distribution of phosphopeptides, which may facilitate the characterization of phosphoproteins with significant biological functions.

Keywords Phosphoproteomics, Mass spectrometry, Alkaline-pH-MS/MS, Phosphorylation, Posttranslational modification

1 Introduction

Protein phosphorylation (O-linked and N-linked) is one of the most ubiquitous and essential posttranslational modifications (PTMs), which is involved in various biological processes [1–3]. The O-linked phosphorylation, which predominantly occurs at serine (S), threonine (T), and tyrosine (Y) residues, is the most common form of protein phosphorylation. In contrast, the N-linked phosphorylation, which takes place at lysine (K), arginine (R), and histidine (H), is present at significantly lower levels in cells [4, 5]. Many proteins simultaneously harbor more than one phosphate group at various sites, which are

Yuqiu Wang and Minchu Tang contributed equally to this work.

Xu Zhang et al. (eds.), *Protein Post-Translational Modifications*, Methods in Molecular Biology, vol. 3018, https://doi.org/10.1007/978-1-0716-5166-7_7,

frequently located within four amino acid residues of each other [6]. Such clustering leads to cooperative interactions, where the presence of one phosphate group modulates the phosphorylation status of neighboring sites, thereby enhancing their overall regulatory impact [7, 8]. Furthermore, these multiple phosphosites enhance the negative charges of proteins, consequently affecting protein–protein interactions, enzymatic activity, and subcellular localization [9–13]. The dysfunction of multiple phosphosites is observed in various diseases, such as Alzheimer's disease and cancer [14–16]. Therefore, comprehensive mapping of multiple phosphosites (O-linked and N-linked) is essential for elucidating the mechanisms of intracellular signaling pathways.

Due to the low stoichiometry and poor ionization efficiency of multiply phosphorylated peptides, the detection of multiple phosphosites in complex mixtures remains challenging. Currently, diverse strategies have been developed to improve the identification to multiply phosphorylated peptides and enhance phosphoproteome coverage. For example, advances in sample preparation lead to more efficient cell lysis and protein digestion, which preserves low-abundance phosphopeptides [17]. Optimized enrichment strategies have also been established to selectively isolate phosphopeptides (including multiply phosphorylated peptides) from complex mixtures, such as immobilized metal affinity chromatography (IMAC) and titanium dioxide (TiO_2) chromatography [18, 19]. In addition, the use of multidimensional fractionation approaches, including high-pH reversed-phase liquid chromatography (RPLC), strong anion exchange (SAX), and strong cation exchange (SCX), significantly reduced sample complexity prior to liquid chromatography tandem mass spectrometry (LC-MS/MS) analysis [20, 21]. However, these fractionation methods are usually adopted as first-dimension fraction, not directly interfaced to the mass spectrometer. Recently, we developed an online alkaline-pH reversed-phase nanoelectrospray-tandem mass spectrometry (alkaline-pH-MS/MS) approach for both proteomic and phosphoproteomic analysis [22, 23]. This alkaline-pH-MS/MS did not influence the chromatographic separation behavior of RP but affected the actual polarity and charge distribution of certain peptides. This characteristic made it complementary to traditional online low-pH reversed-phase nanoelectrospray-tandem mass spectrometry (low-pH-MS/MS) for the identification of both proteins and phosphosites. Furthermore, compared to low-pH-MS/MS, alkaline-pH-MS/MS improved the identification of multiply phosphorylated peptides. Thus, it was a good choice to use alkaline-pH-MS/MS, especially for experiments involving multiple replicates.

In this chapter, we describe a standard protocol for phosphoproteomic analysis using alkaline-pH-MS/MS. After protein extraction, digestion, enrichment, and offline fractionation, phosphopeptides (both O-linked and N-linked) were injected to the RP column packed with C18 beads by a nano-LC system. Then, they were separated by LC with alkaline-pH mobile phase and directly introduced to the mass spectrometer by a nano-electrospray ionization (ESI) source. This detailed protocol is accompanied with practical notes and tips to avoid common pitfalls during the procedure.

2 Materials

All buffers and solutions, except for the MS mobile phase, were prepared with ultrapure water of 18 MΩ cm at 25°C, analytical grade reagents, and highest purity chemicals. MS mobile phase should be prepared with LC-MS grade solvent (*see* **Note 1**).

2.1 Protein Extraction, Reduction, Alkylation, and Digestion

1. Phosphate-buffered saline (PBS).
2. Lysis buffer: 100 mM Tris-HCl, 8 M urea, 1.5% (w/v) sodium deoxycholate (SDC), 5 mM Tris(2-carboxyethyl)phosphine hydrochloride (TCEP), 30 mM chloroacetamide (CAA), 1 mM sodium orthovanadate, 1:10 phosphoSTOP, 1:25 cocktail, 2 mM magnesium chloride, and 10 U/mL DNaseI, pH 8.5.
3. 1.5 mL microcentrifuge tubes.
4. Sonicator (SCIENTZ, JY92-IIDN Ultrasonic Homogenizer, Power 900 W (1%–100% adjustable)).
5. Tissue homogenizer.
6. Rotator.
7. Refrigerated centrifuge.
8. Dilution buffer: 100 mM Tris-HCl, 5 mM TCEP, 30 mM CAA, 1:10 phosphoSTOP, 1:25 cocktail, and 2 mM magnesium chloride, pH 8.5.
9. Benzonase, sequencing grade.
10. Thermomixer.
11. 15 mL tube.
12. Methanol.
13. Chloroform.
14. Digestion buffer: 100 mM Tris-HCl, 1% (w/v) SDC, 5 mM TCEP, and 30 mM CAA, pH 8.5.
15. Trypsin, sequencing grade.

2.2 Offline Fractionation

1. 10% (v/v) formic acid (FA).
2. 1.5 mL microcentrifuge tubes.
3. Refrigerated centrifuge.

4. C18 cartridge (50 mg, Waters tC18 SepPak).
5. Condition buffer: 70% (v/v) acetonitrile (ACN), 0.2% (v/v) trifluoroacetic acid (TFA).
6. Wash buffer: 0.1% (v/v) FA.
7. Phase A: 0.1% (v/v) FA.
8. Phase B: 0.1% (v/v) FA, 80% (v/v) ACN.
9. Vacuum concentrator.

2.3 Phosphopeptide Enrichment

1. 10% (v/v) TFA.
2. 1.5 mL microcentrifuge tubes.
3. High-select Fe-NTA kit with Fe-NTA agarose beads (Thermo Fisher Scientific).
4. Phosphorylation wash buffer: 80% (v/v) ACN, 0.1% (v/v) TFA.
5. Thermomixer.
6. Filter tip (Axygen, TF-200-L-R-S).
7. Refrigerated centrifuge.
8. Phosphorylation elution buffer: 5% (w/w) ammonia solution ($NH_3{\cdot}H_2O$), 50% (v/v) ACN.
9. 1.5 mL low binding microcentrifuge tubes.
10. Vacuum concentrator.

2.4 Alkaline-pH-MS/MS

1. Clean solvent containers for mobile phase.
2. Mobile phase A: 5 mM ammonium formate (NH_4FA) (*see* **Note 2**). Prepare mobile phase A (250 mL): Add 48.13 μL formic acid to 200 mL water. Then, add $NH_3{\cdot}H_2O$ drop by drop until the pH = 8.0. Dilute the mixture with water to 250 mL.
3. Mobile phase B: 5 mM NH_4FA, 80% ACN (*see* **Note 2**). Prepare mobile phase B (250 mL): Add the same amount of FA and $NH_3{\cdot}H_2O$ to 20 mL water. Dilute the mixture with water to 50 mL and finally with ACN to 250 mL.
4. LC-MS/MS system, such as Q Exactive HFX mass spectrometer combined with online nano-flow EASY-nLC1200 system (Thermo Fisher Scientific, USA).
5. Computer with installed software for LC-MS/MS control.
6. In-house packed C18 column or commercial C18 analytical column.

2.5 Data Analysis

1. Computers with installed software, such as MaxQuant, for data analysis.

3 Methods

The complete workflow for phosphoproteomic analysis described in this chapter is illustrated in Fig. 1. It consists of protein extraction, digestion, phosphopeptide enrichment, offline fractionation, and alkaline-pH-MS/MS analysis. The procedures are very simple and straightforward, which can be integrated into already existing methods.

3.1 Protein Extraction, Reduction, Alkylation, and Digestion

1. For cell samples (prokaryotic or eukaryotic cells), carefully remove the culture medium, wash three times with PBS, and add 500 μL lysis buffer. For tissue samples, add 500 μL lysis buffer directly. The volume of lysis buffer depends on the amount of cells and tissues.

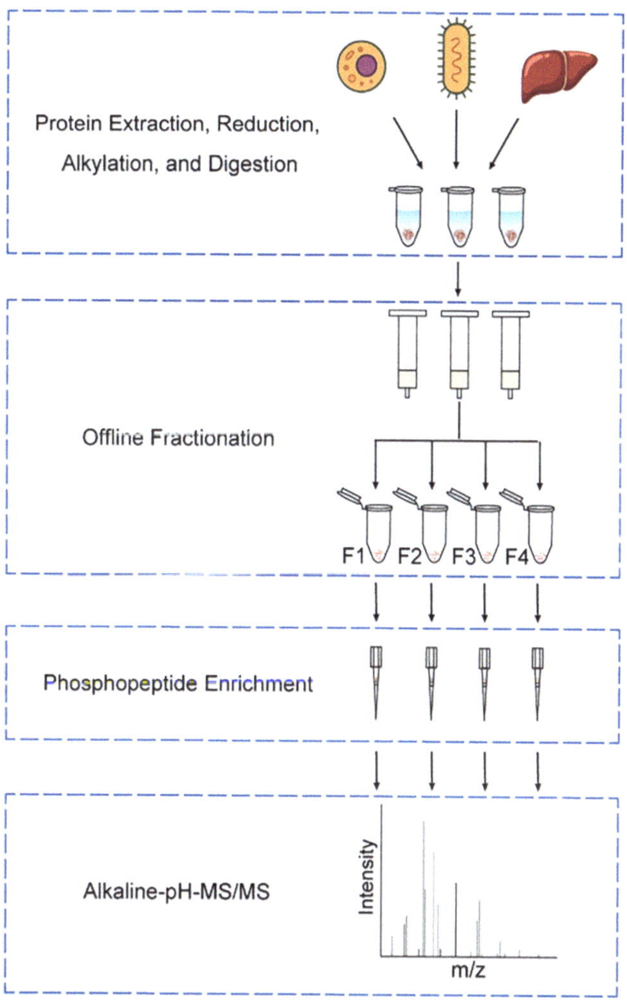

Fig. 1 Schematic overview of phosphoproteomic analysis using alkaline-pH-MS/MS

2. Sonicate the samples for 15 cycles of 10 s ON and 10 s OFF at 30% power on an ice-cold sonication bath. For tissue samples, homogenize them with a tissue homogenizer before sonicating.
3. Incubate the samples for 40 min at 4°C on a rotator.
4. Centrifuge the samples at 4°C, 18,000×*g* for 60 min, and the supernatants are collected.
5. Dilute the supernatants three times with dilution buffer.
6. Add benzonase to achieve a final concentration of 1% and incubate them in a Thermomixer for 120 min at room temperature (RT) at 1200 rpm.
7. Transfer the supernatants to a 15 mL tube, then add methanol at a 4:1 volume ratio, followed by chloroform at a 1:1 ratio, and finally ultrapure water at a 3:1 ratio, mixing thoroughly after each addition.
8. Centrifuge the mixtures at RT, 8000×*g* for 10 min, and carefully discard the supernatants.
9. Add methanol at a 3:1 volume ratio and sonicate for six cycles of 10 s ON and 10 s OFF at 15% power.
10. Centrifuge the mixtures at RT, 8000×*g* for 10 min and carefully discard the supernatants. Air-dry the precipitate (*see* **Note 3**).
11. Resuspend the precipitate in digestion buffer and sonicate for six cycles of 10 s ON and 10 s OFF at 15% power.
12. Determine the protein concentration using the tryptophan-based fluorescence quantification method (*see* **Note 4**).
13. Add trypsin at a 1:50 (enzyme-to-substrate) ratio and incubate all samples in a Thermomixer for 12 h at RT with agitation at 1200 rpm (*see* **Note 5**).
14. Add another aliquot of the same amount of trypsin and incubate in a Thermomixer for 6 h at RT with agitation at 1200 rpm (*see* **Note 5**).

3.2 Offline Fractionation (See Note 6)

1. Acidify all samples with 10% (v/v) FA to pH 3.5 (test with pH strips).
2. Centrifuge the samples at 4°C, 15,000×*g* for 3 min, and the supernatants are collected.
3. Add 1 mL methanol to the C18 cartridge (50 mg), followed by 1 mL condition buffer (*see* **Note 7**).
4. Equilibrate the C18 Cartridge with 1 mL wash buffer twice.
5. Load the samples onto the C18 Cartridge.
6. Wash the C18 Cartridge with 1 mL wash buffer twice.

7. Elute the peptides sequentially with 0.5 mL 5% phase B/95% phase A (v/v) (1), 10% phase B/90% phase A (v/v) (2), 15% phase B/85% phase A (v/v) (3), 20% phase B/80% phase A (v/v) (4), 25% phase B/75% phase A (v/v) (5), 30% phase B/70% phase A (v/v) (6), 35% phase B/65% phase A (v/v) (7), and 90% phase B/10% phase A (v/v) (8).
8. Mix fraction (1) with fraction (5), fraction (2) with fraction (6), fraction (3) with fraction (7), and fraction (4) with fraction (8), respectively.
9. Evaporate the sequentially eluted peptides using a vacuum concentrator.

3.3 Phosphopeptide Enrichment

1. Reconstitute the fractionated peptides in 200 μL phosphorylation wash buffer.
2. Acidify all samples with 10% (v/v) TFA to pH 3.0 (test with pH strips).
3. Incubate each sample with 50 μL Fe-NTA agarose beads for 20 min at 4°C at 1200 rpm (*see* **Note 8**).
4. Transfer the mixtures into the filter tip and centrifuge at 4°C, 50×*g* for 1 min (*see* **Note 9**).
5. Wash the filter tip with 200 μL phosphorylation wash buffer and centrifuge at 4°C, 50×*g* for 1 min; repeat this process three times.
6. Wash the filter tip with 200 μL ultrapure water and centrifuge at 4°C, 50×*g* for 1 min; repeat this process twice.
7. Transfer the filter tip to new 1.5 mL low binding microcentrifuge tubes (*see* **Note 10**).
8. Add 200 μL phosphorylation elution buffer, let it stand for 3 min, and centrifuge at 4°C, 50×*g* for 1 min; repeat this process once.
9. Evaporate the eluted phosphopeptides using a vacuum concentrator.

3.4 Alkaline-pH-MS/MS

1. Purge, flush, and equilibrate the LC system (Fig. 2, using the EASY-nLC 1200 as an example) (*see* **Note 11**).
2. Connect the C18 column to the LC system and wash the column at 55°C and 300 nL/min flow rate using 50/50 mobile phase A/B for 30 min.
3. Set up the gradient elution method according to Table 1 (*see* **Note 12**).
4. Resuspend the phosphopeptides in 7 μL water of LC-MS grade and centrifuge at RT, 15,000×*g* for 10 min (*see* **Note 13**). Transfer 3.5 μL supernatant into an autosampler vial.

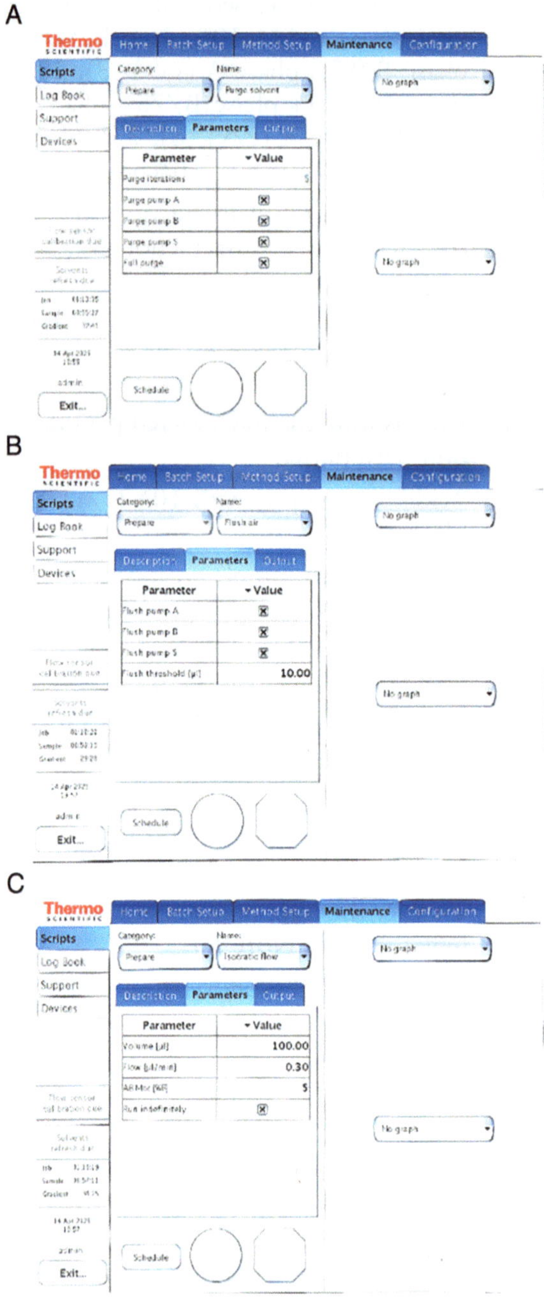

Fig. 2 Prepare the LC system, using the EASY-nLC 1200 as an example. (**A**) Purge; (**B**) flush; and (**C**) equilibrate

5. Inject 3 μL supernatant onto the C18 column using LC system and perform MS analysis. Taking the HFX system as an example (*see* **Note 14**), the parameter settings are as follows: full-scan mass spectra are acquired at positive ion mode, and the

Table 1
Gradient table for alkaline-pH-MS/MS

Times (min)	Flow rate (nL/min)	%A	%B
Initial	300	99.0	1.0
1.00	300	95.0	5.0
79.00	300	75.0	25.0
85.00	300	65.0	35.0
86.00	300	0.0	100.0
90.00	300	0.0	100.0

ion transfer tube temperature is set at 275°C. Data acquisition is programmed in data-dependent acquisition (DDA) mode, in which one full MS scan (350–1650 m/z) is first performed at 60,000 resolution, with a 3e6 automatic gain control (AGC) target and 20 ms maximum injection time. Then, MS2 scans are generated by higher-energy collisional dissociation (HCD) fragmentation with a "Top-20" method and 30 s dynamic exclusion. MS/MS acquisition is performed at 15,000 resolution with a 22 ms maximum injection time, and the AGC target is 1e5. Precursor fragmentation is performed with a 1.6 m/z isolation window and 27% normalized collision energy (NCE). Precursors with charges 1, 7, 8, and >8 are excluded for MS2 analysis.

3.5 Data Analysis

1. All raw files acquired from LC-MS/MS are analyzed by database searching software, such as MaxQuant (*see* **Note 15**). The parameter settings are as follows: carbamidomethylation (C) is set as a fixed modification, while oxidation (M), acetylation (Protein N-term), and phosphorylation (STYKRH) are set as variable modifications. A maximum number of five modifications is allowed for each peptide. Trypsin digestion is defined with cleavage sites at K/R except when followed by P. The maximum missing cleavage site is set as 3, and the minimal peptide length is set at 7. The tolerances of the first search and main search for peptides are set at 20 ppm and 4.5 ppm, respectively. The false discovery rate (FDR) is set to 0.01 at the level of peptides, proteins, and sites.
2. The class I phosphorylation sites (localization probability > 0.75) are considered as highly reliable sites.
3. The phosphopeptides as well as their charge and molecular weight (MW) can be extracted from "Evidence.txt."

4. The phosphosites as well as their intensities can be extracted from "Phospho (STYKRH)Sites.txt."

4 Notes

1. MS mobile phase and organic solutions should be freshly prepared or stored within 2 weeks prior to performing the protocol to avoid changes in buffer composition.
2. Water, ACN, FA, and $NH_3{\cdot}H_2O$ should be of LC-MS grade. We do not recommend using NH_4FA (solid state, LC-MS grade)—the impurities (e.g., metal cation) in NH_4FA (solid state) might contaminate the system and artificially modify the peptides during chromatographic separation or ionization.
3. Alternatively, the precipitate can be evaporated using a vacuum concentrator. Nevertheless, this process should not be overly prolonged, because extended drying times may render the precipitate difficult to dissolve in the digestion buffer.
4. The procedure of tryptophan-based fluorescence quantification: prepare a series of tryptophan standard curve solutions at concentrations of X ng/μL (X=0, 12.5, 25, 37.5, 50, 75, 100, 125, 150, 175, 200, 300, 400). Add 5 μL of each tryptophan standard curve solutions and samples in 1 mL tryptophan dilution buffer (8 M urea, 20 mM Tris-HCl, pH 7.6). Add 200 μL each tryptophan standard curve solutions and samples in a flat-bottom 96-well plate. Measure fluorescence intensity with excitation at 295 nm and emission at 350 nm.
5. To analyze both O-linked and N-linked phosphorylation, we recommend incubating the samples at RT to minimize the degradation of N-linked phosphorylation. For studies focused exclusively on *O*-phosphorylation, incubation at 37°C is recommended.
6. Alkaline-pH-MS/MS has better orthogonality with offline low-pH reversed-phase fractionation rather than offline high-pH reversed-phase fractionation.
7. The 50 mg C18 cartridge is sufficient for loading up to 2.5 mg peptides. For larger amounts of peptides, higher-capacity C18 cartridge (e.g., 200 mg) or HLB cartridge (Agilent Technologies) is recommended.
8. To analyze both O-linked and N-linked phosphorylation, we recommend incubating the samples at 4°C to minimize the degradation of N-linked phosphorylation. For studies focused exclusively on *O*-phosphorylation, incubation at RT is recommended.

9. Do not cover the centrifuge rotor lid and set the centrifuge speed low (<100 ×g) to prevent the beads from falling off the filter tip.
10. It is essential to use 1.5 mL low binding microcentrifuge tubes to prevent the loss of phosphopeptides.
11. Other nano-flow LC systems, for example, EASY-nLC1000, Vanquish Neo, and nanoElute, are also suitable for this protocol.
12. This is the generic gradient for alkaline-pH-MS/MS analysis of phosphopeptides and has been proven as useful for most applications.
13. To analyze both O-linked and N-linked phosphorylation, we recommend using water to minimize the degradation of N-linked phosphorylation. For studies focused exclusively on O-phosphorylation, 0.1% FA is recommended.
14. Other MS systems, for example, Tims-TOF Pro, Orbitrap Astral, and ZenoTOF 7600, are also suitable for this protocol.
15. Other database-searching softwares, for example, pFind, PEAKS studio, and Proteome Discoverer, are also suitable for this protocol.

References

1. Morshed N, Lee MJ, Rodriguez FH et al (2021) Quantitative phosphoproteomics uncovers dysregulated kinase networks in Alzheimer's disease. Nat Aging 1(6):550–565
2. Lu Y, Xu J, Li Y et al (2024) DRAK2 suppresses autophagy by phosphorylating ULK1 at Ser(56) to diminish pancreatic beta cell function upon overnutrition. Sci Transl Med 16(733):eade8647
3. Wang Y, Shu H, Qu Y et al (2024) PKM2 functions as a histidine kinase to phosphorylate PGAM1 and increase glycolysis shunts in cancer. EMBO J 43(12):2368–2396
4. Kalagiri R, Stanfield RL, Meisenhelder J et al (2021) Structural basis for differential recognition of phosphohistidine-containing peptides by 1-pHis and 3-pHis monoclonal antibodies. Proc Natl Acad Sci USA 118(6):e2010644118
5. Adam K, Hunter T (2018) Histidine kinases and the missing phosphoproteome from prokaryotes to eukaryotes. Lab Invest 98(2):233–247
6. Schweiger R, Linial M (2010) Cooperativity within proximal phosphorylation sites is revealed from large-scale proteomics data. Biol Direct 5:6
7. Chrustowicz J, Sherpa D, Li J et al (2024) Multisite phosphorylation dictates selective E2-E3 pairing as revealed by Ubc8/UBE2H-GID/CTLH assemblies. Mol Cell 84(2):293–308.e14
8. Asfaha JB, Ord M, Carlson CR et al (2022) Multisite phosphorylation by Cdk1 initiates delayed negative feedback to control mitotic transcription. Curr Biol 32(1):256–263.e4
9. Raghuraman P, Park S (2025) Exploring the modulation of phosphorylation and SUMOylation-dependent NPR1 conformational switching on immune regulators TGA3 and WRKY70 through molecular simulation. Plant Physiol Bioch 222:109711
10. Deng L, Wan J, Cheng A et al (2022) Duck plague virus US3 protein kinase phosphorylates UL47 and regulates the subcellular localization of UL47. Front Microbiol 13:876820
11. Dengler L, Örd M, Schwab LM, Loog M, Ewald JC et al (2021) Regulation of trehalase activity by multi-site phosphorylation and 14-3-3 interaction. Sci Rep 11(1):962–962. https://doi.org/10.1038/s41598-020-80357-3
12. Jones MH, Gergely ZR, Steckhahn D et al (2024) Kinesin-5/Cut7 C-terminal tail phosphorylation is essential for microtubule sliding force and bipolar mitotic spindle assembly. Curr Biol 34(20):4781–4793.e6
13. Xiao JR, Turner JJ, Koivomägi M et al (2024) Whi5 hypo- and hyper-phosphorylation dynamics control cell-cycle entry and progression. Curr Biol 34(11):2434–2447.e5

14. Stefanoska K, Gajwani M, Tan ARP et al (2022) Alzheimer's disease: ablating single master site abolishes tau hyperphosphorylation. Sci Adv 8(27):eabl8809
15. He P, Bian AW, Miao Y et al (2022) Discovery of a highly potent and orally bioavailable STAT3 dual phosphorylation inhibitor for pancreatic cancer treatment. J Med Chem 65(22): 15487–15511
16. Meng Q, Wang Y, Lu D et al (2023) A dataset resource for clinically associated phosphosites in hepatocellular carcinoma. Proteomics 23(3–4):e2100407
17. Muneer G, Chen CS, Lee TT et al (2024) A rapid one-pot workflow for sensitive microscale phosphoproteomics. J Proteome Res 23(8): 3294–3309
18. Gao L, Tao J, Qi L et al (2022) Synthesis of a metal oxide affinity chromatography magnetic mesoporous nanomaterial and development of a one-step selective phosphopeptide enrichment strategy for analysis of phosphorylated proteins. Anal Chim Acta 1195:339430
19. Yang DS, Ding XY, Min HP et al (2017) Design and synthesis of an immobilized metal affinity chromatography and metal oxide affinity chromatography hybrid material for improved phosphopeptide enrichment. J Chromatogr A 1505:56–62
20. Hardman G, Perkins S, Brownridge PJ et al (2019) Strong anion exchange-mediated phosphoproteomics reveals extensive human non-canonical phosphorylation. EMBO J 38:e100847
21. Cui F, Qian X, Ying W (2021) Integrated strategy for unbiased profiling of the histidine phosphoproteome. Anal Chem 93(47): 15584–15589
22. Gao J, Wang Y, Zhang S et al (2023) Online alkaline-pH reverse phase nanoelectrospray-tandem mass spectrometry complements conventional low-pH method for global proteomic analysis. J Proteome Res 22(5):1446–1454
23. Wang Y, Gao J, Xie W et al (2025) Online alkaline-pH reversed-phase nanoelectrospray-tandem mass spectrometry complements traditional phosphoproteomic analysis via influencing charge state distribution of phosphopeptides. J Proteome Res. doi https://doi.org/10.1021/acs.jproteome.4c01091. Published online 10 April 2025

Chapter 8

Phosphoproteomics Protocol for the Identification of Novel Biomarkers from Extracellular Vesicles of Cancerous Cells

Zoran Minic, Suttinee Poolsup, Yingxi Li, Rochelle D'Mello and Maxim V. Berezovski

Abstract

In this chapter, we describe a general sample preparation workflow for small extracellular vesicles (sEVs) from breast cancer cell lines using techniques such as protein solubilization, enzymatic digestion, and Fe^{3+}-immobilized metal affinity chromatography (IMAC) and titanium dioxide (TiO_2) enrichment prior to mass spectrometry-based proteomic analysis. This workflow can be used to perform global phosphoproteomics of sEVs enabling the study of tumor biology and the discovery of biomarkers for cancer diagnosis and treatment.

Keywords Cells and sEVs preparation, Protein extraction, Tryptic digestion, Titanium dioxide phosphopeptide enrichment, IMAC phosphopeptide enrichment

1 Introduction

Protein phosphorylation refers to the reversible posttranslation modification that is mediated by kinases and phosphatases [1]. This is the most frequent posttranslational modification in humans, playing an essential role in almost all cellular processes. Protein phosphorylation causes structural alterations, assembling or detaching protein complexes, and directing the modulation of protein activity [2]. Understanding the implications of protein phosphorylation can provide a deeper insight into biomolecular processes such as cellular growth, apoptosis, signal transduction, and the major regulatory mechanisms of cell signaling networks [3]. Kinase enzymes covalently bind phosphate groups to serine, tyrosine, and threonine residues [2, 4]. This process can be reversed by phosphatase enzymes that remove phosphate groups from proteins. Both phosphatase and its counterpart, kinase, are essential to cell regulation and cell signaling. Many disorders occur due to mutations in protein kinases and phosphatases that lead to disruptions in the signaling pathways [2, 5].

Xu Zhang et al. (eds.), *Protein Post-Translational Modifications*, Methods in Molecular Biology, vol. 3018,
https://doi.org/10.1007/978-1-0716-5166-7_8,

An entire branch of proteomics, phosphoproteomics, is the large-scale analysis used for the identification and characterization of phosphorylated proteins. However, after the digestion of proteins, the relatively low abundance of phosphorylated peptides can be a limiting factor to their detection. For this reason, it is imperative to use techniques that improve the sensitivity and identification of phosphoproteins/phosphopeptides in phosphoproteomic studies [1, 6]. These enrichment techniques help isolate and concentrate the phosphoproteins/phosphopeptides from the complex sample mixture. This can be done through various approaches, but the most commonly used enrichment techniques, the Fe^{3+}-IMAC and TiO_2 methods, are based on affinity chromatography [1, 6].

Phosphopeptide enrichment using either Fe^{+3}-IMAC or TiO_2 shows different specificities to the phosphopeptides with limited overlap, indicating that no single enrichment method is sufficient for a comprehensive phosphoproteomic analysis [7]. IMAC separates phosphoproteins and phosphopeptides with an affinity for positively charged metal ions [8]. Titanium dioxide (TiO_2) and zirconium dioxide (ZrO_2) are the most used methods for phosphopeptide enrichment. IMAC enriches multiple phosphorylated peptides, while TiO_2 chromatography more strongly selects phosphorylated peptides and is tolerant of most biological buffers [9–12]. Thus, using both techniques, Fe^{+3}-IMAC and TiO_2 can result in increased selectivity and more identified phosphopeptides.

In phosphoproteomic analysis, most phosphorylation sites are identified using mass spectrometry (MS)-based fragmentation technology, including collision-induced dissociation (CID), higher-energy collisional dissociation (HCD), and electron transfer dissociation (ETD) [1]. However, fragmentation of phosphopeptides is not a trivial task due to their low ionization efficiency and a small number of peptide fragment ions for identification caused by the loss of the labile phosphate group. Although CID remains the main fragmentation method used for mass spectrometry, it can cause the preferential loss of phosphate from peptides phosphorylated at serine or threonine (but not tyrosine). HCD [13] works similarly to CID but is slightly more effective for the fragmentation of higher charge state ions [14]. In contrast to CID and HCD, ETD fragmentation does not lead to a preferential loss of phosphoric acid from the phosphorylated peptides. Taken together, strategies combining multiple enrichment and fragmentation methods can lead to the identification of a greater number of phosphopeptides.

Extracellular vesicles (EVs) derived from human cancer cell lines are involved in multiple biological processes in tumor biology, in managing the microenvironment, and in tumor progression [15–21]. EVs can transfer specific biological information by delivering bioactive content (nucleic acids, proteins, and metabolites) from donors to nearby or distant cells [22–24]. Therefore, EVs appear as key regulators of cell–cell communication within multicellular organisms, vital for health and disease. Subtypes of membrane-bound vesicles include exosomes and microvesicles, distinguished by their biogenesis pathway. These subtypes may be roughly categorized by their measured diameter. In this study, vesicles with a measured diameter in the range of 50–300 nm are referred to as small extracellular vesicles (sEVs).

In order to elucidate the molecular basis of cancer or discover new biomarkers, it is important to measure protein expression or posttranslation modification in EVs, such as phosphorylated proteins. Recently, sEVs released from cells have emerged as promising biomarkers for cancer diagnosis and therapies [25]. Several studies have reported the detection of phosphoproteins in biofluids for disease diagnosis [26–28]. However, this research field is still in development and requires more extensive phosphoproteomic studies.

In this chapter, a laboratory protocol describes the sEVs isolation, sample preparation, protein extraction, in-solution digestion with trypsin, and phosphopeptide enrichment with IMAC and TiO_2 using two fragmentation methods (CID and HCD) and nLC-MS/MS phosphoproteomics analyses.

2 Materials

All solutions are prepared using deionized water (with a conductivity of 18 MΩ·cm at 23°C) or MS-grade water.

2.1 Reagents, Solutions, and Buffers

1. 0.1 M TCEP (tris(2-carboxyethyl) phosphine): 0.1 M TCEP in MS-grade water (*see* **Note 1**).
2. MS-grade acetonitrile (ACN) with formic acid (FA): 0.1% FA in ACN.
3. MS-grade water/C18 washing buffer: MS-grade water (optional. 0.1% FA).
4. C18 activation and elution buffer: 70% MS-grade acetonitrile (ACN) containing 0.1% FA.
5. Dithiothreitol (DTT).
6. Bovine serum albumin (BSA).
7. Luna C18 particles, 3 μm, 100 Å (Phenomenex, Torrance, CA, USA).
8. Fused silica capillary using 15 cm LT × 75 μm i.d. × 365 μm o.d. (Polymicro Technologies).
9. 0.5 M iodoacetamide (IAA): 0.5 M IAA in MS-grade water (*see* **Note 2**).
10. Breast cancer epithelial cell lines MDA-MB-231 (ATCC HTB-26) and MCF7 (ATCC HTB-22) and the nontumorigenic

breast epithelial cell line MCF10A (ATCC CRL-10317) (obtained from the American Type Culture Collection [ATCC]).

11. EV-depleted Dulbecco's Modified Eagle's Medium (DMEM)/F-12.
12. Phosphate-buffered saline (PBS).
13. 0.3 μg/μL trypsin/lys-C (Promega, # V5072) (*see* **Note 3**).
14. 10% stock solution of *n*-dodecyl β-D-maltoside (DDM).
15. 50% stock solution of glycerol.
16. 0.2 M HEPES: 0.2 M 4-(2-Hydroxyethyl)piperazine-1-ethanesulfonic acid (HEPES) in MS-grade water (prepared stock at pH = 8).
17. Peptide reconstitution buffer: MS-grade H_2O containing 1% FA.
18. 10% fetal bovine serum.
19. 5% horse serum.
20. Penicillin/Streptomycin.
21. 100 ng/mL cholera toxin.
22. 0.5 μg/mL hydrocortisone.
23. 20 ng/mL EGF protein.
24. 10 μg/mL insulin.
25. 5X lysis buffer for phosphoproteomics (total 10 mL):100 mM HEPES, pH 8.0, 25% glycerol, 0.5% DDM, 1 mM DTT, 8 M urea, and 1/50 (v/v) phosphatase inhibitors (phosphatase inhibitor 2 and 3; Millipore Sigma, # P0044 and # P5726) (*see* **Note 4**).
26. Buffer A: 100 μL MS-grade water, 401 μL ACN, and 2 μL trifluoroacetic acid (TFA).
27. Buffer B: 1 mL Buffer A and 400 μL lactic acid (90%).
28. Elution buffer 1: 20 μL ammonium hydroxide (30%) and 380 μL MS-grade water.
29. Elution buffer 2: 20 μL pyrrolidine and 380 μL MS-grade water.
30. MCF7 and MDA-MB-231 cell culture media: DMEM/F-12 media with 10% EV depleted fetal bovine serum (FBS) and 1% penicillin/streptomycin.
31. MCF10A cell culture media: DMEM/F-12 media with 5% horse serum (HS), 1% penicillin/streptomycin, 20 ng/mL EGF, 0.5 μg/mL hydrocortisone, 100 ng/mL cholera toxin, and 10 μg/mL insulin.

32. Prepare an in-house packed nano column using Luna C18 particles (Luna C18(2), 3 μm, 100 Å, Phenomenex, Torrance, CA, USA) using 15 cm LT × 75 μm i.d. × 365 μm o.d. fused silica capillary (Polymicro Technologies) and fused silica as described on the following website: https://proteomicsresource.washington.edu/docs/protocols05/Packing_Capillary_Columns.pdf (*see* **Note 5**).

2.2 Additional Materials/Equipment

1. Bradford protein assay kit.
2. C18 TopTip microcolumns (*see* **Note 6**).
3. Fe-NTA phosphopeptide enrichment kit (Thermo Fisher Scientific, # 88300).
4. TiO_2-phosphopeptide enrichment kit (Thermo Fisher Scientific, # 88301).
5. 96-well plate.
6. Silicone sealing mat for 96-well PCR microplates.
7. Fisher Scientific™ Vortex Mixers.
8. Savant™ SpeedVac™ SPD120 Vacuum Concentrator.
9. Orbitrap Fusion mass spectrometer.
10. UltiMate 3000 nano-RSLC.
11. Beckman Coulter XL-A Analytical Ultracentrifuge, ultracentrifuge tubes (Polypropylene, Open-Top Thinwall centrifuge tubes, 38.5 mL) and a SW 28 swinging bucket-rotor.
12. ZetaView PMX-110 (Particle Metrix, Meerbusch, Germany). The instrument was calibrated with 102 nm polystyrene beads.
13. Semimicrovolume cuvettes.
14. Spectrophotometer.

3 Methods

Workflow of sample processing for phosphoproteomics analysis of sEVs is presented in Fig. 1.

3.1 Cell Culture

3.1.1 MDA-MB-231 Cell Line

1. Inoculate the cell line starter purchased from ATCC (see Sect. 2.1) in a 100-mm cell culture dish with 20 mL of complete MDA-MB-231 growth medium.
2. Incubate cells at 37°C with 5% CO_2 until reaching 80% confluency.
3. Subculture the cells using trypsinization and perform centrifugation to collect a cell pellet at 200*g* for 5 min at 4°C.

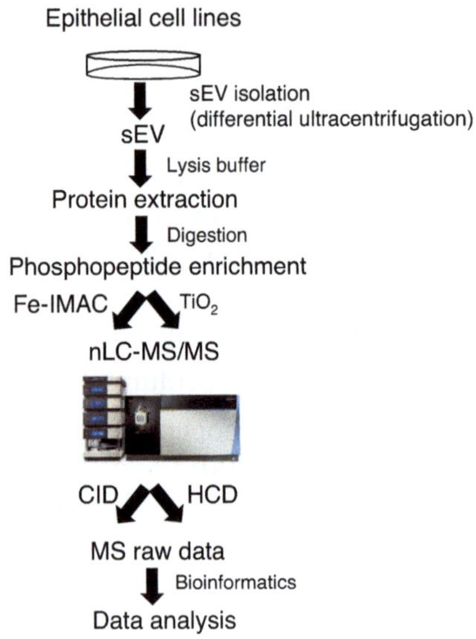

Fig. 1 Workflow illustration of phosphoproteome analysis of sEVs from cancerous cell lines

4. After resuspending the cell pellet with 1 mL fresh growth media, transfer 1 μL of the cell suspension into an empty Eppendorf tube to dilute in trypan blue (1:10 dilution ratio) and determine the total number of cells and percentage of viability (85%–90%) using an automated cell counter. The actual total number of cells is calculated by multiplying the cell density obtained from the automated cell counter by a dilution factor of 10 (*see* **Note 7**).
5. Calculate the volume of cell-suspended media to dilute to the final cell density of 1.0×10^6 cells per 100-mm cell culture dish.
6. Plate 1.0×10^6 cells per 100-mm cell culture dish in 20 mL of complete MDA-MB-231 growth medium.
7. Incubate cells at 37°C with 5% CO_2 for 7 days. Cell density should be 80% confluent on the day of EV-depleted media collection.

3.1.2 MCF10A Cell Line

1. Following the same procedure to determine cell density and viability as MDA-MB-231, plate 1.0×10^6 cells per 100-mm cell culture dish in 20 mL of complete MCF10A growth medium.
2. Incubate cells at 37°C with 5% CO_2 for 7 days. Cell density should be 80% confluent on the day of EV-depleted media collection.
3. Depending on the cell passage and cell culture conditions, the MCF10A cell line may take longer than 7 days to reach 80% confluency.

3.1.3 MCF7 Cell Line

1. Following the same procedure to determine cell density and viability as MDA-MB-231, Plate 1.0×10^6 cells per 100-mm cell culture dish in 20 mL of the complete MCF7 growth media.

Incubate cells at 37°C with 5% CO_2 for 7 days. Cell density should be 80% confluent on the day of EV-depleted media collection.

3.2 sEV Isolation Using Differential Ultracentrifugation (dUC)

1. Harvest conditioned medium (approximately 240 mL) after 7 days of incubation.
2. Centrifuge the collected medium at 300×*g* at 4°C for 10 min.
3. Transfer the supernatant to new centrifuge tubes and centrifuge the supernatant at 2000×*g* at 4°C using a Sigma13190 rotor (MBI) for 20 min to remove apoptotic bodies.
4. Collect the supernatant, then spin it at 16,500×*g* for 1 h at 4°C using an SW28 Ti rotor to deplete microvesicles.
5. Collect the supernatant, then spin it at 100,000×*g* for 3 h at 4°C using the sample rotor to pellet the sEVs.
6. Discard the supernatant and resuspend the sEV pellets with PBS. Combine the resuspended pellets in one UC tube, and centrifuge at 100,000×*g* for 1 h at 4°C using the same rotor.
7. Discard the supernatant, resuspend the sEV pellet in 200 μL PBS, and store at −80°C for later use.

3.3 Quantification of sEVs by Nanoparticle Tracking Analysis (NTA)

1. Prepare a serial dilution of the isolated sEVs in 1X PBS in 1:10, 1:100, 1:1000, and 1:10,000 ratios. The PBS buffer must be filtered through a 0.2-μm syringe filter before use.
2. Measure the average particle diameter and approximate concentration of the isolated sEVs using a ZetaView PMX-110.
3. Calibrate the instrument using 100-nm diameter polystyrene beads. Set the camera shutter speed to 85–40 and the particle diameter focus to 102 nm.
4. Inject 100 μL of the diluted EVs into the sample compartment of the ZetaView PMX-100.
5. The instrument will now measure the average particle diameter of the isolated sEVs using nanoparticle tracking analysis (NTA). A typical result of the particle size distribution is presented in Fig. 2.

3.4 Protein Determination Using Bradford Assay

1. Make 10 mL of 1000 μg/mL BSA working solution (WS) in PBS, pH 7.4.
2. Dilute BSA WS into the following concentrations using MS-grade water: 0, 1, 2.5, 5, 7.5, 10, 12, 15, 20 μg/mL (*see* **Note 8**). All dilution standards should be within BSA's linear dynamic range (0.5–20 μg/mL).

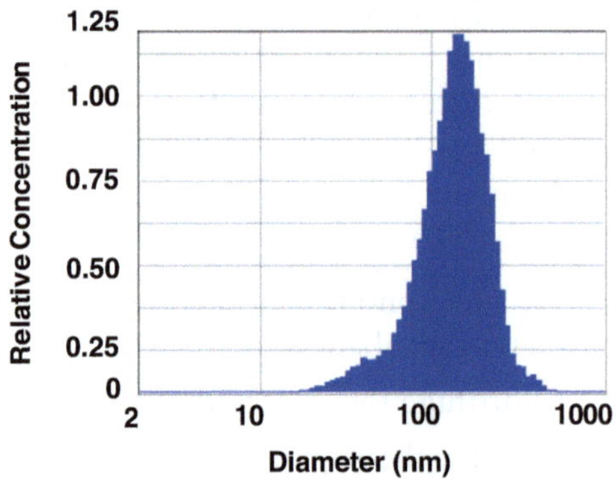

Fig. 2 Typical results of nanoparticle-tracking analysis (NTA) of sEVs from cancerous cell lines

3. Prepare the protein samples using 5–10 μL of sEVs and resuspend to a total volume of 600 μL with MS deionized water.
4. Add 400 μL Bradford buffer to complete the total volume of 1 mL.
5. To make the control sample, repeat Steps 3 and 4, substituting sEVs with PBS.
6. Transfer the mixture of solutions into cuvettes.
7. Measure the absorbance at 595 nm using a spectrophotometer.
8. Create the standard curve to determine the relationship between absorbance and concentration.
9. Calculate the concentration of protein from each sEV sample. These concentrations will be used to determine the amount of each sample used for protein extraction.

3.5 Protein Extraction

1. Resuspend the isolated sEVs containing about 25–50 μg of protein (*see* **Note 9**) in 5X lysis buffer with a volume ratio of 4:1 (fraction/buffer) to give a final concentration of 20 mM HEPES, pH 8.0, 5% glycerol, 0.1% DDM, 0.2 mM DTT, 1.6 M urea, and 1:250 (v/v) phosphatase inhibitors.
2. Gently vortex the mixture for 3 min.
3. Centrifuge the mixture for 1 min at 10,000×*g*.
4. Collect supernatant and continue with in-solution digestion.

3.6 In-Solution Digestion with Trypsin/Lys-C

1. Add 100 μL of sample solution (*see* **Note 10**) containing about 25–50 μg of protein to an Eppendorf tube.
2. Add 4 μL 100 mM TCEP and incubate at room temperature for 30 min.
3. Add 4 μL 500 mM IAA and incubate at room temperature for 40 min in the dark.

4. Add 1.5 μL trypsin/Lys-C solution and incubate at room temperature for 20 h.
5. Add 1.5 μL 100% FA, vortex for 5–10 s and centrifuge at 10,000×*g* for 15 s.
6. Samples are now ready for desalting (*see* **Note 11**).

3.7 TopTip Desalting

1. Add 50 μL of the activation solution and then push the solvent through using a syringe. Repeat this step two additional times. Discard the flow-through.
2. Wash the column with the equilibration solution (MS-grade water containing 0.1% FA) and then push the solvent through. Repeat this step two additional times. Discard the flow-through.
3. Add the sample solution (50–200 μL) on top of the resin bed and slowly push the solvent through to load it onto the column (add a maximum of 70 μL of sample at a time and repeat until all sample solution is loaded) (*see* **Note 12**).
4. Add 50 μL equilibration solution and then push solvent through. Repeat this step two additional times. Discard the flow-through.
5. Place the column into a new Eppendorf tube and add 100 μL activation solution (0.1% FA in 70% ACN) and elute peptides by pushing the solvent through. Repeat this step two times. Collect these three fractions into one tube.
6. Dry the collected samples using a SpeedVac (*see* **Note13**).
7. The dried sample can now be used for Phosphopeptide Enrichment (*see* **Note 14**).

3.8 Phosphopeptide Enrichment with TiO_2

1. Resuspend the dried peptide samples in 150 μL TiO_2 binding buffer (Buffer B) according to the manufacturer's protocol with a slight modification (*see* **Note 15**).
2. Prepare TiO_2 resin spin tips by sequentially rinsing each tip with 20 μL wash buffer (Buffer A) and 20 μL binding buffer (Buffer B).
3. Apply the peptide sample to the spin tip and centrifuge at 1000 ×*g* for 10 min.
4. Wash the column by centrifuging it at 3000 ×*g* for 2 min consecutively, first with 25 μL binding buffer (Buffer B), and then with 25 μL wash buffer (Buffer A), and repeat this Step 4 twice.
5. Release the enriched phosphopeptides by adding 50 μL elution buffer 1 and centrifuge at 1000 ×*g* for 5 min.
6. Elute phosphopeptides by adding 50 μL elution buffer 2 and centrifuge at 1000 ×*g* for 5 min using the same collection tube.

7. Immediately dry the combined eluate using a vacuum concentrator, and then, add 20 µL MS-grade water containing 1% of FA. This sample is now ready for nLC–MS/MS (*see* **Note 16**).

3.9 Phosphopeptide Enrichment by IMAC

1. Resuspend the dried peptide samples obtained at the end of Sect. 3.7 Step 7 in 200 µL binding/wash buffer according to the Pierce Fe-NTA phosphopeptide enrichment kit protocol (*see* **Note 17**).
2. Prepare the Fe-NTA spin column by removing the storage solution by centrifuging it at 1000 ×*g* for 30 s, and then rinse twice by adding 200 µL binding/wash buffer and centrifuging at 1000 ×*g* for 30 s.
3. Resuspend the dried peptide sample in 200 µL binding/wash buffer, vortex for 15 s and then centrifuge at 3000×*g* for 1 min.
4. Add the peptide sample to the equilibrated spin column and incubate for 30 min with gentle mixing every 10 min at 4–8°C.
5. Wash the column by adding 200 µL binding/wash buffer and centrifuge it at 500 ×*g* for 30 s. Repeat this Step 3 times.
6. Rinse the column by adding 200 µL of MS-grade water and centrifuge it at 500 ×*g* for 30 s.
7. Place the column in a new microcentrifuge tube for the elution step.
8. Release the phosphopeptides by adding 100 µL elution buffer and then centrifuge it at 1000 ×*g* for 30 s. Repeat this step and then immediately dry the combined eluate using a vacuum concentrator (*see* **Note 18**).
9. Add 20 µL MS-grade water containing 1% of FA to the dried eluate. This sample is now ready for nLC–MS/MS.

3.10 NanoRPLC–MS/MS Analysis of Phosphopeptides

1. Put 20 µL of enriched phosphopeptide samples obtained in Step 7 of Sect. 3.8 or Step 9 of Sect. 3.9 into a 96-well plate covered with a silicone sealing mat. Put the loaded 96-well plate in the autosampler of UltiMate 3000 nanoRSLC.
2. Pre-equilibrate the UltiMate 3000 nanoRSLC column with 98% mobile phase A for 30 min before peptide loading.
3. Load 3 µL of sample onto the column.
4. Elute the peptides at a flow rate of 300 nL/min for 65 min with the following gradient: 0–10 min, 2—2% ACN; 10–40 min, 2—38% ACN; 40–45 min, 38—98% ACN; 45–50 min, 98—98% ACN; 50–55 min, 98—2% ACN; 55–65 min, 2—2% ACN.
5. Process the samples with the Orbitrap Fusion mass spectrometer (*see* **Note 19**).

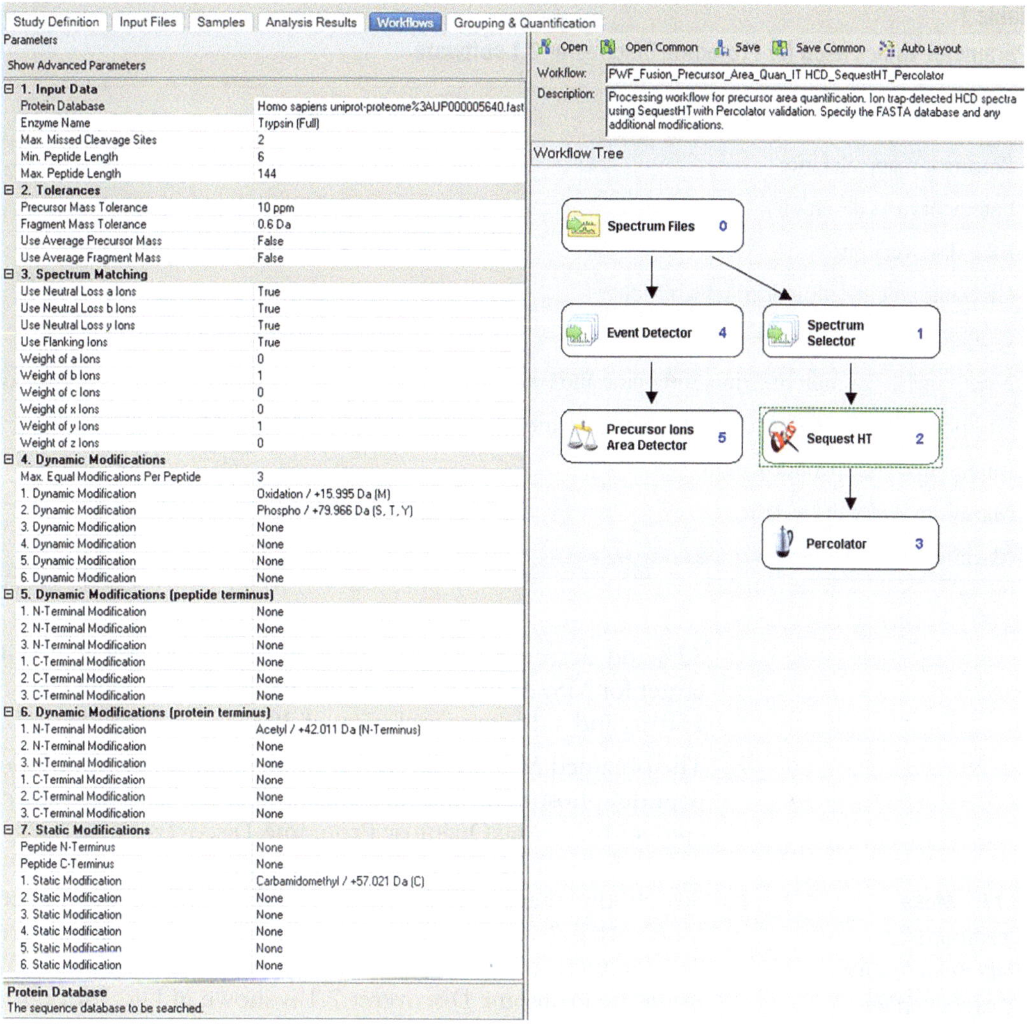

Fig. 3 Workflow and settings for processing phosphopeptide-enriched samples using Proteome Discoverer

6. The ESI+ parameters for MS processing are as follows: top speed mode, ion source temperature 250°C, ion spray voltage 2.1 kV, and a full-scan MS (m/z 350–2000) resolution of 60,000.
7. For CID, set the automatic gain control (AGC) target to 5×10^5 for full scans and the intensity threshold to 0.5×10^3 for MS/MS scans. Filter precursor ions from +2 to +7 charge states within a 2 m/z isolation window. CID in the linear ion trap is performed at a normalized collision energy of 35%.
8. For HCD, set the AGC target for precursor ions to 5×10^5 with an ion filling time of 150 ms. Isolate and fragment the highest intensity ions using a normalized collision energy of

Table 1
Parameter values used for Proteome Discoverer 2.1 software

Parameter	Value
Precursor mass tolerance	10 ppm
Fragment mass deviation	0.5 Da
False discovery rate	0.01
Carbamidomethylation of cysteine residues	Fixed modification
Oxidation of methionine residues	Variable modification
N-terminal acetylation of serine, threonine and tyrosine	Variable modification
Phosphorylation of serine, threonine, and tyrosine	Variable modification
Minimum tryptic peptide length	6 amino acids
Maximum molecular weight	4600 Da
Maximum missed cleavages	2

32% and detect at a mass resolution of 15,000. Set the AGC target for MS/MS to 5×10^4 with a maximum injection time of 150 ms and a dynamic exclusion of 30 s.

9. The obtained MS raw files can now be processed for phosphopeptide/phosphoprotein identification using different search engines (e.g., MaxQuant or Proteome Discoverer).

3.11 Mass Spectrometry Data Searches for Phosphorylation Analysis

1. Search the mass spectrometry raw files against the human UniProtKB/SwissProt database using Proteome Discoverer 2.1 (Sequest HT) (*see* **Note 20**). The processing workflow using the Proteome Discoverer 2.1 is shown in Fig. 3 for phosphopeptide-enriched samples.
2. The settings found in Table 1 can be used to preprocess the phosphoproteomic data.
3. Further downstream statistical and differential analysis of the data can be conducted using R packages (*see* **Note 21**).

4 Notes

1. After preparing the 0.1 M TCEP stock solution, distribute into 100 μL aliquots and store at −20°C until use.
2. Prepare, aliquot, and store IAA stock solution at −20°C. Note that IAA is light-sensitive so minimize exposure to light at all times.

3. Prepare trypsin stock solution in 50 mM Tris–HCl (pH 8.0) and store at −20°C for up to 3 months.
4. Ensure that the buffer used for in-solution digestion is at pH 8.0 for optimal trypsin activity. In-solution digestion can also be done using other buffers, such as 50 mM ammonium bicarbonate.
5. For the separation of peptides, it is also possible to use Thermo Scientific™ Acclaim™ PepMap™ RSLC C18 column (P/N ES800A), 15 cm × 75 μm ID, 3 μm, 100 Å.
6. Glygen TopTip microcolumns can also be purchased from PolyLCinc. Additionally, samples can also be desalted using disposable Pierce C-18 tips (Thermo Fisher Scientific, # 87784) with the addition of two C-18 resins from Pierce C-18 Spin Columns (Thermo Fisher Scientific, # 89870).
7. Here, MDA-MB-231, MCF7, and MCF10A cells are used as examples. If other cell types are used, culture conditions should be adjusted accordingly. The dilution factor varies upon the initial cell density, which may require additional dilution if the density of the cells exceeds the readable range on the automated cell counter.
8. The standard curve requires 600 μL of Bradford buffer, 100 μ L of each diluted protein sample, and an additional 300 μL of MS-grade water.
9. A minimum of 25 μg of protein is recommended for sufficient phosphopeptide enrichment. For optimal results, it is recommended to isolate as high quantity of sEVs as possible.
10. Maximum volume should not exceed 200 μL.
11. Samples can now be frozen at −20°C until further processing.
12. To ensure complete binding, recover flow-though and repeat this step.
13. Ensure that all solvent is evaporated.
14. Samples can be stored at −20°C or −80°C.
15. Samples can be desalted using TopSpin columns. Verify that pH of resuspended samples is <3 using pH strip. The manufacturer's protocol is slightly modified, especially for washing columns, by adding 150 μL of buffer B. The reason is to efficiently remove nonphosphorylated peptides.
16. Eluate cannot be stored in elution buffer.
17. Manufacturer's protocol is slightly modified, especially for washing columns, by adding 200 μL binding/wash buffer. The reason is to efficiently remove nonphosphorylated peptides. In addition, a smaller velocity for centrifugation was used for washing steps to avoid drying the columns.

18. Eluate cannot be stored in elution buffer.
19. Any high-resolution mass spectrometer can be used.
20. Any appropriate search engine, such as MaxQuant, can be used.
21. Statistical analysis depends on the research objective. One example is described in [25].

References

1. Minic Z, Dahms TES, Babu M (2018) Chromatographic separation strategies for precision mass spectrometry to study protein-protein interactions and protein phosphorylation. J Chromatogr B Analyt Technol Biomed Life Sci 1102-1103:96–108. https://doi.org/10.1016/j.jchromb.2018.10.022
2. Cohen P, Snelling T (2025) Diseases caused by altered specificity of a protein kinase for its allosteric activators. Trends Biochem Sci 50(1):61–70. https://doi.org/10.1016/j.tibs.2024.10.008
3. Pang K, Wang W, Qin JX, Shi ZD, Hao L, Ma YY, Xu H, Wu ZX, Pan D, Chen ZS, Han CH (2022) Role of protein phosphorylation in cell signaling, disease, and the intervention therapy. MedComm (2020) 3(4):e175. https://doi.org/10.1002/mco2.175
4. Pearlman SM, Serber Z, Ferrell JE Jr (2011) A mechanism for the evolution of phosphorylation sites. Cell 147(4):934–946. https://doi.org/10.1016/j.cell.2011.08.052
5. Gerritsen JS, White FM (2021) Phosphoproteomics: a valuable tool for uncovering molecular signaling in cancer cells. Expert Rev Proteomics 18(8):661–674. https://doi.org/10.1080/14789450.2021.1976152
6. Dudley E, Bond AE (2014) Phosphoproteomic techniques and applications. Adv Protein Chem Struct Biol 95:25–69. https://doi.org/10.1016/b978-0-12-800453-1.00002-6
7. Bodenmiller B, Mueller LN, Mueller M, Domon B, Aebersold R (2007) Reproducible isolation of distinct, overlapping segments of the phosphoproteome. Nat Methods 4(3):231–237. https://doi.org/10.1038/nmeth1005
8. Yang C, Zhong X, Li L (2014) Recent advances in enrichment and separation strategies for mass spectrometry-based phosphoproteomics. Electrophoresis 35(24):3418–3429. https://doi.org/10.1002/elps.201400017
9. Ficarro SB, McCleland ML, Stukenberg PT, Burke DJ, Ross MM, Shabanowitz J, Hunt DF, White FM (2002) Phosphoproteome analysis by mass spectrometry and its application to Saccharomyces cerevisiae. Nat Biotechnol 20(3):301–305. https://doi.org/10.1038/nbt0302-301
10. Jensen SS, Larsen MR (2007) Evaluation of the impact of some experimental procedures on different phosphopeptide enrichment techniques. Rapid Commun Mass Spectrom 21(22):3635–3645. https://doi.org/10.1002/rcm.3254
11. Pinkse MW, Uitto PM, Hilhorst MJ, Ooms B, Heck AJ (2004) Selective isolation at the femtomole level of phosphopeptides from proteolytic digests using 2D-NanoLC-ESI-MS/MS and titanium oxide precolumns. Anal Chem 76(14):3935–3943. https://doi.org/10.1021/ac0498617
12. Larsen MR, Thingholm TE, Jensen ON, Roepstorff P, Jørgensen TJ (2005) Highly selective enrichment of phosphorylated peptides from peptide mixtures using titanium dioxide microcolumns. Mol Cell Proteomics 4(7):873–886. https://doi.org/10.1074/mcp.T500007-MCP200
13. Olsen JV, Macek B, Lange O, Makarov A, Horning S, Mann M (2007) Higher-energy C-trap dissociation for peptide modification analysis. Nat Methods 4(9):709–712. https://doi.org/10.1038/nmeth1060
14. Zhang Z, Wu S, Stenoien DL, Paša-Tolić L (2014) High-throughput proteomics. Annu Rev Anal Chem (Palo Alto Calif) 7:427–454. https://doi.org/10.1146/annurev-anchem-071213-020216
15. Kruger S, Abd Elmageed ZY, Hawke DH, Wörner PM, Jansen DA, Abdel-Mageed AB, Alt EU, Izadpanah R (2014) Molecular characterization of exosome-like vesicles from breast cancer cells. BMC Cancer 14: 44. https://doi.org/10.1186/1471-2407-14-44
16. Palazzolo G, Albanese NN, DI Cara G, Gygax D, Vittorelli ML, Pucci-Minafra I (2012) Proteomic analysis of exosome-like vesicles derived from breast cancer cells. Anticancer Res 32(3):847–860
17. Hurwitz SN, Rider MA, Bundy JL, Liu X, Singh RK, Meckes DG Jr (2016) Proteomic profiling of NCI-60 extracellular vesicles uncovers common protein cargo and cancer type-specific biomarkers. Oncotarget 7(52):86999–87015. https://doi.org/10.18632/oncotarget.13569

18. Demory Beckler M, Higginbotham JN, Franklin JL, Ham AJ, Halvey PJ, Imasuen IE, Whitwell C, Li M, Liebler DC, Coffey RJ (2013) Proteomic analysis of exosomes from mutant KRAS colon cancer cells identifies intercellular transfer of mutant KRAS. Mol Cell Proteomics 12(2):343–355. https://doi.org/10.1074/mcp.M112.022806
19. Liang B, Peng P, Chen S, Li L, Zhang M, Cao D, Yang J, Li H, Gui T, Li X, Shen K (2013) Characterization and proteomic analysis of ovarian cancer-derived exosomes. J Proteomics 80:171–182. https://doi.org/10.1016/j.jprot.2012.12.029
20. Risha Y, Minic Z, Ghobadloo SM, Berezovski MV (2020) The proteomic analysis of breast cell line exosomes reveals disease patterns and potential biomarkers. Sci Rep 10(1):13572. https://doi.org/10.1038/s41598-020-70393-4
21. Risha Y, Susevski V, Hüttmann N, Poolsup S, Minic Z, Berezovski MV (2021) Breast cancer-derived microvesicles are the source of eunctional metabolic enzymes as potential targets for cancer therapy. Biomedicines 9(2):107. https://doi.org/10.3390/biomedicines9020107
22. Huang X, Yuan T, Tschannen M, Sun Z, Jacob H, Du M, Liang M, Dittmar RL, Liu Y, Liang M, Kohli M, Thibodeau SN, Boardman L, Wang L (2013) Characterization of human plasma-derived exosomal RNAs by deep sequencing. BMC Genomics 14: 319. https://doi.org/10.1186/1471-2164-14-319
23. Carvalho AS, Baeta H, Silva BC, Moraes MCS, Bodo C, Beck HC, Rodriguez MS, Saraswat M, Pandey A, Matthiesen R (2020) Extra-cellular vesicles carry proteome of cancer hallmarks. Front Biosci (Landmark Ed) 25(3):398–436. https://doi.org/10.2741/4811
24. Brzozowski JS, Jankowski H, Bond DR, McCague SB, Munro BR, Predebon MJ, Scarlett CJ, Skelding KA, Weidenhofer J (2018) Lipidomic profiling of extracellular vesicles derived from prostate and prostate cancer cell lines. Lipids Health Dis 17(1):211. https://doi.org/10.1186/s12944-018-0854-x
25. Minic Z, Hüttmann N, Poolsup S, Li Y, Susevski V, Zaripov E, Berezovski MV (2022) Phosphoproteomic analysis of breast cancer-derived small extracellular vesicles reveals disease-specific phosphorylated enzymes. Biomedicines 10(2):408. https://doi.org/10.3390/biomedicines10020408
26. Chen IH, Xue L, Hsu CC, Paez JS, Pan L, Andaluz H, Wendt MK, Iliuk AB, Zhu JK, Tao WA (2017) Phosphoproteins in extracellular vesicles as candidate markers for breast cancer. Proc Natl Acad Sci U S A 114(12):3175–3180. https://doi.org/10.1073/pnas.1618088114
27. Iliuk A, Wu X, Li L, Sun J, Hadisurya M, Boris RS, Tao WA (2020) Plasma-derived extracellular vesicle phosphoproteomics through chemical affinity purification. J Proteome Res 19(7):2563–2574. https://doi.org/10.1021/acs.jproteome.0c00151
28. Wu X, Li L, Iliuk A, Tao WA (2018) Highly efficient phosphoproteome capture and analysis from urinary extracellular vesicles. J Proteome Res 17(9):3308–3316. https://doi.org/10.1021/acs.jproteome.8b00459

Chapter 9

Mapping Cellular Protein Lysine Methylation Using Targeted-Mass Spectrometry

Anand Chopra, Matthew Hoekstra, William G. Willmore and Kyle K. Biggar

Abstract

Lysine methylation is a dynamic posttranslation modification (PTM) involved in the regulation of numerous biological processes. Depending on the context, site- and state-specific lysine methylation has been shown to regulate key features of proteins, such as stability, activity, and interactions. Assessing site- and state-specific lysine methylation within the proteome is difficult without access to antibodies that are able to detect the modification site. As the commercially available options for reliable antibodies that are specific to a methylation site and state are limited, mass spectrometry (MS)-based methods are more appealing as they enable detection of site- and state-specific methylation events with broad applicability. Here, we describe the use of immunoprecipitation and targeted-MS to detect and perform relative quantification of site-specific lysine methylation events occurring on endogenous proteins of interest. As an example, detection and relative quantification of hypoxia-responsive PGC1α-K224me2 and RNF20-K610me3/K614me3 lysine methylation, occurring in hypoxia-treated human colorectal cancer cells, are described.

Keywords Posttranslational modification, Lysine methylation, Targeted-mass spectrometry, Parallel reaction monitoring, Hypoxia

1 Introduction

Lysine methylation is a reversible PTM occurring on the ε-amino group of lysyl residues. Unlike other PTMs, lysine methylation exists in three distinct forms; mono-, di-, and tri-methylation; whereby one, two, or three methyl (-CH3) groups are added, respectively [1]. The roles of the enzymatic "writers" (lysine methyltransferases; KMT) and "erasers" (lysine demethylase; KDM) of lysine methylation have increasingly been shown to be relevant to a plethora of biological processes, of which several of these enzymes are key therapeutic targets for disease [2–4]. Of increasing relevance to biological function, lysine methylation has been shown to be modulated by atmospheric oxygen tension. This relationship is mostly driven by the oxygen-dependent catalytic activity of the largest known class of

Xu Zhang et al. (eds.), *Protein Post-Translational Modifications*, Methods in Molecular Biology, vol. 3018,
https://doi.org/10.1007/978-1-0716-5166-7_9,

demethylases, Jumonji C demethylases, which consists of several families, including the KDM2, KDM3, KDM4, KDM5, and KDM6 families, as well as several other individual JmjC KDMs [5]. Furthermore, the number of known methylation sites has been rapidly increasing, as well as the number of modification sites associated with methyl-regulating enzymes, and thus the tool kit for lysine methylation is in need for methodologies applicable to a broad range of methylation sites.

Since the early 2000s, the rapid growth in the number of known lysine methylation sites has been primarily driven by shotgun liquid chromatography–tandem mass spectrometry (LC–MS/MS) approaches, typically following an identification workflow that incorporates affinity enrichment of methylation sites [6–11]. However, as individual methylation sites continue to demonstrate biological significance, there is a need to detect and quantify site- and state-specific methylation events on specific proteins of interest; methylation sites that may not have commercially available antibodies. Further, the number of non-MS based methodologies that enable detection of cellular site- and state-specific methylation is limited and cumbersome in nature; often involving custom and expensive methyl-specific antibodies, methyl-reader domains, mutant constructs, as well as substrate and enzyme overexpression. In this manner, the same methodologies or tools may not be applicable from target-to-target.

Rather than relying on traditional "shotgun" MS methodologies, the methodology described here outlines the use of targeted-MS, specifically parallel reaction monitoring mass spectrometry (PRM-MS), as a label-free approach to detect and quantify lysine methylation events on endogenous proteins in mammalian cell extracts. The major steps of this pipeline include (1) affinity enrichment of the target protein by immunoprecipitation, (2) proteolytic digestion, (3) MS sample preparation, and (4) the use of targeted-MS software (i.e., Skyline [12]) to guide MS experiments, (5) performing the mass spectrometry experiments, and (6) analyzing acquired data. Here, we describe the method of utilizing PRM-MS for detecting di- and tri-methylation of key target proteins, isolated from cells. This method focuses on the preparation and data analysis of samples for PRM-MS as mass spectrometry can typically be performed at local or institutional facilities. In essence, we have successfully applied this workflow previously to detect and perform relative quantification of Nab3 mono-methylation at lysine-363 (i.e., Nab3-K363me1) in *Saccharomyces cerevisiae* [13]. Herein, as our prototypical example of the use of PRM-MS for detection of lysine methylation events, we demonstrate detection of peroxisome proliferator-activated receptor gamma coactivator 1 alpha (PGC1α)-K224me2 and ring finger protein 20 (RNF20)-K610me3/K614me3 lysine methylation events occurring on endogenous proteins in human colorectal carcinoma cells (Fig. 1). Furthermore, we demonstrate that relative

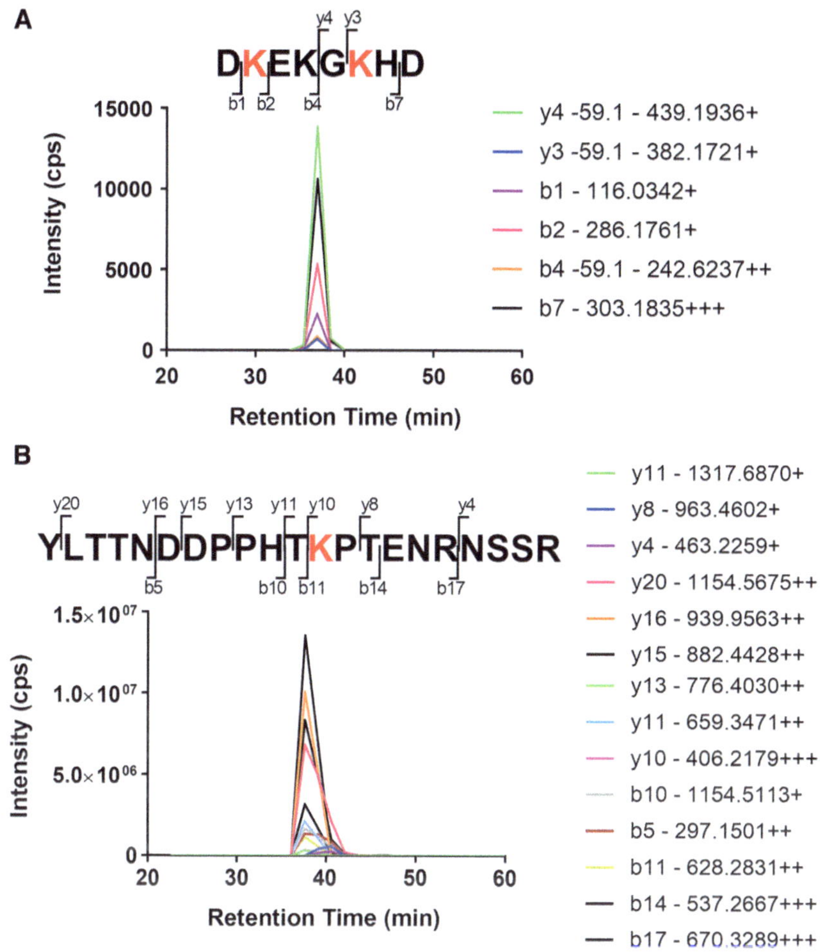

Fig. 1 Detection of (**A**) RNF20-K610me3/K614me3 and (**B**) PGC1α-K224me2 methylation by PRM-MS. Individual fragment ion intensities from (**A**) RNF20-K610me3/614me3 ($m/z = 347.5293$, $z = +3$) and (**B**) PGC1α-K224me2 ($m/z = 824.4018$, $z = +3$) precursors are shown. The peptide sequences are displayed above each chromatogram, and the methylation site is indicated with red font. Proteins were immunoprecipitated from human colorectal carcinoma cells (i.e., HCT116)

quantification is enabled as chromatographic peak areas of select peptides may be extracted using Skyline software. Specifically, we show that hypoxia increases relative levels of both PGC1α-K224me2 and RNF20-K610me3/K614me3 lysine methylation events (Fig. 2). Previously, PGC1α mono-methylation levels at this residue have been shown to increase in hypoxia due to inhibition of its demethylase (i.e., lysine demethylase 3A) [14]. Thus, we have outlined a method for detection and quantification of site- and state-specific lysine methylation events on multiple endogenous proteins from mammalian cells using a label-free targeted-MS approach.

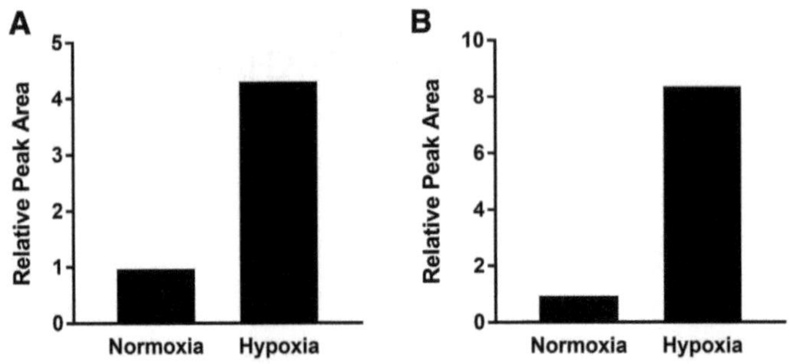

Fig. 2 Relative quantification (i.e., fold change) of (**A**) RNF20-K610me3/K614me3 and (**B**) PGC1α-K224me2 methylation in normoxia- versus hypoxia-treated HCT116 cells. Charts indicating relative total peak areas of chromatograms (as shown in Fig. 1) normalized by relative total peak areas of internal peptide controls (i.e., fold change). The results shown here are a representation of a single replicate

2 Materials

2.1 Cell Lysis and Immunoprecipitation

1. 0.5 M EDTA, pH 8: weigh 46.53 g ethylenediaminetetraacetic acid (EDTA; MW = 372.24 g/mol) and transfer to a beaker. Add 200 mL of water, put on a magnetic stir plate to mix. Gradually add sodium hydroxide (NaOH) pellets until the solution is dissolved. Continue to add NaOH pellets to adjust the pH to 8, then top up the volume to 250 mL with water.
2. 100 mM PMSF: weigh 0.0174 g phenylmethylsulfonyl fluoride (PMSF) and transfer to a 1.5 mL Eppendorf tube. Dissolve in 1 mL DMSO and store at −20°C.
3. Immunoprecipitation (IP) buffer: 20 mM Tris–Cl, pH 7.9, 150 mM KCl, 0.2 mM EDTA, 20% glycerol, 0.1% v/v NP-40. Weigh 0.4846 g Tris and 2.2365 g KCl and transfer to a beaker. Add 100 mL of water, 40 mL of glycerol, and 80 μL of 0.5 M EDTA pH 8 solution. Add stir bar and put on a magnetic stir plate to mix, adjust pH to 7.9, top up the volume to 200 mL with water, and autoclave on a liquid cycle with 20 min of sterilization. After the solution has cooled, add 0.2 mL of NP-40 and store at 4°C (*see* **Note 1**).
4. Lysis buffer: IP buffer supplemented with 1 mM PMSF and 0.4% v/v NP-40. Just before use, thaw the PMSF solution. Transfer 1 mL of IP buffer to a 1.5 mL Eppendorf tube. Add 10 μL of 100 mM PMSF and 4 μL NP-40 and vortex to mix. Use immediately.
5. Vibra cell Sonicator from Sonics & Materials.

6. Protein A bead solution: rProtein A Beads 4FF (BioBasic, MSA015005).
7. RNF20 antibody: RNF20 Recombinant Rabbit Monoclonal Antibody (Invitrogen, 701879).
8. PGC1αantibody: PGC-1αAntibody (Santa Cruz Biotechnology, sc-13067).

2.2 Protein Digestion

1. Digestion buffer: 50 mM $(NH_4)HCO_3$, pH 7.8. Weigh 0.7960 g $(NH_4)HCO_3$ and transfer to a beaker. Dissolve in 150 mL of water, if needed adjust the pH to 7.8, and top up to a final volume of 200 mL. Store at room temperature.
2. Trypsin stock: 200 ng/μL trypsin. Reconstitute entire 20 μg contents of Sequencing Grade Modified Trypsin, Porcine (Promega, V511A) in 100 μL of digestion buffer. Immediately aliquot and store at −80°C.
3. Trypsin working solution: 20 ng/μL trypsin. In a 1.5-mL Eppendorf tube, dilute 10 μL of Trypsin stock tenfold by addition of 90 μL water. Store at −80°C until use.
4. AspN stock: 100 ng/μL AspN. Reconstitute entire 50 μg contents of endoproteinase AspN (NEB, P8104S) with water. Immediately aliquot and store at −80°C.
5. AspN digestion buffer: 2× endoproteinase AspN reaction buffer (NEB, B8104S).
6. AspN working solution: 20 ng/μL AspN. In a 1.5-mL Eppendorf tube, dilute 10 μL of AspN stock 5-fold by addition of 40 μL water. Store at −80°C until use.
7. 50 mM TCEP solution: In a 1.5-mL Eppendorf tube, dilute 50 μL of 0.5 M Tris(2-carboxyethyl)phosphine hydrochloride (TCEP) stock (Sigma, 646547) 10-fold by the addition of 450 μL water.

2.3 C18 Sample Desalting

1. ZipTip with 0.6 μL C18 resin (Millipore Sigma, ZTC18S096).
2. Formic acid solution: 10% v/v formic acid. In a 1.5-mL Eppendorf tube, mix 10 μL of formic acid with 90 μL of water. Invert to mix.
3. Acetonitrile with 0.1% v/v formic acid.
4. MS-grade water with 0.1% formic acid.
5. Elution buffer: 60% acetonitrile, 0.1% formic acid. To a 1.5-mL Eppendorf tube transfer 600 μL acetonitrile and 400 μL MS-grade water and invert to mix.
6. Savant SC100 SpeedVac.

2.4 Software Tool

Skyline software [12] (accessible at http://proteome.gs.washington.edu/software/skyline).

3 Methods

This procedure begins using cell pellets as starting material, obtained from standard mammalian tissue culturing techniques. Specifically, this procedure was used for one replicate of human colorectal carcinoma cell, HCT116, pellets collected from 10-cm dishes that were incubated in either normoxia (21% oxygen) or hypoxia (1% oxygen) for 24 h.

3.1 In Silico Digestion and Isolation List Preparation

1. Open Skyline software (version number 20.1.0.76 for this demonstration) [12].
2. On the lefthand side, under "Targets" paste the sequence of the protein of interest.
3. Under "Settings," select "Peptide Settings."
4. In the pop-up window, select the "Digestion" tab.
5. Select the desired protease (*see* **Note 2**) under the "Enzyme" drop-down menu and set the number of "Max missed cleavages" allowed (*see* **Note 3**).
6. In the "Modifications" tab, select "Carbamidomethyl (C)," if iodoacetamide will be used in downstream sample preparation steps.
7. Under "Targets," view the in silico digestion patterns and locate peptides containing the methylation site of interest (*see* **Note 2**).
8. Right-click on the peptide sequence and in the drop-down menu select "Modify."
9. In the pop-up window, on the left-hand side under "Structural," select the lysine residue of interest and select the desired modification (*see* **Note 4**).
10. Under "Settings," select "Transition Settings."
11. In the pop-up window, select the "Filter" tab and adjust settings as desired (*see* **Note 5**).
12. Once finalized, under "File," select "Export" then "Isolation list."

3.2 Cell Lysis

Perform all steps either on ice or at 4°C.

1. Place fresh or frozen cell pellets on ice.
2. Add 200 μL of lysis buffer to each cell pellet (*see* **Note 6**). Pipette up and down until the solution is homogenous.

3. Sonicate each pellet for three cycles of 10 s "on" and 10 s "off" with the % duty cycle set to 60 and the output control set to four.
4. Incubate the lysates for 30 min with end-to-end rotation.
5. Centrifuge the lysate for 15 min at 16,000×*g*.
6. Transfer the supernatant to a new 1.5 mL Eppendorf tube (*see* **Notes 7 and 8**).

3.3 Immuno-precipitation and Digestion

Perform all steps either on ice or at 4°C unless otherwise stated.

1. Transfer 500 μg (*see* **Notes 9 and 10**) soluble cell extract to a 1.5 mL Eppendorf tube.
2. Top up the volume to 500 μL with IP buffer.
3. Thaw 100 mM PMSF solution at room temperature, add 5 μL to each immunoprecipitation reaction, and invert the tubes to mix.
4. Add 1 μg of antibody to each immunoprecipitation reaction (*see* **Note 11**).
5. Incubate the soluble cell extract-antibody mixture overnight with end-to-end rotation.
6. Transfer 40 μL of Protein A bead solution (i.e., 20 μL dry bed volume) to a new 1.5 mL Eppendorf tube for each immunoprecipitation reaction.
7. Add 500 μL IP buffer to each tube containing Protein A beads and invert to mix.
8. Centrifuge for 5 min at 100×*g* (*see* **Note 12**).
9. Carefully remove and discard the supernatant.
10. Repeat steps 7–9 twice.
11. Transfer the entire volume of soluble extract-antibody mixture to a tube containing equilibrated Protein A beads.
12. Incubate for 3 h with end-to-end rotation.
13. Centrifuge for 5 min at 100×*g*.
14. Carefully remove the supernatant and transfer to a new 1.5 mL Eppendorf tube (*see* **Note 13**).
15. Repeat steps 7–9 twice (*see* **Note 14**).
16. Add 500 μL digestion buffer to each tube containing Protein A beads and invert to mix.
17. Repeat steps 8 and 9.
18. Repeat steps 16 and 17.
19. Resuspend the beads in 50 μL digestion buffer (*see* **Note 15**).

20. Add 0.5 µL of 20 ng/µL Trypsin working solution; alternatively, add 50 µL 2X AspN digestion buffer and 0.5 µL AspN working solution.
21. Incubate overnight with end-to-end rotation at 37°C.
22. Proceed to C18 cleaning or store digests at −80°C (*see* **Notes 16 and 17**).

3.4 C18 Sample Desalting

Perform all steps at room temperature.

1. To 50 µL of digest add 0.5 µL 10% formic acid solution.
2. Transfer 1.5 mL of acetonitrile with 0.1% formic acid and 1.5 mL of MS-grade water with 0.1% formic acid into two separate 1.5 mL Eppendorf tubes.
3. Attach a µ-C18 ZipTip to an appropriately sized pipette set to 10 µL.
4. Depress the pipette plunger to the "first stop."
5. To avoid introducing bubbles in your sample, keep the plunger depressed to the "first stop" and move the ZipTip into the acetonitrile with 0.1% formic acid.
6. Draw up the liquid.
7. Expel the liquid within the ZipTip by depressing the pipette plunger to the "first stop."
8. Repeat steps 6 and 7 twice.
9. Keep the plunger depressed to the "first stop" and move the ZipTip into the MS-grade water with 0.1% formic acid.
10. Repeat steps 6 and 7 twice.
11. Keep the plunger depressed to the first stop and move the ZipTip into the digest.
12. Repeat steps 6 and 7 ten times.
13. Keep the plunger depressed to the first stop and move the ZipTip into the MS-grade water with 0.1% formic acid.
14. Repeat steps 6 and 7 six times.
15. Keep the plunger depressed to the first stop and move the ZipTip into the elution buffer.
16. Draw up the liquid, move the ZipTip into a new 1.5 mL Eppendorf tube and pipette the solution up-and-down ten times.
17. Repeat step 16 by drawing up an additional volume of elution buffer and transferring into the same 1.5 mL Eppendorf tube.
18. Repeat steps 1–17 for all samples.
19. Evaporate the elution buffer by drying the samples in a SpeedVac.

20. Resuspend dried peptides in 20 µL of MS-grade water with 0.1% formic acid (*see* **Note 18**).
21. Proceed to subject the sample to PRM-MS analysis. If not performing MS on the same day, then store the sample at −80°C until ready.

3.5 Targeted Mass Spectrometry

In our case, a Q-Exactive Plus hybrid quadrupole-orbitrap mass spectrometer was used with a Thermo EASY-nLC II system. MS facilities will have standard procedures for PRM-MS analysis; however, the specific details will vary in relation to the equipment that is available (e.g., mass spectrometry, separating columns, ionspray, mass analyzers). Thus, a general procedure is outlined below, and more specific details are mentioned in the Notes section.

1. Load resuspended peptides (prepared in the previous section) onto the separating column for 100 min LC gradient at a flow rate of 0.25 µL/min (*see* **Note 19**).
2. Separate peptides using linear LC gradient of 6%–43% acetonitrile (*see* **Note 20**).
3. Electrospray compounds that begin to elute during the previous step are directly into the mass spectrometry (*see* **Note 21**).
4. Conduct PRM scans using the isolation list prepared in Sect. 3.1 (*see* **Note 22**).

3.6 Data Analysis

1. Open the developed isolation list (prepared in Sect. 3.1) in Skyline Software.
2. Under "File," select "Import" then "Results."
3. Select and load the raw data files (e.g., ".raw" file extension from Thermo instruments) obtained from MS analysis (*see* **Note 23**).
4. Locate precursor ions for the peptide of interest and manually select each individual fragment ion and visually verify whether the peak is chromatographic in nature. In other words, fragment ions that display a symmetrical elution profile.
5. Identify those chromatographic peaks containing at least three fragment ions that elute with the same retention time (Fig. 3) (*see* **Note 24**).
6. Peak boundaries are automatically defined by the Skyline software, if necessary, boundaries can be manually changed to flank the chromatographic peaks by dragging the vertical dotted lines with the cursor.
7. Measure peak areas by navigating to "View" and selecting "Peak Areas," then "Replicate Comparisons."

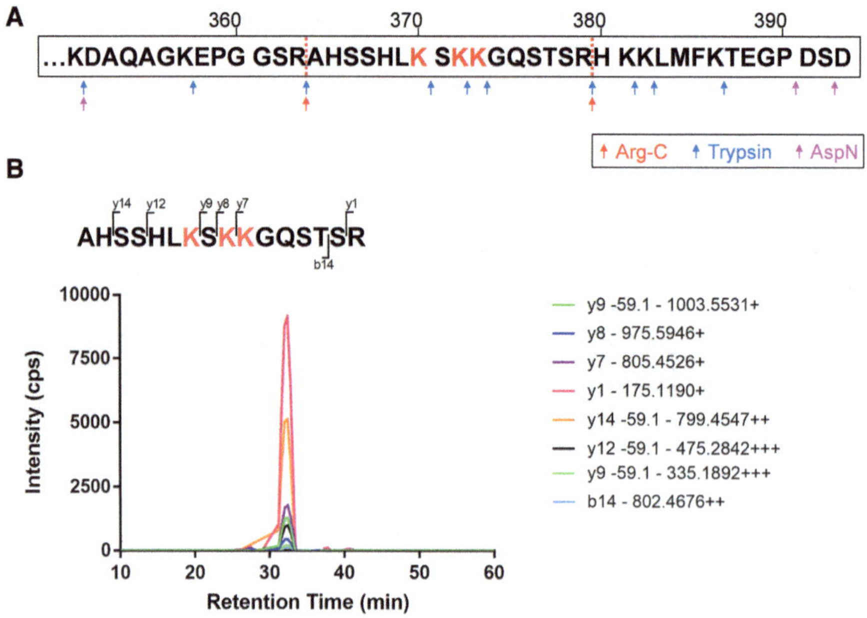

Fig. 3 Common protease sites on $p53_{351-393}$ sequence and detection of p53 methylation by PRM-MS. (**A**) Common protease cleavage sites on C-terminal end sequence of p53. The lysine methylation sites of interest (red text) may be encompassed in a peptide (flanked by red dotted lines) resultant from proteolytic digestion using Arg-C. Trypsin, Arg-C, and AspN cleavage sites are shown as blue, red, and pink, respectively. (**B**) Individual fragment ion intensities from p53-K370me3/372me3/K373me3 ($m/z = 622.3620$, $z = +3$) are shown. The peptide sequences are displayed above each chromatogram, and the methylation site is indicated with red font. Proteins were immunoprecipitated from human colorectal carcinoma cells (i.e., HCT116) and digested with Arg-C protease

8. Right-click on the graph in the pop-up window, select "Copy Data" in the drop-down menu, and paste in the desired data processing software.
9. Repeat steps 4–8 for each methylated peptide of interest as well as several internal unmodified peptides (*see* **Note 25**).
10. To determine relative methylation levels between multiple conditions, first summate the individual ion intensities to obtain the total peak area for each individual peptide.
11. To obtain a relative fold change, divide the total peak area of the experimental conditions by the total peak area of the control.
12. To normalize for protein abundance, divide the relative values of each methylated peptide by the relative areas of the internal unmodified peptide (Fig. 2).

4 Notes

1. Nearly, the same lysis buffer and immunoprecipitation buffer used in this procedure have been used previously for coimmunoprecipitation of RNF20 by In et al. [15]. We have successfully applied the same buffer system to study PGC1α; however, the buffer conditions may need to be adjusted if the immunoprecipitation does not sufficiently enrich in the target protein. We recommend beginning with buffer conditions that have been either established or published in the literature previously.
2. The protease selected will vary depending on the sequence in which the methylation site resides. In the main examples provided here, trypsin and AspN were used for PGC1α-K224 and RNF20-K610/K614, respectively. Ideally, the modification site should reside on a peptide between 8 and 25 residues (under "Settings" > "Peptide Settings" > "Filter," set the "Min length" and "Max length" to 8 and 25, respectively). If the methylation site cannot be covered by trypsin under these conditions, consider using other proteases by performing additional in silico digestions with the Skyline software. As an example, we have previously been interested in detecting triply-trimethylated p53 at lysine residues 370, 372, and 373 on a single peptide. In this case, Trypsin and AspN would not enable generation of a peptide harboring all three of these residues (would not satisfy our ideal criterion of an 8–25 residue long peptide), thus Arg-C was used as an alternative protease (Fig. 3).
3. We find that digestions are not always 100% efficient, and thus it must be anticipated the possibility that missed cleavages may occur. We recommend setting the Max missed cleavages to "1" to accommodate for this possibility.
4. Since lysine methylation can exist in three distinct forms, the modification can be set to either mono-, di-, or tri-methylation (resulting in increased monoisotopic masses of 14.01565, 28.03130, and 42.04695 Da, respectively). To note, a characteristic mass loss of −59.0735 is often observed with tri-methylation. If interested in studying different methylation states of the same site, under the "Targets" window on the right-hand side, individual peptides may be copied and pasted. After this, the modification type of the copied peptide may be changed to a different methylation state.
5. We recommend generally using the following settings: precursor charges: 2, 3. Ion charges: 1, 2, 3, 4. Ion types: y, b, p.

To ensure all y- and b-ions are selected under "Product ion selection," set "From" to "ion 1" and "To" to "last ion."

6. The volume of lysis buffer may be adjusted; however, we recommend ensuring the volume is as low as possible such that the final protein solution collected is sufficiently concentrated. We use 200 μL of lysis buffer for a cell pellet collected from a 10-cm dish and 100 μL for a 6-cm dish.
7. The supernatant contains soluble proteins. The protein may be snap frozen in liquid nitrogen and stored at −80°C for long-term storage. However, we recommend performing the immunoprecipitation procedure the same day as the cell lysis procedure to avoid subjecting the proteins to freeze–thaw cycles.
8. Ensure to measure protein concentration with an appropriate quantification method. We use Pierce BCA Protein Assay Kit, however, a standard Bradford assay will be compatible with the components of the lysis buffer.
9. The amount of cell extract used for the immunoprecipitation procedure may need to be adjusted depending on the abundance of the protein of interest in the cell. We have found that generally 0.5–1 mg of soluble cell extract is sufficient for most proteins. Generally, we validate the immunoprecipitation procedure by a standard western blot to confirm three aspects: (1) that the target protein is expressed in the given cell line, (2) that the antibody is capable of binding to the target protein, and (3) that the immunoprecipitation procedure enriches in the target protein and captures relatively equal amounts of precipitated protein between conditions (Fig. 4).
10. If the study involved multiple conditions, we recommend diluting the soluble cell extracts to the same protein concentration with lysis buffer prior to beginning the Immunoprecipitation and Digestion procedure. This will ensure that the detergent concentration will be equal between samples during the immunoprecipitation.
11. The amount of antibody used may be adjusted. Generally, we have had success using 0.5–1 μg of antibody for immunoprecipitations using 500–1000 μg of soluble cell extract.
12. Look carefully at the supernatant to ensure all beads have collected at the bottom of the tube. If some beads are still suspended, recentrifuge the tube or allow the beads to settle for a few minutes on ice.
13. At this point, the supernatant represents the unbound fraction. This fraction may be saved and used for troubleshooting as this solution may be subjected to western blotting.

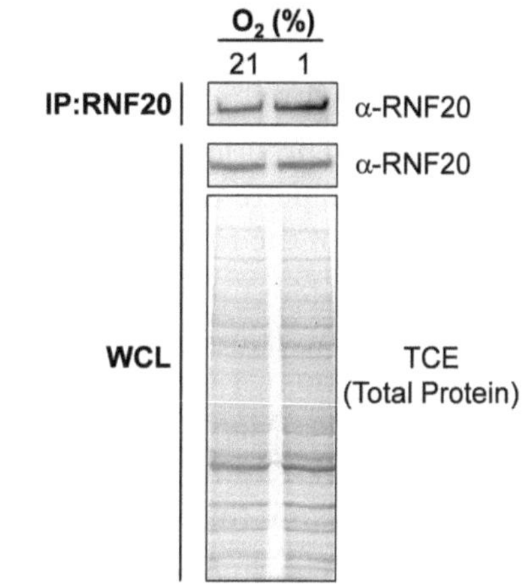

Fig. 4 Example of standard western blot analysis of soluble cell extracts and immunoprecipitation (IP) procedure using RNF20 IP. HCT116 cells were exposed to normoxia (21% O_2) or hypoxia (1% O_2) for 24 h prior to harvesting. After soluble cell extracts were collected, 30 μg of whole cell lysate (WCL) were subject to SDS-PAGE and western blot analysis to ensure even loading with 2,2,2-trichloroethanol (TCE) and observe RNF20 abundance. Immunoprecipitations of RNF20 from 500 μg WCL were assessed by western blotting. The results shown here are a representation of a single replicate

14. Rather than proceeding to step 16 in the Immunoprecipitation and Digestion procedure, the beads may be resuspended in a standard Laemmli buffer and subject to western blotting to confirm the efficacy of the Immunoprecipitation portion of the procedure. To avoid having to repeat the full Immunoprecipitation procedure for the downstream MS, approximately 10% of beads may be transferred to a separate tube and resuspended in Laemmli buffer. In our experience, continuing the procedure with the remaining 90% of beads does not compromise PRM-MS detection and analysis.
15. Note that the digestion reaction conditions (i.e., digestion buffer composition, protease amount, and temperature) may need to be adjusted depending on the protease used. We advise the use of manufacturer's recommendations.
16. In our experience, digested proteins (i.e., peptides) may be stored at −80°C for several months to years without detrimental effects on sample quality.
17. If the peptides of interest contain a cysteine(s), we recommend performing a reduction and alkylation step prior to sample clean-up in downstream steps. Transfer a portion of the digest to a new 1.5 mL Eppendorf tube (e.g., 15 μL). Add 1.6 μL of 50 mM TCEP (5 mM final concentration) and incubate for 10 min at room temperature with shaking. Add

1.85 µL of 100 mM iodoacetamide (10 mM final concentration) and incubate for 30 min at room temperature in the dark.

18. The dried peptides will not form a visible pellet. Ensure to pipette the solution up-and-down, as well as along the sides of the tube near the bottom, to resuspend the peptides.
19. Here, peptides were loaded onto a Thermo Scientific™ Acclaim PepMap™ 100 precolumn (2 cm × 100 x 100 µm ID, C18, 5 µm, 100 Å) and a Thermo Scientific™ Acclaim™ PepMap™ RSLC C18 analytical column (15 cm × 75 x 75 µm ID, 3 µm, 100 Å) attached to a Thermo EASY-nLC II system. Note that the columns may vary between instruments.
20. The linear gradients of MS-grade water with 0.1% formic acid (defined as "A") and acetonitrile with 0.1% formic acid (defined as "B") were performed as follows: 0—6% B for 2 min, 6—43% B for 70 min, 43—100% B for 4 min. To clear the column of residual peptides prior to the loading of the next sample, washes with 100% B were then performed for 10 min, followed by a gradient of 100—10% B for 4 min, and washes with 100% A for 10 min.
21. A Thermo Scientific™ EASY spray was used for positive electrospray ionization (ESI). The ion source temperature was 250°C and the voltage was 2.1 kV. Note that the specific equipment and parameters for ionization will vary between different MS systems.
22. The following parameters were used for PRM scanning: an orbitrap resolution of 35,000 (at m/z 200), automatic gain control (AGC) target of 1e5, a maximum injection time of 200 ms, an isolation window of 2 m/z, and normalized collision energy (NCE) of 27.
23. The file extension of raw data files should be ".raw". Once the data files are selected, the software will begin to process the data using the isolation list loaded in software. On the right-hand side, data will be shown for those peptides or precursor ions that are selected in the "Targets" window.
24. In addition to using at least three fragment ions for quantification, for unambiguous detection of site-specific methylation events, it must be ensured that the modification site is encompassed by these ions. In the example data for PGC1α-K224 and RNF20-K610/K614, it can be seen that several fragment ions containing the methylation site(s) were detected (Fig. 1).
25. Quantification of the internal unmodified peptide is used for normalization, thereby controlling for slight differences in protein abundance between samples. For relative

quantification, we highly recommend ensuring near-equal protein abundance between samples by subjecting soluble cell extracts and immunoprecipitation reactions to standard western blotting prior to MS (Fig. 4).

Acknowledgments

This work was supported by Discovery Grants from the Natural Science and Engineering Council (NSERC) of Canada to KKB (Grant No. RGPIN-2016-06151) and WGW (Grant No. RGPIN-2017-06414). A.C. previously held a Canada Graduate Scholarship—Doctoral (CGS-D) from the NSERC of Canada and is now supported by PDF grants from the Azrieli Foundation (Azrieli International Postdoctoral Research Fellowship) and the NSERC of Canada. Mass Spectrometry was performed at the John L. Holmes Mass Spectrometry Facility at the University of Ottawa.

References

1. Hyun K, Jeon J, Park K et al (2017) Writing, erasing and reading histone lysine methylations. Exp Mol Med 49:e324
2. Bhat KP, Ümit Kaniskan H, Jin J et al (2021) Epigenetics and beyond: targeting writers of protein lysine methylation to treat disease. Nat Rev Drug Discov 20:265–286
3. Faundes V, Newman WG, Bernardini L et al (2018) Histone lysine methylases and demethylases in the landscape of human developmental disorders. Am J Hum Genet 102:175–187
4. Black JC, Whetstine JR (2013) Tipping the lysine methylation balance in disease. Biopolymers 99:127–135
5. Chopra A, Adhikary H, Willmore WG et al (2019) Insights into the function and regulation of Jumonji C lysine demethylases as hypoxic responsive enzymes. Curr Protein Pept Sci 21:642–654
6. Wang R, Huang M, Li L et al (2018) Affinity purification of methyllysine proteome by site-specific covalent conjugation. Anal Chem 90:13876–13881
7. Lund PJ, Lehman SM, Garcia BA (2019) Quantitative analysis of global protein lysine methylation by mass spectrometry. Methods Enzymol 626:475–498
8. Liu H, Galka M, Mori E et al (2013) A method for systematic mapping of protein lysine methylation identifies functions for HP1β in DNA damage response. Mol Cell 50:723–735
9. Carlson SM, Moore KE, Green EM et al (2013) Proteome-wide enrichment of proteins modified by lysine methylation. Nat Protoc 9:37–50
10. Cao XJ, Arnaudo AM, Garcia BA (2013) Large-scale global identification of protein lysine methylation in vivo. Epigenetics 8:477–485
11. Cao XJ, Garcia BA (2016) Global proteomics analysis of protein lysine methylation. Curr Protoc Protein Sci 86:24.8.1–24.8.19
12. MacLean B, Tomazela DM, Shulman N et al (2010) Skyline: an open source document editor for creating and analyzing targeted proteomics experiments. Bioinformatics 26:966–968
13. Lee KY, Chopra A, Burke GL et al (2020) A crucial RNA-binding lysine residue in the Nab3 RRM domain undergoes SET1 and SET3-responsive methylation. Nucleic Acids Res 48:2897–2911
14. Qian X, Li X, Shi Z et al (2019) KDM3A senses oxygen availability to regulate PGC-1α -mediated mitochondrial biogenesis. Mol Cell 76:885–895
15. In S, Kim YI, Lee JE et al (2019) RNF20/40-mediated eEF1BδL monoubiquitylation stimulates transcription of heat shock-responsive genes. Nucleic Acids Res 47:2840–2855

Chapter 10

Enrichment and Metaproteomic Analysis of Lysine Acetylation in Fecal Microbiome Samples

Xu Zhang, Zhibin Ning and Daniel Figeys

Abstract

Among the various posttranslational modifications (PTMs) found in microbiome samples, lysine acetylation is known to be abundant and plays an important role in regulating microbial short-chain fatty acid (SCFA) metabolism. The latter is a crucial microbiome function that significantly impacts human intestinal health. This chapter describes a detailed protocol for lysine acetylomic profiling of microbial proteins in human fecal microbiome samples. The protocol consists of stool sample preprocessing, microbiome protein extraction and digestion, immunoaffinity enrichment of lysine acetylated peptides, and high-resolution mass spectrometry analysis for the identification and quantification of lysine-acetylated proteins.

Keywords Lysine acetylation, Microbiome, Metaproteomics, Immunoaffinity enrichment, Bioinformatics workflow

1 Introduction

The human microbiome is increasingly recognized as a crucial environmental factor influencing health and diseases [1–3]. Over the past two decades, microbiome research has evolved from species and gene cataloging to a greater focus on functionality. This shift leverages functional meta-omics tools, such as metaproteomics, to characterize the expression and posttranslational modifications (PTMs) of microbiome proteins [4].

Mass spectrometry (MS)-based metaproteomics is so far the most important method to characterize microbiome proteomes, including those from the human gut. The sensitivity and throughput of this approach have been dramatically increased by recent advances in high-resolution mass spectrometers (e.g., Orbitrap Astral, timsTOF Ultra) and more efficient data-independent acquisition (DIA) MS workflows [5, 6]. Consequently, the modern DIA–MS-based metaproteomics enables deeper, more robust, and higher-throughput

Xu Zhang et al. (eds.), *Protein Post-Translational Modifications*, Methods in Molecular Biology, vol. 3018,
https://doi.org/10.1007/978-1-0716-5166-7_10,

analysis of microbiomes, making it more readily applicable to clinical microbiome studies [7–10].

Moreover, the capability of metaproteomics is far more than measuring protein expression levels [11]. It can be used for diverse objectives, including the analysis of PTMs, protein synthesis/dynamics through stable isotopic probing (SIP) [12], and active enzymes through activity-based protein profiling (ABPP) [13]. Metaproteomics also allows to identify and characterize host proteins, dietary components, proteins from eukaryotic organisms present in the samples, in addition to the microbial proteins [14]. These unique capabilities make metaproteomics an exceptionally powerful tool for translational and clinical microbiome studies.

PTMs are known to be important factors in both human and environmental microbiomes in regulating key activities, including short-chain fatty acid (SCFA) metabolism and environmental adaption [15, 16]. Unlike in mammalian cells, where phosphorylation is among the most commonly studied PTMs, abundant modifications in microbiomes have been reported to be methylation, deamination, acetylation, hydroxylation, and oxidation [15]. We have demonstrated, for example, that lysine acetylation is widespread in the human gut microbiome and plays an important role in regulating SCFA metabolism—a key function of the commensal microbiota [16]. Subsequent metaproteomic analysis of the microbiome lysine acetylation in patients with new onset Crohn's diseases revealed that gut microbial protein lysine acetylation plays a crucial role in the functional dysbiosis in patients compared to healthy controls [16].

In this chapter, we will detail the protocols used for metaproteomic analysis of lysine acetylation with high-resolution mass spectrometry method. The protocol includes steps starting from stool sample preprocessing with differential centrifugation, protein extraction and purification with acetone precipitation, in-solution proteolytic digestion, peptide desalting, and immunoaffinity enrichment with a commercially available Anti-acetyl-lysine motif kit, to tandem mass spectrometry analysis and downstream bioinformatic data processing (Fig. 1).

2 Materials

2.1 Reagents and Buffers

1. 1 M Tris–HCl buffer (pH 8.0) stock solution: dissolve 12.1 g Tris base in 70 mL ddH_2O in a beaker. While mixing using a magnetic stirrer, add 4 mL concentrated HCl and finely adjust the pH to 8.0 by adding concentrated HCl. Fill to a final volume of 100 mL with ddH_2O and store at room temperature for up to 3 months.

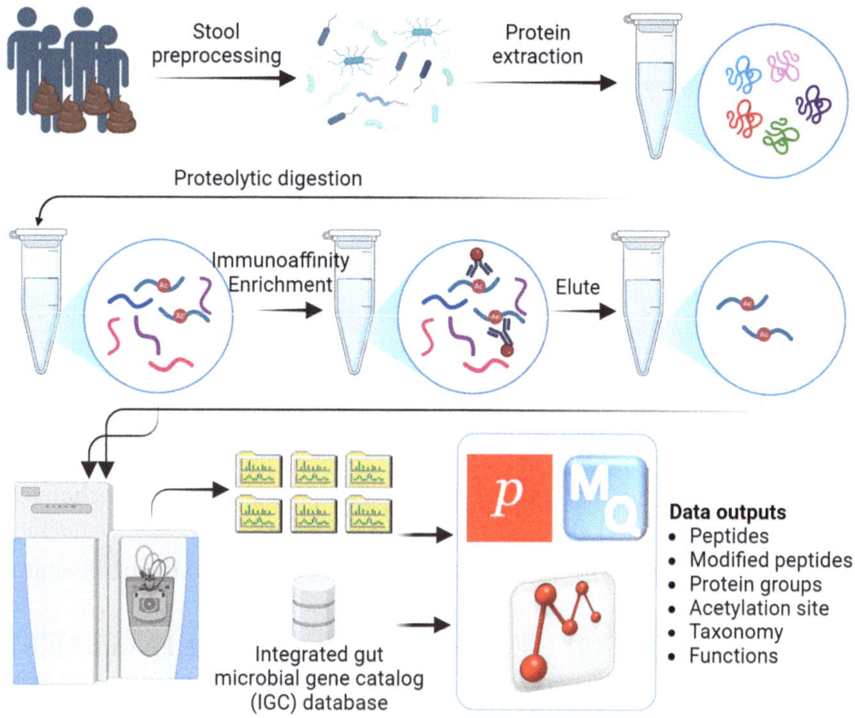

Fig. 1 Overview of the experimental and bioinformatic workflow for lysine acetylomic analysis of human microbiomes. Created in BioRender. https://BioRender.com/j85o927

2. 20% (w/v) sodium dodecyl sulfate (SDS) stock solution: dissolve 20 g SDS in 80 mL ddH_2O by mixing on magnetic stirrer in a beaker (warm at 50°C to dissolve if needed). Fill to a final volume of 100 mL with ddH_2O and store at room temperature.
3. 1 M ammonium bicarbonate (ABC) stock solution: dissolve 7.9 g ammonium bicarbonate in 80 mL ddH_2O, and fill to a final volume of 100 mL with ddH_2O. Check pH and it should be around pH 8. Store this stock solution at 4°C for up to 3 months.
4. Protein precipitation buffer: mix 250 mL acetone, 250 mL anhydrous ethanol, and 0.5 mL acetic acid. Store the precipitation buffer at −20°C at least overnight before use.
5. Protein lysis buffer (4% SDS in 100 mM Tris–HCl): for 10 mL lysis buffer, mix 2 mL 20% (w/v) SDS stock solution, 1 mL 1 M Tris–HCl buffer (pH 8.0), and 7 mL ddH_2O. Add one tablet cOmplete™ Protease Inhibitor and mix well prior to use.
6. Protein digestion buffer (6 M urea in 100 mM ABC): add 3.6 g urea, 1 mL 1 M ABC stock solution, and 6.5 mL ddH_2O.

Dissolve urea by shaking and hand warm at 37°C, if needed (this solution needs to be freshly prepared).

7. 100 mM ABC buffer: mix 1 mL 1 M ABC and 9 mL ddH_2O (calculate the amount needed and adjust the total volumes; this solution needs to be prepared freshly).
8. 250 mM 1,4-dithiothreitol (DTT): dissolve 38.58 mg DTT in 1 mL ddH_2O; use fresh solution, or make aliquots and store at −80°C.
9. 500 mM iodoacetamide (IAA): dissolve 92.48 mg IAA in 1 mL ddH_2O. Use fresh solution, or make aliquots and store at −80°C. Wrap IAA with aluminum foil as it is light sensitive.
10. Lysyl-endopeptidase (Lys-C; Wako Pure Chemical Corp., Osaka, Japan).
11. TPCK-treated trypsin (Worthington Biochemical Corp., Lakewood, NJ, USA)
12. 0.1% (v/v) formic acid (FA): mix 0.01 mL formic acid with 10 mL MS-grade water.
13. 10% (v/v) formic acid (FA): mix 1 mL FA with 9 mL MS-grade water.
14. Peptide elution buffer (80% ACN, 0.1% FA): mix 8 mL acetonitrile (ACN), 0.01 mL FA, and 2 mL MS-grade water (this solution needs to be freshly prepared).
15. PTMScan® Acetyl-Lysine Motif [Ac-K] Kit (Catalog #13416, Cell Signaling Technology, Inc.).
16. DC Protein Assay Kit (Catalog #5000112, Bio-Rad).
17. 0.1% (v/v) trifluoroacetic acid (TFA): mix 0.01 mL TFA with 10 mL MS-grade water.
18. 1× immunoaffinity purification (IAP) buffer: add 1 mL 10× IAP buffer (provided in the enrichment kit) to 9 mL of ddH_2O and mix well. Store 1× IAP buffer at 4°C for up to 1 month.

2.2 Equipment and Software Tools

1. Q Exactive HF-X mass spectrometer (ThermoFisher Scientific Inc.) coupled with an Eksigent Ekspert NanoLC 400 (AB SCIEX Inc.).
2. Tube rotator: can hold 1.5 mL or 2.0 mL centrifuge tubes.
3. Refrigerated centrifuge: can cool to 4°C and reach a speed of at least 16,000×*g*.
4. HyperSep™ C18 Cartridges (200 mg) (Catalog #60108-303, Thermo Scientific Inc.).
5. Tip-based C18 desalting column: to make one C18 desalting column, one 20 μL filter tip (cut the tip end), one 0.6 mL tube (cut the lid and bottom), and one 2.0 mL tube (remove lid) are needed. Insert the above 20 μL filter tip into the 0.6 mL

Eppendorf tube holder, and then put on the 2.0 mL tube to assemble the column. For each column, add 5 mg 10 μm AQ C18 beads (GmbH. R10 aq0010) resuspended in 100 μL ACN (*see* **Note 1**).

6. SpeedVac Vacuum Concentrator.
7. Q700 Ultrasonicator with cup-horns (Qsonica L.L.C) (*see* **Note 2**).
8. MaxQuant (version 1.6.3.4): https://www.maxquant.org.
9. MetaLab (version 2.3): https://imetalab.ca/.
10. pFind 3: https://github.com/pFindStudio/pFind3.

3 Methods

3.1 Stool Preprocessing with Differential Centrifugation

1. Add 1.5 mL ice-cold PBS, 3–5 glass beads (2 mm diameter) to 2 mL tubes with ~0.2 g stool samples, vortex with maximum speed till there are no visible large particles (*see* **Note 3**).
2. Centrifuge at 300 ×*g* for 5 min at 4°C, carefully transfer supernatant to a 10 mL tube;
3. Resuspend the pellets with 1.5 mL PBS, vortex with maximum speed, centrifuge at 300 ×*g* for 5 min at 4°C, and carefully transfer supernatant to the same 10 mL tube;
4. Repeat Step 3 for one more time to collect a total of ~4 mL supernatant per sample; discard the pellets (biohazards, treatment with bleach or PREempt for decontamination) (*see* **Note 4**).
5. Clear the supernatant by centrifuge at 300 ×*g*, 4°C for 5 min, and carefully transfer supernatant to a new tube; repeat the clearing step for two more times.
6. Centrifuge the cleared supernatant at 14,000 ×*g*, 4°C for 20 min to harvest microbial cells (discard supernatant).
7. Wash the microbial cells by resuspending pellets with 1.5 mL ice-cold PBS, spin down at 14,000 ×*g*, 4°C for 20 min, and discard the supernatant;
8. Repeat the wash step two more times, and then store the pellets in −80°C or use for protein extraction immediately.

3.2 Protein Extraction and Purification

1. Add ~1 mL protein lysis buffer directly to the frozen or freshly prepared microbial pellets, and let solutions sit at room temperature for 5–10 min while shaking until the pellet is well lifted and resuspended (*see* **Note 5**).
2. Use the QSonica sonicator with parameter set as follows for ultrasonication facilitated microbial cell lysis: 10 min

processing time with 10 s on and 10 s off; 50% amplitude, and Chiller temperature of 8°C to avoid crystallization (*see* **Note 6**).

3. After sonication, centrifuge the lysates at 16,000 ×*g*, 8°C for 10 min to remove debris.
4. Carefully transfer supernatant into a new tube, and add five volumes of protein precipitation buffer, and mix by inversion up and down for three to five times.
5. Incubate at −20°C overnight to precipitate proteins.
6. In the next day, centrifuge the tubes directly at 16,000 ×*g*, 4°C for 25 min to pellet the proteins.
7. Carefully remove supernatant and resuspend the protein pellets using 1 mL ice-cold acetone with sonication or vortexing until there are no visible chunks (*see* **Note 7**).
8. Spin down at 16,000 ×*g*, 4°C for 10 min, and wash the pellets two more times (*see* **Note 8**).
9. Add 1 mL protein digestion buffer to the pellets, and resuspend by pipetting or vortexing at room temperature.
10. Measure protein concentrations using DC protein assay and aliquot 10 mg protein lysate for proteolytic digestion (*see* **Note 9**).

3.3 In-Solution Digestion

1. Transfer 10 mg protein lysate for each sample and add protein digestion buffer to a total of 5 mL to make a final protein concentration of 2 mg/mL (*see* **Note 10**).
2. Add DTT stock solution to a final concentration of 10 mM DTT in the samples, mix well and incubate at room temperature for 1 h with agitation (*see* **Note 11**).
3. Add IAA stock solution to a final concentration of 20 mM IAA in the samples, mix well and incubate for 40 min at room temperature in the dark (*see* **Note 12**).
4. Dilute samples with nine volumes of 100 mM ABC buffer to make the final concentration of urea to 0.6 M for digestion.
5. Add 1 mAU of Lys-C per 50 μg protein for 4 h followed by the addition of 1 μg of trypsin per 50 μg protein for overnight at room temperature with agitation. Keep digestion time between 18 and 20 h (*see* **Note 13**).
6. Quench the reaction by gradually adding FA or TFA with mixing to pH 2–3 (avoid rapidly generating bubbles).

3.4 Desalting

1. To desalt 10 mg peptides, two HyperSep C18 Cartridges (200 mg) will be needed.
2. Add 3 mL of 100% ACN to the cartridges and centrifuge at 100 ×*g* for 1 min. Repeat for a total of three times to fully activate the column (*see* **Note 14**).

3. Add 3 mL 0.1% (v/v) FA to the cartridges and centrifuge at 300 ×*g* for 2 min. Repeat for a total of two times to equilibrate (*see* **Note 15**).
4. Load 3 mL of sample to the cartridge at a time and centrifuge at 300 ×*g* for 2 min to discard flow through. Continue this step until all samples are loaded.
5. Add 3 mL 0.1% (v/v) FA to the cartridges and centrifuge at 300 ×*g* for 2 min; repeat the wash one more time (*see* **Notes 15 and 16**).
6. Transfer the C18 cartridge to a new 15 mL tube, add 1 mL of peptide elution buffer, and centrifuge at 100×g for 1 min. Repeat the elution step one more time to collect all eluents and ensure beads are dry after the last elution.
7. Make an aliquot of eluents corresponding to ~50 μg peptides for LC–MS/MS analysis (total proteome) and keep the rest for immunoaffinity enrichment.
8. Dry all the eluents with a SpeedVac vacuum concentrator.

3.5 Immunoaffinity Enrichment of Lysine Acetylated Peptides

1. Add 1.4 mL IAP buffer to ~10 mg dried peptides, resuspend the peptides by vortexing and/or pipetting up and down, if needed (*see* **Note 17**).
2. Clear the peptide solution by spinning at 10,000 ×*g*, 4°C for 5 min. Keep the cleared peptide solution on ice while preparing the beads.
3. Centrifuge a vial of antibody-bead slurry at 2000 ×*g*, 4°C for 30 s and remove all buffer from the beads (*see* **Note 18**).
4. Add 1 mL PBS to the beads and mix gently by inverting the tubes up and down. Centrifuge at 2000 ×*g*, 4°C for 30 s and carefully remove the supernatant. Repeat the wash for three more times.
5. After the last PBS wash, resuspend the beads in 40 μL PBS.
6. Transfer the peptide solution into the tube containing antibody beads (*see* **Note 19**).
7. Tighten and seal the cap with parafilm to avoid leakage, and incubate on a tube rotator for 2 h at 4°C.
8. Centrifuge at 2000 ×*g*, 4°C for 30 s and transfer the supernatant to a new tube to save at −80°C for future use (*see* **Note 20**).
9. Add 1 mL 1× IAP buffer to the beads, mix by inverting up and down five to six times, and centrifuge at 2000 ×*g*, 4°C for 30 s; discard the supernatant.
10. Add 1 mL ice-cold MS-grade water to the beads, mix by inverting up and down five to six times, centrifuge at 2000 ×*g*, 4°C for 30 s, and discard the supernatant. Repeat this wash

step two more times, and remove the supernatant as much as possible in the last wash.

11. Add 55 μL of 0.15% TFA elution buffer to the beads, tap the bottom of the tube several times (do not vortex), and let it stand at room temperature for 10 min (tap the bottom of the tube gently to mix every 2–3 min).
12. Centrifuge at 2000 ×*g*, 4°C for 30 s and transfer the supernatant to a new 1.5 mL Eppendorf tube.
13. Add another 55 μL of 0.15% TFA to the beads to repeat the elution step. Combine both eluents for desalting using a 20 μL filter tip-based C18 cartridge.
14. Add 200 μL of 100% ACN to the cartridges and centrifuge at 100 ×*g* for 1 min. Repeat for a total of three times to fully activate the column.
15. Add 200 μL 0.1% FA to the cartridges and centrifuge at 300 ×*g* for 2 min. Repeat for a total of two times to equilibrate.
16. Load all samples to the cartridge and centrifuge at 200 ×*g* for 2 min to discard flow through.
17. Add 200 μL 0.1% FA to the cartridges and centrifuge at 200 ×*g* for 2 min (*see* **Note 21**).
18. Transfer the column to a new 1.5 mL tube, add 130 μL of peptide elution buffer, and centrifuge at 100 ×*g* for 1 min. Repeat the elution step one more time to collect all eluents.
19. Dry the eluents with a SpeedVac vacuum concentrator and resuspend in 20 μL 0.1% FA for LC–MS/MS analysis.

3.6 LC–MS/MS Analysis and Bioinformatics

1. Load both unenriched and enriched peptide samples for LC–MS/MS analysis with typical parameters for proteomics on the instrument/platform in your lab. In our case, we utilized a Q Exactive HF-X mass spectrometer with a 2-h HPLC gradient from 5% to 35% (v/v) ACN at a flow rate of 300 nL/min. For the unenriched samples, peptides corresponding to 250 ng of protein were loaded for MS analysis. For enriched samples, peptides were resuspended in 20 μL of 0.1% FA, and 4 μL was injected for MS analysis.
2. Peptide identification: upload the generated MS raw files to pFind 3 for database search using the human gut microbial gene catalog database (IGC; https://db.cngb.org/microbiome/genecatalog/genecatalog_human/). Perform a separate search for enriched and unenriched raw files, respectively, with the same search parameters as follows: enzyme, trypsin KR_C; full-specific; up to three missed cleavages; 20 ppm for both precursor and fragment tolerance; checked open search with no selected fixed and variable modifications; no quantitation; and keep the remaining parameters as default.

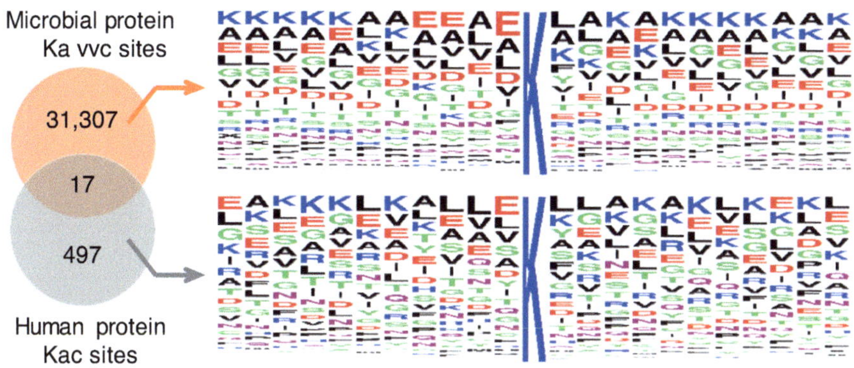

Fig. 2 Number of identified human and gut microbial protein lysine acetylation sites in adult stool samples. The uniquely assigned microbial or human protein sites were analyzed using WebLogo (https://weblogo.berkeley.edu/logo.cgi) to demonstrate the distribution of amino acids surrounding the identified lysine acetylation sitesAdapted from Fig. 1 and Supplementary Fig. 2 of Zhang et al [16]

3. Reduced database generation: generate a reduced protein database from the output of pFind database search (pFind-Filtered. spectra). Briefly, all identified proteins from the pFind searches of both enriched and unenriched samples will be collected, their sequences extracted from the original IGC database to generate a reduced database in FASTA format (*see* **Note 22**).
4. Protein and peptide quantification: use MaxQuant for a second database search to identify and quantify lysine acetylation (Kac) sites using the reduced gut microbial database generated above (*see* **Note 23**). The same database search parameters should be used for both enriched and unenriched data sets to allow for integrated data analysis. Up to four missed cleavages are allowed with the potential modifications, including methionine oxidation, lysine acetylation, and protein N-terminal acetylation. Keep default for all other parameters. The identified peptides and protein groups in unenriched samples can be obtained from the modificationSpecificPeptides.txt and proteinGroups. txt files, respectively. The identified lysine acetylated peptides and sites in enriched samples were obtained from the modificationSpecificPeptides.txt and Acetyl (K) Sites files, respectively. Only Kac sites with a localization probability of >0.75 should be used for further analysis, such as using WebLogo (https://weblogo.berkeley.edu/logo.cgi) for analysis of amino acid distribution surrounding identified sites (Fig. 2).
5. Taxonomic and functional annotation: use MetaLab v2.3 for taxonomic and functional annotations using peptides and protein group outputs, respectively. In MetaLab v2.3, select "Taxonomy analysis" or "Functional annotation" from the left panel and select Result type of "MaxQuant" to upload either peptides.txt or proteinGroups.txt file for taxonomic

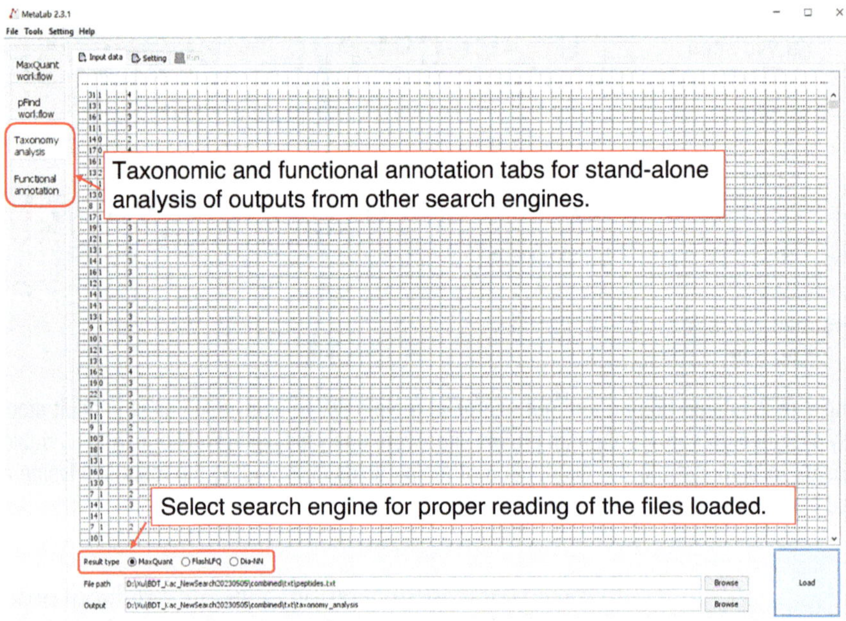

Fig. 3 MetaLab interface for taxonomic and functional annotations using MaxQuant output from the second step database search

or functional annotation (Fig. 3). A report will be generated along with detailed result data tables in the MetaLab output folder for further downstream.

4 Notes

1. To make the C18 beads evenly distributed across different columns, mix the bead slurry by pipetting up and down or vortexing while adding C18 bead suspension to the multiple columns. Add C18 beads to the column when you are ready to desalt and do not let beads dry out. We used a home-made tip-based C18 column for a small amount of peptide sample desalting; however, there are commercially available premade C18 desalting plates or tips available that can be used as alternatives.
2. A probe sonicator can be used as an alternative. The parameter used in this protocol is optimized for this specific model. Adjust the parameters according to the sonicator used.
3. All steps involve pipetting fecal slurries should be performed on ice and in a biosafety cabinet. The vortexing step is very important for the efficiency of microbial cell harvesting. Make

sure there are no visible chucks and vortex to the maximum speed to ensure high and consistent microbial cell harvesting.

4. Usually, during the third step microbial extraction, a clear supernatant can be observed indicating sufficient extraction of microbial cells from the stool pellets. If the supernatant still looks cloudy, perform the microbial extractions for more times or using more PBS for each extraction.
5. Pipette tips can be used to stir the pellet to help resuspending the cells in lysis buffer; however, SDS can easily generate foams, so pipette the solution gently.
6. Alternative methods can be used for protein extraction; however, our previous evaluations suggest that ultrasonication yielded the best performance in extracting stool microbial proteins [17].
7. It is critical that the pellets are disrupted completely to efficiently remove SDS from proteins. The pellets can be manually disrupted using gel-loading tips—using more of mortar and pestle action not pipetting action—followed by water bath sonication.
8. The proteins can be kept in acetone at −20°C for long-term storage, if digestion is not performed in the same day. If protein samples are to be shipped to another lab for processing, remove acetone and air-dry the protein pellets. The dry protein pellets can then be shipped at room temperature or on dry ice for long-distance shipping.
9. We use the DC protein assay because it is detergent (e.g., SDS and urea) compatible; however, other protein assays or methods can be used as long as if they can provide an accurate quantity estimation. Perform the assay according to the manufacturer's instructions and use a protein standard (e.g., BSA) to generate a standard curve for quantification.
10. We use 2 mg/mL protein concentration to perform reduction and alkylation, and dilute tenfold for in-solution digestion. Higher concentration of proteins can be used to reduce the volume of digestion reaction system; however, that needs to be evaluated for the performance of digestion.
11. We use room temperature for the DTT reduction step instead of 56°C, which is generally used for trypsin digestion. This modification is to avoid protein carbamylation when urea is present in the digestion buffer [18]. Protein carbamylation will generate lysine carbamylated peptides, and these peptides are structurally similar to lysine acetylation, which can be co-enriched during immunoaffinity enrichment.
12. The IAA reaction needs to be performed in the dark, so wrap the samples using foil paper and place the samples in a drawer without shaking.

13. Digestion time affects the efficiency of proteolytic digestion, so always keep the same digestion time for the same project. Room temperature was used in this study for digestion due to the same reason as described in **Note 11**.
14. A 15 mL falcon tube can be used as receiving tube to collect flow-through waste buffer. Empty the receiving tube every two to three spins to avoid contaminating the cartridge.
15. 0.1% FA may not easily flow through the column depending on the samples; therefore, the speed can be increased up to 400 g and for 5 min. However, do not over-dry the beads on the column.
16. Make sure to fill the cartridge with wash buffer to the level where the loaded samples were present to efficiently remove any residual salts in the samples.
17. If the peptides are over-dried, it could be difficult to dissolve them in the IAP buffer. If that's the case, keep the peptides in IAP buffer at 4°C for 30–60 min followed by vortexing. After dissolving the peptide, check the pH of the peptide solution with a pH strip, and the pH should be close to neutral, or no lower than 6.0.
18. The experiment can be scaled down, if using high-sensitive MS or preliminary test shows higher amount of acetylation in the samples as long as if maintaining the ratio of 10 mg peptides to one vial of antibody beads. The new version of the PTMscan HS kit using the magnetic beads technology can be used if less protein per sample is available (requires ~1 mg peptide input). Tandem Mass Tag (TMT)-based multiplexing approach can also be used for further lowering the need of sample input for each sample as well as reducing the cost for reagent kits [19].
19. Pipette the sample directly on top of the beads at the bottom of the tube and immediately mix by gently pipetting up and down (avoid generating excessive bubbles).
20. The supernatant collected in this step can be used for enrichment of peptides with other modifications, such as propionylation and succinylation [20], using PTMscan kits or for glycopeptide enrichment with hydrophilic interaction liquid chromatography (HILIC) enrichment as reported in our previous studies [19].
21. One wash is needed for the desalting of enriched lysine acetylated peptides to reduce sample loss during washing and due to the fact that eluted peptides from the beads are relatively clean and in a small volume.

22. To streamline the fasta sequence extraction, we have constructed and made available a Perl script (can be downloaded from: https://drive.google.com/file/d/17RzSp51dEUMY-IBHJrM-mgkrit1qBWgm/view?usp=drive_link).
23. Metaproteomics has the advantage of identifying both microbial and host proteins from the same generated MS data. A human proteome database can be concatenated with the reduced gut microbial database for a combined database search, allowing identification of both human and microbial peptides.

Acknowledgments

This work was supported by the Government of Canada through Health Canada, Genome Canada, the Natural Sciences and Engineering Research Council of Canada, and partly by the funding from Genomics Research and Development Initiative (GRDI) of Canada (Phase VIII).

References

1. Gilbert JA, Blaser MJ, Caporaso JG, Jansson JK, Lynch SV, Knight R (2018) Current understanding of the human microbiome. Nat Med 24(4):392–400
2. The Integrative Human Microbiome Project (2019) Nature 569(7758):641–648
3. Joos R, Boucher K, Lavelle A, Arumugam M, Blaser MJ, Claesson MJ, Clarke G, Cotter PD, De Sordi L, Dominguez-Bello MG, Dutilh BE, Ehrlich SD, Ghosh TS, Hill C, Junot C, Lahti L, Lawley TD, Licht TR, Maguin E, Makhalanyane TP, Marchesi JR, Matthijnssens J, Raes J, Ravel J, Salonen A, Scanlan PD, Shkoporov A, Stanton C, Thiele I, Tolstoy I, Walter J, Yang B, Yutin N, Zhernakova A, Zwart H, Dore J, Ross RP (2025) Examining the healthy human microbiome concept. Nat Rev Microbiol 23(3):192–205
4. Zhang X, Li L, Butcher J, Stintzi A, Figeys D (2019) Advancing functional and translational microbiome research using meta-omics approaches. Microbiome 7(1):154
5. Meier F, Brunner AD, Frank M, Ha A, Bludau I, Voytik E, Kaspar-Schoenefeld S, Lubeck M, Raether O, Bache N, Aebersold R, Collins BC, Rost HL, Mann M (2020) diaPASEF: parallel accumulation-serial fragmentation combined with data-independent acquisition. Nat Methods 17(12):1229–1236
6. Zhang F, Ge W, Huang L, Li D, Liu L, Dong Z, Xu L, Ding X, Zhang C, Sun YAJ, Gao J, Guo T (2023) A comparative analysis of data analysis tools for data-independent acquisition mass spectrometry. Mol Cell Proteomics 22(9):100623
7. Sun Y, Xing Z, Liang S, Miao Z, Zhuo LB, Jiang W, Zhao H, Gao H, Xie Y, Zhou Y, Yue L, Cai X, Chen YM, Zheng JS, Guo T (2024) metaExpertPro: a computational workflow for metaproteomics spectral library construction and data-independent acquisition mass spectrometry data analysis. Mol Cell Proteomics 23(10):100840
8. Wang A, Fekete EEF, Creskey M, Cheng K, Ning Z, Pfeifle A, Li X, Figeys D, Zhang X (2024) Assessing fecal metaproteomics workflow and small protein recovery using DDA and DIA PASEF mass spectrometry. Microbiome Res Rep 3(3):39
9. Wu E, Xu G, Xie D, Qiao L (2024) Data-independent acquisition in metaproteomics. Expert Rev Proteomics 21(7–8):271–280
10. Zhang X, Ning Z, Mayne Y, Figeys D (2025) Clinical microbiome analysis by mass spectrometry–based metaproteomics. Ann Rev Anal Chem 18:149–172
11. Kleiner M (2019) Metaproteomics: much more than measuring gene expression in microbial communities. mSystems 4(3):e00115-19
12. Smyth P, Zhang X, Ning Z, Mayne J, Moore JI, Walker K, Lavallee-Adam M, Figeys D

(2020) Studying the temporal dynamics of the gut microbiota using metabolic stable isotope labeling and metaproteomics. Anal Chem 92(24):15711–15718
13. Thuy-Boun PS, Wang AY, Crissien-Martinez A, Xu JH, Chatterjee S, Stupp GS, Su AI, Coyle WJ, Wolan DW (2022) Quantitative metaproteomics and activity-based protein profiling of patient fecal microbiome identifies host and microbial serine-type endopeptidase activity associated with ulcerative colitis. Mol Cell Proteomics 21(3):100197
14. Valdes-Mas R, Leshem A, Zheng D, Cohen Y, Kern L, Zmora N, He Y, Katina C, Eliyahu-Miller S, Yosef-Hevroni T, Richman L, Raykhel B, Allswang S, Better R, Shmueli M, Saftien A, Cullin N, Slamovitz F, Ciocan D, Ouyang KS, Mor U, Dori-Bachash M, Molina S, Levin Y, Atarashi K, Jona G, Puschhof J, Harmelin A, Stettner N, Chen M, Suez J, Honda K, Lieb W, Bang C, Kori M, Maharshak N, Merbl Y, Shibolet O, Halpern Z, Shouval DS, Shamir R, Franke A, Abdeen SK, Shapiro H, Savidor A, Elinav E (2025) Metagenome-informed metaproteomics of the human gut microbiome, host, and dietary exposome uncovers signatures of health and inflammatory bowel disease. Cell 188(4):1062–1083.e1036
15. Zhang W, Sun J, Cao H, Tian R, Cai L, Ding W, Qian P Y. (2016) Post-translational modifications are enriched within protein functional groups important to bacterial adaptation within a deep-sea hydrothermal vent environment. Microbiome 4(1):49
16. Zhang X, Ning Z, Mayne J, Yang Y, Deeke SA, Walker K, Farnsworth CL, Stokes MP, Couture JF, Mack D, Stintzi A, Figeys D (2020) Widespread protein lysine acetylation in gut microbiome and its alterations in patients with Crohn's disease. Nat Commun 11(1):4120
17. Zhang X, Li L, Mayne J, Ning Z, Stintzi A, Figeys D (2018) Assessing the impact of protein extraction methods for human gut metaproteomics. J Proteomics 180:120–127
18. Sun S, Zhou JY, Yang W, Zhang H (2014) Inhibition of protein carbamylation in urea solution using ammonium-containing buffers. Anal Biochem 446:76–81
19. Fekete EE, Wang A, Creskey M, Cummings SE, Lavoie JR, Ning Z, Li J, Figeys D, Chen R, Zhang X (2024) Multilevel proteomic profiling of colorectal adenocarcinoma Caco-2 cell differentiation to characterize an intestinal epithelial model. J Proteome Res 23(7):2561–2575
20. Zhang X, Cheng K, Ning Z, Mayne J, Walker K, Chi H, Farnsworth CL, Lee K, Figeys D (2021) Exploring the microbiome-wide lysine acetylation, succinylation, and propionylation in human gut microbiota. Anal Chem 93(17):6594–6598

Chapter 11

Generation of High-Quality Succinyl Spectral Libraries for Improved Proteome-Wide Succinylome Analysis Using Data-Independent Acquisition

Alice Zhang, Birgit Schilling and Joanna Bons

Abstract

Protein posttranslational modifications (PTMs) are crucial and dynamic modulators of protein functions, interactions, and localizations, as well as biological pathways and cellular signaling. Lysine succinylation analysis remains challenging, but sophisticated workflows combining succinylated peptide enrichments and quantitative mass spectrometry approaches have revolutionized proteome-wide succinylome analysis. The implementation of data-independent acquisition (DIA)-mass spectrometry has greatly advanced the detection of low-abundance succinylated peptides, succinylome coverage, identification reproducibility, and quantification accuracy. However, the complexity of DIA data requires dedicated data processing algorithms and tools, which typically rely on spectral libraries. These reference libraries can be generated from experimental data-dependent acquisition (DDA) acquisitions of representative study samples submitted to DDA database search engines for confident succinylated peptide identification and precise PTM site localization. Here, we describe how to build DDA PTM spectral libraries using various software tools, specifically Spectronaut, SpectroMine, and MSFragger. The generated libraries were imported into Skyline for the analysis of previously published DIA succinylome data of Sirtuin-5 knocked-out vs wild-type mouse brains in order to accurately quantify and visualize PTM-containing peptides.

Keywords Spectral library, Succinylation, Data-dependent acquisition, Posttranslational modifications, Data-independent acquisition, Quantitative proteomics

1 Introduction

Posttranslational modifications (PTMs) occur after protein biosynthesis and involve the addition of functional covalent groups, the modification of amino acids, or the cleavage of peptide bonds, through enzymatic or nonenzymatic processes. They are crucial for dynamically regulating protein function, activity, stability, subcellular localization, and interaction networks [1]. Succinylation is a reversible modification that is dynamically regulated within the cells and impacts proteins through the addition of a succinyl group to the ε -amine group of lysine residues [2]. Sirtuin-5 (SIRT5), a member

Xu Zhang et al. (eds.), *Protein Post-Translational Modifications*, Methods in Molecular Biology, vol. 3018,
https://doi.org/10.1007/978-1-0716-5166-7_11,

of the sirtuin family of NAD$^+$-dependent deacetylases, has been identified as a key regulator of protein succinylation [3–5]. SIRT5 removes succinyl groups from lysine residues, thereby regulating the succinylome and impacting cellular metabolism through reactive oxygen species (ROS) detoxification, the TCA cycle, the electron transport chain, the fatty acid cycle, nitrogenous waste management, and ketone body formation [4, 6–8]. SIRT5 has also been found to play a role in various cancers, neurodegenerative conditions, and cardiovascular function [9–12].

Protein succinylation analysis is challenging due to the low abundance of the modified peptides/proteins as well as the substoichiometric nature and low site occupancy of the modified residues, which complicates correct PTM identification, site localization, and accurate quantification [1]. Fortunately, in recent years, there have been significant advancements in succinyl enrichment strategies, fast scanning and high resolution mass spectrometry (MS) instrumentation, and PTM identification and site localization bioinformatic methodologies [13], making liquid chromatography–tandem mass spectrometry (LC–MS/MS) the leading method for comprehensive PTM analysis [14–16]. LC–MS/MS offers high sensitivity and the ability to pinpoint PTMs to the amino acid residue-level. Large-scale quantitative PTM analysis has been successful in uncovering the regulatory mechanisms governed by PTMs and several studies have demonstrated the success of data-independent acquisition (DIA) to systematically profile thousands of PTMs in biological [17–22]. DIA relies on the simultaneous collection of fragment ion spectra for all detectable precursor ions isolated within a defined wide *m/z* range, generating highly complex spectra with fragment ions from multiple coeluting peptides and allowing for comprehensive, reproducible, and unbiased quantitative proteome analysis [23]. Deconvoluting DIA MS/MS spectra is, however, computationally intensive, requiring complex, sophisticated, and dedicated analysis tools. One approach is to use the information contained in a spectral library generated from data-dependent acquisition (DDA) data to perform DIA data extraction, peak group scoring, and PTM site localization probability determination for peptide and PTM identification and quantification [15, 24, 25]. A spectral library generated from high-quality DDA data contains experimentally observed MS/MS spectra confidently assigned to a peptide sequence, leading to high identification confidence, sensitive and accurate quantification, and enhanced probabilities of correctly identifying and precisely localizing the modifications, which is necessary to understand their biology. Additionally, spectral library-based approaches reduce the computational burden, as the search space for the targeted extraction of the DIA data is limited to the modified peptides present in the library, leading to fast and efficient DIA data processing. Several sophisticated and efficient computational tools have been developed and/or optimized to generate study-specific spectral libraries from experimental DDA data for DIA PTM analysis, such as OpenSwath [26], Skyline [27], MaxQuant [28], Spectronaut [29] (Biognosys), PeakView (Sciex), ProteomeDiscoverer (Thermo Fisher Scientific), MSFragger [30], DIA-NN [31], and SpectroMine (Biognosys). They rely on various search engines, validation tools, and PTM site localization scoring algorithms to match peptide sequences to the fragmentation ion spectra and further build spectral libraries. In addition, some are commercially available—e.g., Spectronaut, SpectroMine, PeakView, and Proteome Discoverer, and other tools are free and/or open-source—e.g., OpenSwath, Skyline, MSFragger, MaxQuant, and DIA-NN.

Here, we provide step-by-step protocols to guide researchers on how to use vendor agnostic spectral library generation tools, especially the commercial Spectronaut and SpectroMine (both from Biognosys), that use the Pulsar search engine, as well as the open-source MSFragger [30] through the FragPipe platform (https://fragpipe.nesvilab.org/) tools to generate DDA spectral libraries for comprehensive and unbiased protein succinylation analysis by DIA (Fig. 1). To note, the aim of these protocols was not to compare the

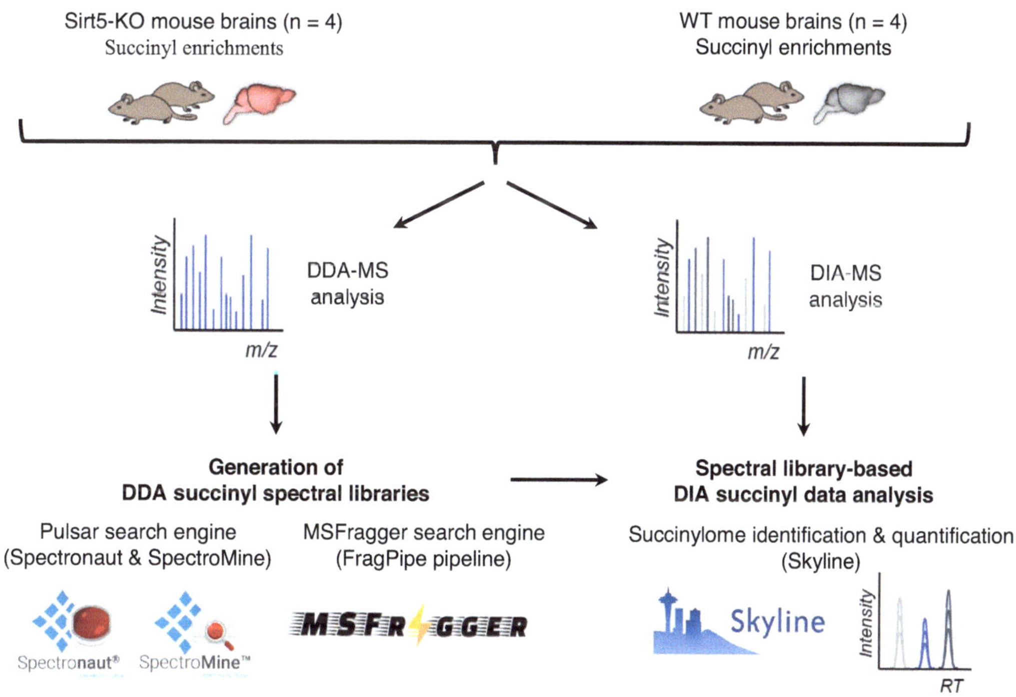

Fig. 1 Experimental workflows for succinylome analysis by data-independent acquisition using data-dependent succinyl spectral libraries. Succinyl enrichments of Sirtuin-5-knocked out (SIRT-KO) and wild-type (WT) mouse brains were analyzed by data-dependent acquisition (DDA) and data-independent acquisition (DIA) on an Orbitrap Eclipse Tribrid platform. DDA data were used to generate high-quality spectral libraries with confident PTM site localization using Pulsar search engine embedded in Spectronaut and SpectroMine, as well as MSFragger search tool through the FragPipe pipeline. The spectral libraries were then imported into Skyline to analyze DIA data for succinylated peptide/protein identification and quantification

different software tools—for that, we point the readers to extensive benchmarking studies performed by Gotti et al. [32], Fröhlich et al. [33], and Lou et al. [34]—but to describe several available workflows to generate DDA-based spectral libraries with high confidence for succinylation identification, quantification, and PTM site localization using a previously described succinylome dataset on SIRT5-KO and wild-type (WT) brain tissues collected from 18-month female mice [19]. Briefly, whole brain protein lysates from SIRT5-KO and WT mice ($n=4$ each) were digested with trypsin, and proteolytic peptides were enriched for succinylation using antisuccinyl antibody conjugated to agarose beads (Succinyl-Lysine Motif Kit, Cell Signaling Technology). Each sample was then individually acquired in both DDA and DIA modes on an Orbitrap Eclipse Tribrid platform. DDA datasets with IDs MSV000096592 and PXD058580 are uploaded and available at ftp://massive.ucsd.edu/v07/MSV000096592/ and DIA datasets with IDs MSV000089606 and PXD034275 are available at ftp://massive.ucsd.edu/v04/MSV000089606/). For guidance on acylated peptide enrichment for mass spectrometry analysis, please refer to Bons et al. [19] and Zhang et al. [35].

2 Materials

2.1 Liquid Chromatography–Tandem Mass Spectrometry (LC–MS/MS)

1. LC–MS grade formic acid (FA).
2. LC–MS grade acetonitrile (ACN).
3. Resuspension buffer: 0.2% FA in water.
4. LC solvent A: 2% ACN, 0.1% FA in water.
5. LC solvent B: 98% ACN, 0.1% FA in water.
6. iRT peptides (Biognosys).
7. Acclaim PepMap 100 C_{18} trap column (75 μm × 50 cm, 3 μm particle size; Thermo Fisher Scientific).
8. Acclaim PepMap 100 C_{18} analytical column (75 μm × 50 cm, 3 μm particle size; Thermo Fisher Scientific).
9. Dionex UltiMate 3000 system coupled to an Orbitrap Eclipse Tribrid mass spectrometer (both from Thermo Fisher Scientific).

2.2 Data Analysis

1. Spectronaut (Biognosys), Spectronaut (Biognosys), or MSFragger [30] software for DDA succinyl spectral library generation.
2. Skyline [27] software for spectral library-based DIA data processing, graphical representation, and statistical data analysis.

3 Methods

3.1 DDA Data Generation with LC–MS/MS

1. Reconstitute dried succinylated peptide enrichments in 20 μL of 0.2% FA and thoroughly mix the solution. Centrifuge at 12,000 × *g* at room temperature for 2 min.
2. Add 1 μL of iRT peptides prepared according to manufacturer's instructions.
3. Transfer the sample to an autosampler vial and inject 4 μL peptide for LC–MS/MS analysis.
4. Load the peptides onto the trap column over 10 min at 2 μL/min with 100% solvent A (*see* **Note 1**).
5. Elute the peptides with the analytical column heated at 35°C and at a flow rate of 300 nL/min over 215 min using the following gradient of solvent B: 2% for 10 min, linear from 2% to 20% in 125 min, linear from 20% to 32% in 40 min, up to 80% in 1 min, 80% for 9 min, and down to 2% in 1 min. Equilibrate the column with 2% solvent B for 29 min.
6. Acquire the data in DDA mode using the following settings. Collect full MS spectra at 240,000 resolution (AGC target: 1.2e6 ions, maximum injection time: Auto, 350–1500 *m/z*). Select precursor ions with a charge state 2–5+ for HCD fragmentation at NCE 28 in the ion trap using a dynamic exclusion of 60 s. Collect MS2 spectra with scan rate set to Turbo (AGC target: 3e4 ions, maximum injection time: 25 ms, scan range: Auto) in a 1-s cycle time data-dependent mode (*see* **Note 2**).

3.2 DDA Succinyl Data Analysis for Spectral Library Generation

1. Search the DDA data against the *Mus musculus* proteome database (58,430 protein entries from UniProtKB-TrEMBL) and the Biognosys-11 iRT peptide sequences used for retention time calibration using the following settings: Trypsin/P as digestion enzyme; up to four missed cleavages allowed; carbamidomethylation of cysteine residues as fixed modification; methionine oxidation, protein N-terminal acetylation, and lysine succinylation as variable modifications; PTM localization probability at 0.75; 1% false discovery rate (FDR) at the precursor, peptide, and protein level.
2. Apply the detailed settings specifically described below for each software tool, Spectronaut (version 17.6, Sect. 3.3), SpectroMine (version 4.4, Sect. 3.4), and MSFragger (version 22.0, Sect. 3.5) (*see* **Note 3**).

3.3 DDA Succinyl Spectral Library Generation with Spectronaut

1. Ensure that the lysine succinylation modification is defined. If not, in the *Databases* tab, navigate to the *Modifications* tab, create *New* and set *Composition* to "+C4+H4+O3," *Position* to "Anywhere," *Specificity* to "Residue," and *Site* to "K."

2. In the *Library* tab, select *Generate Library from Pulsar/Search Archives* and apply default settings unless specified below.
3. Add the DDA raw data files and assign the relevant organism FASTA file.
4. To configure the **database search with Pulsar**, apply the *BGS Factory Settings* with some modifications as in the following steps.
5. In *Peptides*, set the *Enzyme/Cleavage Rules* to "Trypsin/P," *Digest Type* to "Specific," and *Missed Cleavages* to "4."
6. In *Modifications*, set *Max Variable Modifications* to "8," *Fixed Modifications* to "Carbamidomethyl (C)," and *Variable Modifications* to "Acetyl (Protein N-term)," "Oxidation (M)," and "Succinyl."
7. In *MS2*, set *Ion Types* to "b" and "y."
8. Perform protein inference on the same FASTA file as the originally imported one.
9. Select the gene annotation file relevant to the experiment organism.
10. To configure the **spectral library generation** settings, select the *BGS Factory Settings (Default)*.
11. In *Tolerances*, set *Calibration Search* and *Main Search* to "Dynamic" and *MS1 Correction Factor* and *MS2 Correction Factor* to "1."
12. In *Identification*, set *Peptide FDR*, *Protein FDR*, and *PSM FDR* to "0.01." Check *PTM Localization Filter* and set *Min Localization Threshold* to "0.75."
13. In *Spectral Library Filters*, set *Ion AA Length* to "3," *Max m/z* to "3000," *Min m/z* to "200", *Min Relative* Intensity to "1," *Max Best N Fragments per Peptide* to "6" and *Min Best N Fragments per Peptide* to "3."
14. In *iRT Calibration*, set *iRT Reference Strategy* to "Deep Learning Assisted iRT Regression."
15. In *Workflow*, set *Fragment Ion Selection Strategy* to "Intensity Based" and *Use DNN Predicted Ion Mobility* to "Auto."
16. Press *Finish* to start the search (*see* **Note 3**).
17. To export the generated spectral library, navigate to the **Library** tab, click on *Spectral Libraries*, select the spectral library of interest, and click *Export Spectral Library*.
18. Save as a "Tab separated xls File (*.xls)" (*see* **Note 4**).

3.4 DDA Succinyl Spectral Library Generation with SpectroMine

1. Ensure that the lysine succinylation modification is defined. If not, define the modification according to Sect. 3.3, Step 1.
2. In the *Analysis* tab, select *Set up a SpectroMine Experiment* and apply default settings unless specified below.

3. Add the DDA raw data files and assign the relevant FASTA file.
4. To configure the **database search with Pulsar**, select the *BGS Factory Settings* with some modifications as in the following steps.
5. In *Peptides*, set the *Enzyme/Cleavage Rules* to "Trypsin/P," *Digest Type* to "Specific," and *Missed Cleavages* to "4."
6. In *Modifications*, set *Max Variable Modifications* to "8," *Fixed Modifications* to "Carbamidomethyl (C)," and *Variable Modifications* to "Acetyl (Protein N-term)," "Oxidation (M)," and "Succinyl."
7. In *MS1*, set *Number of Searches* to "15."
8. In *MS2*, set *Ion Types* to "b" and "y."
9. Assign runs to block: each run is automatically assigned to its own block.
10. Perform protein inference on the same FASTA file as the originally imported one.
11. Select the gene annotation file relevant to the experiment organism.
12. To configure the **spectral library generation** settings, select the *BGS Factory Settings.*
13. In *Tolerances*, set *Calibration Search* and *Main Search* to "Dynamic" and *MS1 Correction Factor* and *MS2 Correction Factor* to "1."
14. In *Output*, select "Spectral Library" and "Search Archive."
15. In *Identification*, set *Peptide FDR*, *Protein FDR*, and *PSM FDR* to "0.01."
16. In *PTM Workflow*, select *PTM Localization Filter* and set *Min localization threshold* to "0.75."
17. In *Spectral Library Filters*, set *Ion AA Length* to "3," *Max m/z* to "3000," *Min m/z* to "200", *Min Relative* Intensity to "1," and *Max Best N Fragments per Peptide* to "6" and *Min Best N Fragments per Peptide* to "3."
18. In *iRT Calibration*, check "Auto-assign iRT source" and set *iRT Reference Strategy* to "Deep Learning Assisted iRT Regression."
19. In *Workflow*, set *Fragment Ion Selection Strategy* to "Intensity Based" and *Use DNN Predicted Ion Mobility* to "Auto."
20. Press *Finish* to start the search (*see* **Note 3**).
21. To export the generated spectral library, navigate to the **Archive** tab, click on *Spectral Libraries*, select the spectral library of interest, and click *Export Spectral Library.*

22. Save as a "Tab separated xls File (*.xls)."
23. To export a report, go to the **Analysis** tab, load the completed SpectroMine experiment, go to the **Report** tab and in *PTM Site Report* to select the report template (*see* **Note 4**).

3.5 DDA Succinyl Spectral Library Generation with MSFragger Through FragPipe

1. In the **Workflow** tab, load the workflow "Default" and apply default settings unless specified below.
2. Add the DDA raw data files and select "Regular MS."
3. In the **Database** tab, import the relevant organism FASTA file supplemented with the Biognosys-11 iRT peptide sequences. Click "Add decoys."
4. To configure the **database searchwith MSFragger**, in the **MSFragger** tab, select "Run MSFragger."
5. In *Peak Matching*, set *Precursor mass tolerance* to "−20 PPM" – "20 PPM" and *Fragment mass tolerance* to "20 PPM."
6. In *Protein Digestion*, set *Cleavage* to "ENZYMATIC," *Enzyme name 1* to "stricttrypsin," and *Missed Cleavages 1* to "4." Ignore enzyme 2 if only trypsin is used for protein digestion. Set *Peptide length* to "7" and "52."
7. In *Modifications*, set *Max Variable Modifications* to "5" (this is the maximum). In *Variable Modifications*, enable "[^" for N-terminus acetylation (+42.0106 Da), "M" for methionine oxidation (+15.9949 Da), and "K" for lysine succinylation (+100.016 Da). In *Fixed modifications*, enable "C (Cysteine)" (+57.02146 Da) and the other sites with no delta mass.
8. In the *Validation* tab, ensure Crystal-C is not checked (recommended for open searches only).
9. In *PSM Validation*, select "Run PSM Validation" and "Run PeptideProphet" with the "Closed Search" defaults loaded.
10. In *PTM Site Localization*, select "Run PTMProphet," press "Load defaults" and "Load Mods from MSFragger Settings," and set *MINPROB* to "0.75" in the command line.
11. In *Protein Inference*, select "Run ProteinProphet" and keep the default command line.
12. In *FDR Filter and Report*, select "Generate reports" and keep the default—*prot* filter of "0.01."
13. To configure the **spectral library generation** settings, in the **Spectral Library** tab, select "Generate spectral library from Search Results," set *Filetype to Convert* to "psm.tsv," *RT calibration* to "Biognosys_iRT," *Fragment types* to "b" and "y," and select "Make library Skyline compatible."

14. In the **Run** tab, name the "Output dir," conduct a "Dry Run" to ensure that the modules run properly.
15. Press *Run* to start the search (*see* **Note 3**).
16. Locate all intermediate and output files stored in the "Output dir" after completion of the spectral library generation (*see* **Note 4**).

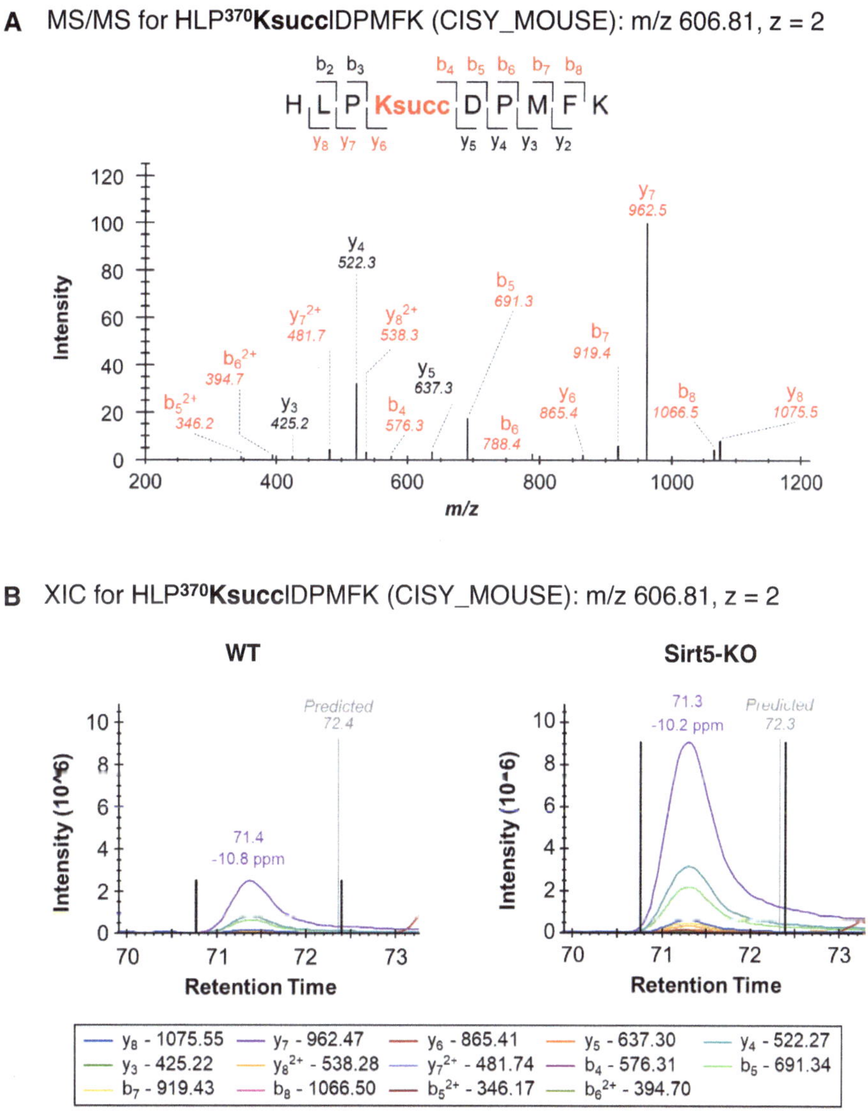

Fig. 2 Example for the succinylated peptide HLP[370]K_{succ}DPMFK from the mouse mitochondrial citrate synthase analyzed using the various workflows. (**A**) DDA MS/MS spectrum for HLP[370]K_{succ}DPMFK from the mouse mitochondrial citrate synthase (UniProt ID: Q9CZU6) with a succinyl group on the lysine K-370 (precursor ion at *m/z* 606.81, z = 2). Red fragment ions contain the succinyl group and are direct evidence of its localization site. (**B**) Representative extracted ion chromatograms (XICs) of HLP[370]K_{succ}DPMFK obtained from a WT and SIRT-KO replicate. DIA data were analyzed in Skyline using the DDA succinyl spectral library.

3.6 Succinylation Quantification and Visualization in Skyline

1. Succinylated peptide and protein quantification from the DIA data were performed in Skyline, leveraging its compatibility with spectral libraries generated from various search engines and tools, such as Spectronaut, SpectroMine, and MSFragger (*see* **Note 5**). Each DDA succinyl spectral library was individually imported into Skyline-daily (version 24.0.9.184) using Biognosys-11 as iRT standard peptides, and three separate Skyline documents were generated, one for each of the libraries. All peptides matching the library were picked. Ion match tolerance was set to 0.5 *m/z* and the ten most intense product ions from filtered charges were extracted from the DIA succinyl data. All matching scans were included.
2. Ion chromatograms extracted from the DIA data can be visualized in Skyline and relative quantification of succinylation levels performed across conditions, for instance, comparing Sirt5-KO vs WT mice (Fig. 2).

4 Notes

1. Trap and analytical columns with other specifications can be employed with eventually slight adjustments of the loading and eluting conditions. As an example, peptides can be loaded on a 5-μm particle size trap column over 5 min at 5 μL/min with 100% solvent A.
2. Other high-resolution mass spectrometry platforms can be used with adjusted acquisition parameters to equally perform high-quality DDA-PTM acquisitions and subsequently generate spectral libraries.
3. We summarized the characteristics of the three spectral library generation tools, Spectronaut, SpectroMine, and MSFragger, in Table 1. Spectronaut and SpectroMine are commercially

Table 1
Characteristics of the spectral library generation tools

	Spectronaut	SpectroMine	MSFragger
Open source	No	No	Yes (GitHub)
Vendor agnostic	Yes	Yes	Yes
Distributer	Biognosys	Biognosys	Nesvizhskii Lab
Run time	~2 h	~2.5 h	~3.5 h
PTM site localization in spectral library	Yes	Yes	Yes

available from Biognosys, while MSFragger, run through FragPipe, is an open source software maintained by the Nesvizhskii Lab [30]. They, however, are all MS vendor agnostic. Support is easily available through the User Manuals for Spectronaut and SpectroMine or the GitHub page maintained by the Nesvizhskii Lab for MSFragger. From initializing the pipeline to exporting the spectral library, Spectronaut required ~2 h, SpectroMine required ~2.5 h, and FragPipe required ~3.5 h. Finally, PTM site localization information (e.g., residue number, site probability) was provided for each modified peptide contained in the libraries.

4. Generated spectral libraries were exported as *.xlsx* files for Spectronaut and SpectroMine and *psm.tsv* file for MSFragger. The export files were imported into R to remove peptides identified with less than three fragment ions and succinylated peptides with a succinyl group on the C-terminal lysine residue and to ensure a minimum PTM site localization probability of 0.75. The R scripts are available at ftp://massive.ucsd.edu/v07/MSV000096592/ (MassIVE ID: MSV000096592 and ProteomeXchange ID: PXD058580). PTM site-specific localizations and probabilities were available in SpectroMine through the PTM Report tab (Sect. 3.4, Step 23) and MSFragger through PTMProphet [36] (Sect. 3.5, Step 16).

5. For a comparison between the spectral library-free and spectral library-based approaches for proteome-wide succinylome analysis, please see the evaluation in Bons et al. [19], which presents the dynamic remodeling of the mouse brain succinylome upon SIRT5 deletion. Briefly, a sample specific spectral library was generated in Spectronaut from the DDA acquisitions of the eight individual succinylation enrichments and further employed to analyze the succinyl DIA data with the same tool. This strategy was compared to the library-free directDIA one, which revealed that similar identification performances were achieved with both approaches.

Acknowledgments

We acknowledge the support of instrumentation for the Orbitrap Eclipse Tribrid system from the NCRR shared instrumentation grant 1S10 OD028654 (PI: Birgit Schilling).

Competing Interests The authors have no conflicts of interest to declare that are relevant to the content of this chapter.

References

1. Olsen JV, Mann M (2013) Status of large-scale analysis of post-translational modifications by mass spectrometry. Mol Cell Proteomics 12(12):3444–3452
2. Zhang Z, Tan M, Xie Z et al (2011) Identification of lysine succinylation as a new post-translational modification. Nat Chem Biol 7(1):58–63
3. Du J, Zhou Y, Su X et al (2011) Sirt5 is a NAD-dependent protein lysine demalonylase and desuccinylase. Science 334(6057):806–809
4. Park J, Chen Y, Tishkoff DX et al (2013) SIRT5-mediated lysine desuccinylation impacts diverse metabolic pathways. Mol Cell 50(6):919–930
5. Peng C, Lu Z, Xie Z et al (2011) The first identification of lysine malonylation substrates and its regulatory enzyme. Mol Cell Proteomics 10(12):M111.012658
6. Kumar S, Lombard DB (2018) Functions of the sirtuin deacylase SIRT5 in normal physiology and pathobiology. Crit Rev Biochem Mol Biol 53(3):311–334
7. Silva Barbosa AC, Pfister KE, Chiba T et al (2024) Dicarboxylic acid dietary supplementation protects against AKI. J Am Soc Nephrol 35(2):135–148
8. Zhang R, Fang J, Xie X et al (2024) Regulation of urea cycle by reversible high-stoichiometry lysine succinylation. Nat Metab 6(3):550–566
9. Xu Y, Shi Z, Bao L (2022) An expanding repertoire of protein acylations. Mol Cell Proteomics 21(3):100193
10. Sadhukhan S, Liu X, Ryu D et al (2016) Metabolomics-assisted proteomics identifies succinylation and SIRT5 as important regulators of cardiac function. Proc Natl Acad Sci U S A 113(16):4320–4325
11. Abril YLN, Fernandez IR, Hong JY et al (2021) Pharmacological and genetic perturbation establish SIRT5 as a promising target in breast cancer. Oncogene 40(9):1644–1658
12. Bringman-Rodenbarger LR, Guo AH, Lyssiotis CA et al (2018) Emerging roles for SIRT5 in metabolism and cancer. Antioxid Redox Signal 28(8):677–690
13. Yang F, Wang C (2020) Profiling of post-translational modifications by chemical and computational proteomics. Chem Commun (Camb) 56(88):13506–13519
14. Brandi J, Noberini R, Bonaldi T et al (2022) Advances in enrichment methods for mass spectrometry-based proteomics analysis of post-translational modifications. J Chromatogr A 1678:463352
15. Hunter CL, Bons J, Schilling B et al (2023) Perspectives and opinions from scientific leaders on the evolution of data-independent acquisition for quantitative proteomics and novel biological applications. Aust J Chem 76(8):379–398
16. Kitata RB, Yang JC, Chen YJ (2022) Advances in data-independent acquisition mass spectrometry towards comprehensive digital proteome landscape. Mass Spectrom Rev 42(6):2324–2348
17. Basisty N, Meyer JG, Wei L et al (2018) Simultaneous quantification of the acetylome and succinylome by 'one-pot' affinity enrichment. Proteomics 18(17):e1800123
18. Bekker-Jensen DB, Bernhardt OM, Hogrebe A et al (2020) Rapid and site-specific deep phosphoproteome profiling by data-independent acquisition without the need for spectral libraries. Nat Commun 11(1):787
19. Bons J, Rose J, Zhang R et al (2023) In-depth analysis of the Sirtuin 5-regulated mouse brain malonylome and succinylome using library-free data-independent acquisitions. Proteomics 23(3-4):e2100371
20. Di Y, Li W, Salovska B et al (2023) A basic phosphoproteomic-DIA workflow integrating precise quantification of phosphosites in systems biology. Biophys Rep 9(2):82–98
21. Hadjineophytou C, Loh E, Koomey M et al (2024) Combining FAIMS based glycoproteomics and DIA proteomics reveals widespread proteome alterations in response to glycosylation occupancy changes in Neisseria gonorrhoeae. Proteomics 24(14):e2300496
22. Skowronek P, Thielert M, Voytik E et al (2022) Rapid and in-depth coverage of the (phospho-)proteome with deep libraries and optimal window design for dia-PASEF. Mol Cell Proteomics 21(9):100279
23. Gillet LC, Navarro P, Tate S et al (2012) Targeted data extraction of the MS/MS spectra generated by data-independent acquisition: a new concept for consistent and accurate proteome analysis. Mol Cell Proteomics 11(6):O111.016717
24. Ludwig C, Gillet L, Rosenberger G et al (2018) Data-independent acquisition-based SWATH-MS for quantitative proteomics: a tutorial. Mol Syst Biol 14(8):e8126
25. Schilling B, Gibson BW, Hunter CL (2017) Generation of high-quality SWATH((R)) acquisition data for label-free quantitative proteomics studies using TripleTOF((R)) mass spectrometers. Methods Mol Biol 1550:223–233
26. Rost HL, Rosenberger G, Navarro P et al (2014) OpenSWATH enables automated, targeted analysis of data-independent acquisition MS data. Nat Biotechnol 32(3):219–223
27. MacLean B, Tomazela DM, Shulman N et al (2010) Skyline: an open source document editor for creating and analyzing targeted proteomics experiments. Bioinformatics 26(7):966–968

28. Cox J, Mann M (2008) MaxQuant enables high peptide identification rates, individualized p.p.b.-range mass accuracies and proteome-wide protein quantification. Nat Biotechnol 26(12):1367–1372
29. Bruderer R, Bernhardt OM, Gandhi T et al (2015) Extending the limits of quantitative proteome profiling with data-independent acquisition and application to acetaminophen-treated three-dimensional liver microtissues. Mol Cell Proteomics 14(5):1400–1410
30. Kong AT, Leprevost FV, Avtonomov DM et al (2017) MSFragger: ultrafast and comprehensive peptide identification in mass spectrometry-based proteomics. Nat Methods 14(5):513–520
31. Demichev V, Messner CB, Vernardis SI et al (2020) DIA-NN: neural networks and interference correction enable deep proteome coverage in high throughput. Nat Methods 17(1):41–44
32. Gotti C, Roux-Dalvai F, Joly-Beauparlant C et al (2021) Extensive and accurate benchmarking of DIA acquisition methods and software tools using a complex proteomic standard. J Proteome Res 20(10):4801–4814
33. Frohlich K, Brombacher E, Fahrner M et al (2022) Benchmarking of analysis strategies for data-independent acquisition proteomics using a large-scale dataset comprising inter-patient heterogeneity. Nat Commun 13(1):2622
34. Lou R, Cao Y, Li S et al (2023) Benchmarking commonly used software suites and analysis workflows for DIA proteomics and phosphoproteomics. Nat Commun 14(1):94
35. Zhang R, Bons J, Rose JP et al (2024) Protocol for mass spectrometric profiling of lysine malonylation by lysine acetyltransferase in CRISPRi K562 cell lines. STAR Protoc 5(2):103074
36. Shteynberg DD, Deutsch EW, Campbell DS et al (2019) PTMProphet: fast and accurate mass modification localization for the trans-proteomic pipeline. J Proteome Res 18(12):4262–4272

Part III

Proteoform Level Analysis

Chapter 12

Combining SDS-PAGE with Capillary Zone Electrophoresis–Tandem Mass Spectrometry for Top-Down Proteomics Analysis of Intact Histone Proteoforms

Fei Fang and Liangliang Sun

Abstract

Mass spectrometry (MS)-based top-down proteomics (TDP) analysis of histone proteoforms provides critical information about combinatorial posttranslational modifications (PTMs), which is vital for pursuing a better understanding of epigenetic regulation of gene expression. Sodium dodecyl-sulfate polyacrylamide gel electrophoresis (SDS-PAGE) separates histone proteins (H1, H2, H3, and H4) based on their molecular weight, while capillary zone electrophoresis (CZE) provides additional separation of proteoforms of each histone protein based on their electrophoretic mobility, which is affected by PTMs, e.g., acetylation and phosphorylation. Here, we describe the combination of SDS-PAGE-based protein fractionation with CZE–tandem MS (MS/MS) for high-resolution characterization of histone proteoforms. Over 200 histone proteoforms from a commercial calf thymus histone sample were identified with good reproducibility.

Keywords Capillary electrophoresis-mass spectrometry, Histone proteoform, Posttranslational modifications, SDS-PAGE, Top-down proteomics

1 Introduction

Histone proteins mediate epigenetic regulation through a variety of posttranslational modifications (PTMs), including methylation, phosphorylation, acetylation, ubiquitylation, SUMOylation, glycosylation, and ADP-ribosylation [1, 2]. Given that histone PTMs rarely function independently and often exist in a staggering number of combinatorial patterns, mass spectrometry (MS)-based top-down proteomics (TDP) has proven valuable for characterizing intact histone proteoforms and uncovering complex combinatorial PTMs [3–5].

Due to the increased diversity from the histone variants and their heavily decorated PTMs, high-resolution separations of different histone proteins (H2A, H2B, H3, H4) before MS measurement

Xu Zhang et al. (eds.), *Protein Post-Translational Modifications*, Methods in Molecular Biology, vol. 3018,
https://doi.org/10.1007/978-1-0716-5166-7_12,

are critical for better sensitivity and coverage. The main prevailing methods used for histone proteoform separation are reverse-phase liquid chromatography (LC) [6–9], weak cation exchange hydrophilic interaction liquid chromatography (WCX-HILIC) [10, 11], and capillary zone electrophoresis (CZE) [12–15].

Multidimensional separations before MS and MS/MS analysis are essential for deciphering the diverse and dynamic histone proteoforms. Sodium dodecyl-sulfate polyacrylamide gel electrophoresis (SDS-PAGE) is one of the most widely used techniques for protein-related biological research, offering high-resolution separation of proteins according to their mass [16]. In addition, the protein fractions from SDS-PAGE could be passively extracted for subsequent MS and MS/MS analysis [17–21].

We describe a two-dimensional approach for TDP of histone proteoforms via integration of SDS-PAGE and CZE–MS/MS. SDS-PAGE offered efficient isolation and purification of each histone protein, that is, H1, H2, H3, and H4. CZE provided further separation of proteoforms of each histone protein based on their charge-to-size ratios, which could be influenced by the PTMs, for example, acetylation and phosphorylation. The methodology produced reproducible and high-resolution separations of histone proteoforms, enabling the identification of over 200 histone proteoforms.

2 Materials

Prepare all solutions using ultrapure water (prepared by purifying deionized water, to attain a sensitivity of 18 MΩ-cm at 25°C) and analytical grade reagents (unless indicated otherwise). Prepare and store all reagents at room temperature (unless indicated otherwise). Diligently follow all waste disposal regulations when disposing of waste materials. We do not add sodium azide to reagents.

2.1 SDS Polyacrylamide Gel

1. Resolving gel buffer: 1.5 M Tris-HCl, pH 8.8. Add about 100 mL of water to a 1 L graduated cylinder or a glass beaker (*see* **Note 1**). Weigh 181.7 g Tris-HCl and transfer to the cylinder. Add water to a volume of 900 mL. Mix and adjust pH to 8.8 with hydrochloric acid (HCl) (*see* **Note 2**). Make up to 1 L with water. Store at 4°C.
2. Stacking gel buffer: 0.5 M Tris–HCl, pH 6.8. Weigh 60.6 g Tris–HCl and prepare a 1 L solution as in the previous step. Store at 4°C.
3. Acrylamide mixture: 30% acrylamide/Bis solution (37.5:1). Store at 4°C.
4. 10% Sodium dodecyl-sulfate (SDS, wt/vol): Weigh 50 mg of ammonium persulfate and dissolve in 500 μL water.

5. 10% Ammonium persulfate (APS, wt/vol): Weigh 50 mg of ammonium persulfate and dissolve in 500 µL water (*see* **Note 3**).
6. *N*, *N*, *N*, *N*′-Tetramethyl-ethylenediamine (TEMED): Store at 4°C.
7. SDS-PAGE running buffer: 0.025 M Tris–HCl, pH 8.3, 0.192 M glycine, 0.1% SDS (wt/vol).
8. SDS sample loading buffer (4×): 250 mM Tris-HCl, pH 6.8, 40% glycerol (vol/vol), 8% SDS (wt/vol), and 20% 2-mercaptoethanol (vol/vol). To prepare the loading buffer, weigh 1.514 g Tris-HCl and dissolve it in 10 mL of water, adjusting the pH to 6.8 with HCl. Add 20 mL glycerol to the solution and vortex. Weigh 4 g SDS and 200 mg bromophenol blue and add to the solution. Add 10 mL 2-mercaptoethanol to the solution. Adjust the final volume to 50 mL with water. Store the buffer at −20°C (*see* **Note 4**).
9. SDS-PAGE aqueous staining solution: Weigh 70 mg of Coomassie Brilliant Blue G-250 and dissolve in 1 L of water. Stir the solution and gently vortex until the Coomassie Brilliant Blue is completely dissolved. Adjust the pH to 2.0 with HCl.
10. Precision plus protein dual color standards: Store at −20°C.

2.2 Protein Extraction

1. Protein extraction buffer: 0.1% SDS (wt/vol) in 50 mM ammonium bicarbonate (NH_4HCO_3), pH 9.0. To prepare 50 mM NH_4HCO_3, weigh 0.395 g of NH_4HCO_3 and dissolve in 100 mL of LC-MS grade water, adjusting the pH to 9.0 with ammonium hydroxide. Weigh 10 mg SDS and dissolve in 10 mL of 50 mM NH_4HCO_3 (pH 9.0).
2. Trichloroacetic acid (TCA) solution (100%, wt/vol): To prepare the solution, add 227 mL of LC-MS grade water to the bottle containing 500 g of TCA crystals. Store at 4°C.

2.3 Preparation of Linear Polyacrylamide (LPA)-Coated Separation Capillaries

1. 1 M HCl: Take 4.93 mL of 37% HCl and dilute it to a final volume of 50 mL with water. Solutions are stable for up to 1 month at room temperature.
2. 1 M sodium hydroxide (NaOH): Weigh 2 g NaOH and dissolve in 50 mL of water. Solutions are stable for up to 1 month at room temperature.
3. Insulin syringe: Med vet international exel syringe and needle, 1 mL, 28 G × 1/2" (*see* **Note 5**).
4. Syringe pump: Kd scientific.
5. Vacuum pump: Fisherbrand™ MaximaDry™ Diaphragm Vacuum Pump.

6. Acrylamide solution (4%, w/v): Weigh 40 mg of acrylamide and dissolve in 1 mL of water (*see* **Note 6**).
7. APS solution (5%, w/v): Weigh 1 mg of ammonium persulfate and dissolve it in 20 μL of water.
8. Hydrofluoric acid (HF): ACS reagent, 48%–51% solution in water (*see* **Note** 7).
9. Saturated NaOH solution: Add 500 mL water to a beaker, slowly add NaOH pellets in small portions to the water while stirring continuously. Continue adding NaOH until no more dissolves, and some solid NaOH remains at the bottom. Cool down the solution to room temperature.

2.4 CZE–MS/MS Analysis of the Histone Protein Fractions

1. Background electrolyte (BGE): 5% acetic acid. To prepare the BGE, add 2.5 mL of LC–MS grade acetic acid to 47.5 mL of LC-MS grade water. Thoroughly mix the solution using a vortex mixer for 10 s, followed by degassing in an ultrasonic bath for 10 min.
2. Sheath buffer: 10% methanol and 0.2% formic acid. To prepare the sheath buffer, add 5 mL of LC-MS grade methanol and 0.1 mL of LC–MS grade formic acid to 44.9 mL of LC–MS grade water. Thoroughly mix the solution using a vortex mixer for 10 s, followed by degassing in an ultrasonic bath for 10 min.
3. Sample buffer: 50 mM ammonium bicarbonate (NH_4HCO_3), pH 9.0.
4. CESI 8000 Plus CE system: Beckman Coulter.
5. Orbitrap Exploris 480 Mass Spectrometer: Thermo Fisher Scientific.
6. Electrokinetically Pumped Sheath-Flow CE-MS Nanospray Interface: EMASS-II, CMP Scientific.
7. CESI Vials and CESI Caps: Sciex.
8. Sample vials: PCR Microvials, Sciex.

3 Methods

Carry out all procedures at room temperature unless otherwise specified.

3.1 18% Sodium Dodecyl Sulfate-Polyacrylamide Gel Electrophoresis

1. Prepare the 18% separating gel by mixing 3 mL of resolving gel buffer, 7.2 mL of acrylamide mixture, and 1.5 mL of water in a 50 mL conical flask. Degas with helium for 10 min. Add 120 μL of SDS, 120 μL of APS, and 12 μL of TEMED and cast gel within a 7.25 cm × 10 cm × 1.0 mm gel cassette. Allow space for stacking gel and gently overlay with ethanol.

2. Prepare the 4% stacking gel by mixing 1.25 mL of stacking gel buffer, 0.67 mL of acrylamide mixture, and 3 mL of water in a 50 mL conical flask. Degas with helium for 10 min. Add 50 μL of SDS, 50 μL of APS, and 5 μL of TEMED. Insert a 10-well gel comb immediately without introducing air bubbles.
3. Add 4 μL of SDS sample loading buffer (4×) to 10 μL of histone proteins (15 μg) and shake at 1500 rpm for 3 min. Heat

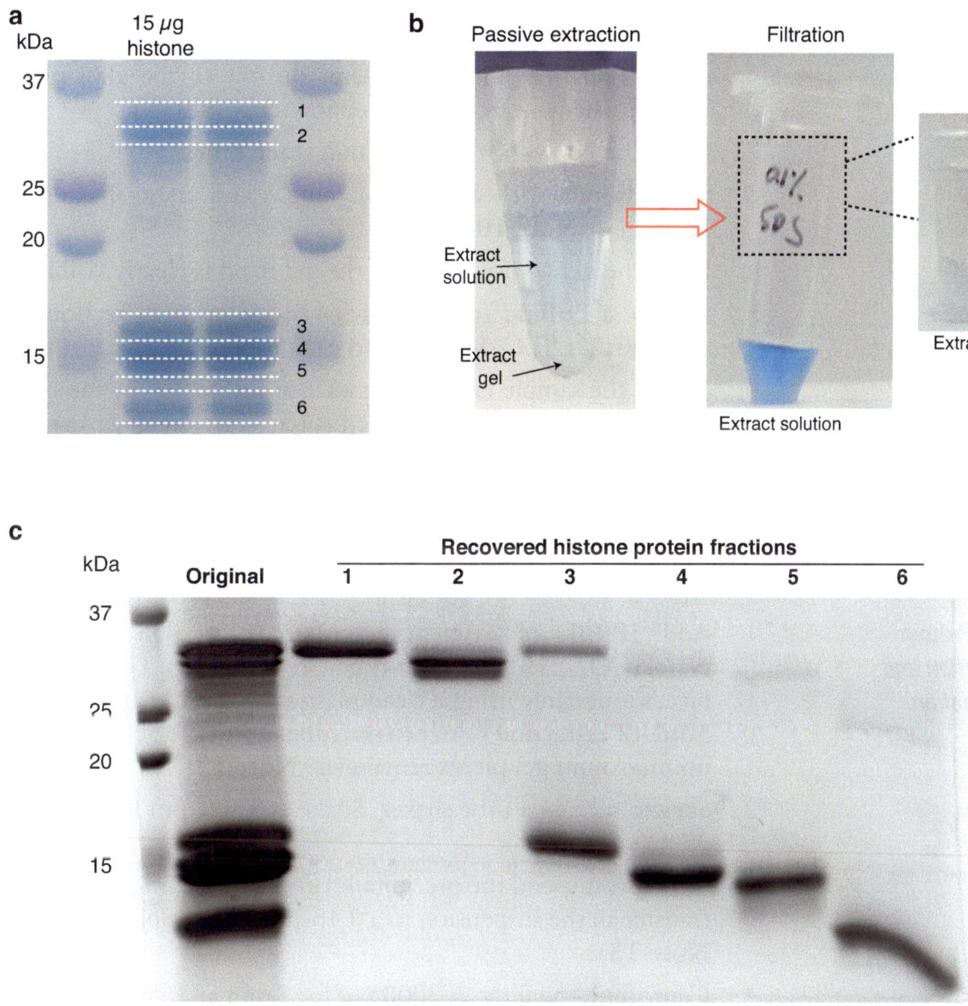

Fig. 1 Passive extraction of intact histone protein fractions separated with SDS–PAGE gels. (**a**) Gel electrophoresis. SDS–PAGE of bovine histone protein samples was conducted using 18% (wt/vol) acrylamide gels, and the gel was stained with aqueous CBB after electrophoresis. (**b**) Passive extraction. Collected gel pieces were finely crushed, mixed with extract solution (0.1% (wt/vol) SDS/100 mM NH_4HCO_3 pH 9), and subjected to shaking using a tube mixer. Then, 10 min later, the gel became clear, and the proteins in the gel were released into the extract solution with CBB. (**c**) A representative SDS–PAGE image of the original bovine histone protein (15 μg) and all the fractions. The protein bands were stained with aqueous CBBPanel c is adapted from [14], CC BY 4.0

the sample at 90°C for 10 min. Centrifuge the heated samples at 3000 × *g* for 30 s to bring down the condensate (*see* **Note 8**).

4. Load the histone sample into one well and load protein standards in both neighboring wells. Perform SDS-PAGE electrophoresis at a constant voltage of 90 V until the sample fully enters the separating gel (~15–20 min). Then increase the voltage to 140 V and continue running until the 10 kDa molecular weight (MW) marker just passes the bottom of the gel, ensuring not to overrun (Fig. 1a) (*see* **Note 9**).

3.2 Aqueous Staining

1. Following electrophoresis, pry the gel plates open with the use of a spatula. The gel remains on one of the glass plates. Transfer the gel carefully to a container with water.
2. Rinse the gel with water for 5 min three times.
3. Stain the gel with aqueous staining solution by shaking for 10 min (*see* **Note 10**).
4. Rinse the gel with water until the gel band is clearly visible.
5. Prepare a LoBind tube for each histone protein fraction and add 500 μL of LC–MS water to each tube.
6. Divide the sample lane. Using a clean scalpel to cut the bands corresponding to each histone protein fraction (Fig. 1a). Cut each band into small cubes (ca. 2 × 2 mm).
7. Transfer the gel pieces into a LoBind tube containing 500 μL of LC–MS water and spin them down on a bench-top microcentrifuge. Carefully remove the water with a pipette.

3.3 Protein Extraction and Purification

1. Add 100 μL of extraction buffer (0.1% SDS in 50 mM NH_4HCO_3, pH 9) to each tube (*see* **Note 11**). Grind the gel pieces uniformly using a clean plastic pestle. Add an additional 50 μL of extraction buffer to the tube, grind the gel pieces until no unground gel pieces remain (*see* **Note 12**).
2. Set the tube in a tube shaker. Shake at 1500 rpm for 10 min at 37°C.
3. Gently resuspend the precipitated gels with a 1-mL pipette and transfer all the suspension to a 0.45-μm Costar Spin-X filter (*see* **Note 13**).
4. Centrifuge the filter at 3000 × *g* for 3 min at 15°C, collect the filtered solution (Fig. 1b).
5. Add 50 μL of 100% TCA to the 150 μL of the filtered solution drop by drop (*see* **Note 14**). Invert the tube several times, incubate the protein sample on ice for 30 min, and centrifuge the sample at 10,000 × *g* for 10 min at 4°C.

6. Carefully remove the supernatant, wash the pellet with 500 µL of ice-cold acetone. Centrifuge the sample at 16,000× *g* for 5 min at 4°C.
7. Repeat the acetone wash and centrifugation steps once more. Carefully remove the supernatant, air-dry the histone pellet for 20 min at room temperature (*see* **Note 15**).
8. Dissolve the obtained histone fraction pellet with 100 µL of LC–MS grade H_2O and lyophilize the sample with a vacuum concentrator equipped with a cold trap (*see* **Note 16**).

3.4 Preparation of Linear Polyacrylamide (LPA)-Coated Separation Capillaries for CZE–MS/MS

1. Cut 105 cm of capillary (50 µm i.d., 365 µm o.d.). Flush the capillary with 100 µL 1 M HCl, water, 1 M NaOH, water, and methanol through the capillary with insulin syringe.
2. Dry the capillary with nitrogen gas for at least 2 h.
3. Pump 50% (v/v) 3-(trimethoxysilyl) propyl methacrylate in methanol through the capillary for 10 min under a 5 µL flow rate controlled by syringe pump (*see* **Note 17**).
4. Seal both ends of the capillary and incubate it at room temperature for at least 24 h (*see* **Note 18**).
5. Flush the capillary with 100 µL methanol through the capillary to flush away 50% (v/v) 3-(trimethoxysilyl) propyl methacrylate in methanol.
6. Dry the capillary with nitrogen gas for at least 2 h.
7. Prepare polyacrylamide solution in a 1.5 mL Eppendorf tube by mixing 500 µL of acrylamide solution with 3.5 µL ammonium persulfate solution.
8. Put the polymer solution under nitrogen gas flow for 10 min and move the capillary around to remove the oxygen from the solution thoroughly.
9. Suck the polyacrylamide solution into the capillary by using a vacuum pump until the capillary is filled with polyacrylamide solution.
10. Seal both ends of the capillary and place it in a 50°C water bath for 1 h.
11. Cut both ends of the capillary and flush the polymer solution out with distilled water until the polymer inside the capillary is pushed out (*see* **Note 19**).
12. Burn out 2–3 cm outside the coating of the capillary at 1 cm away from one end of the capillary with a lighter.
13. Take a 0.6 mL centrifuge tube and poke a hole with a size slightly smaller than 365 µm (o.d. of the capillary) at the bottom of the tube.

14. Thread the end of the capillary that is close to the burnt portion through the hole until the burnt portion is in the tube.
15. Stabilize the tube on a tube holder in the hood, add 150 μL hydrogen fluoride (HF) solution to the tube, and let the HF solution reach about halfway up the burnt portion on the capillary (*see* **Note 20**).
16. Leave the capillary in the HF solution at room temperature for 70 min.
17. Trash the HF solution into a saturated NaOH solution.
18. Pull the etched capillary out of the tube gently. Rinse the etched part of the capillary under the running deionized water for 3 min.

Table 1
MS1 and MS/MS parameters used for CZE–MS/MS

Parameter	High-high mode	
MS1 resolution (at m/z 200)	120,000	
Microscans (MS1)	3	
m/z range (MS1)	400–1500	
AGC target (MS1)	1e6	
Max injection time (MS1)	50	
DDA method (Top N)	Top 5	
Isolation window (MS/MS)	2 m/z	
HCD collision energy (NCE)	F1	12
	F2	12
	F3	16
	F4	16
	F5	14
	F6	18
MS/MS resolution (at m/z 200)	60,000	
Microscans (MS/MS)	1	
m/z range (MS1)	200–2000	
AGC target (MS/MS)	1e6	
Max injection time (MS/MS)	400 ms	
Precursor charge states	>6	
Intensity threshold	>2E5	
Dynamic exclusion	50 s	

19. Cutoff the capillary at the thinnest point of the etched part with a capillary cutter.

3.5 CZE–MS/MS Analysis of the Histone Protein Fractions

1. Analyze the samples on a Thermo Scientific Q Exactive HF mass spectrometer in line with a Sciex CESI 8000 Plus CE system via a CMP Scientific third-generation electro-kinetically pumped sheath flow CE–MS interface.
2. Thread the etched end of the LPA-coated capillary through the tee of the CE–MS interface and introduce it to the electrospray emitter (*see* **Note 21**). Set the electrospray voltage at 2.0 kV for electrospray ionization.
3. Resuspend each histone fraction pellet with 10 μL sample buffer.
4. Set up buffers in the tray. Place the Sample vial in the inlet sample tray in position SI:A1. Place the BGE vial and Sample buffer vial in the inlet buffer tray in position BI:A1 and BI:B1, respectively.
5. For the histone sample with gel fractionation, inject 50 nL of each fraction. Apply a 30 kV voltage to the injection end for 1800 s for separation. Apply 30 kV voltage and 10 psi pressure for 600 s to push out the residue analytes in the capillary.
6. The MS parameters were set as follows in Table 1.
7. Check the raw files with Xcalibur software (see Fig. 2).

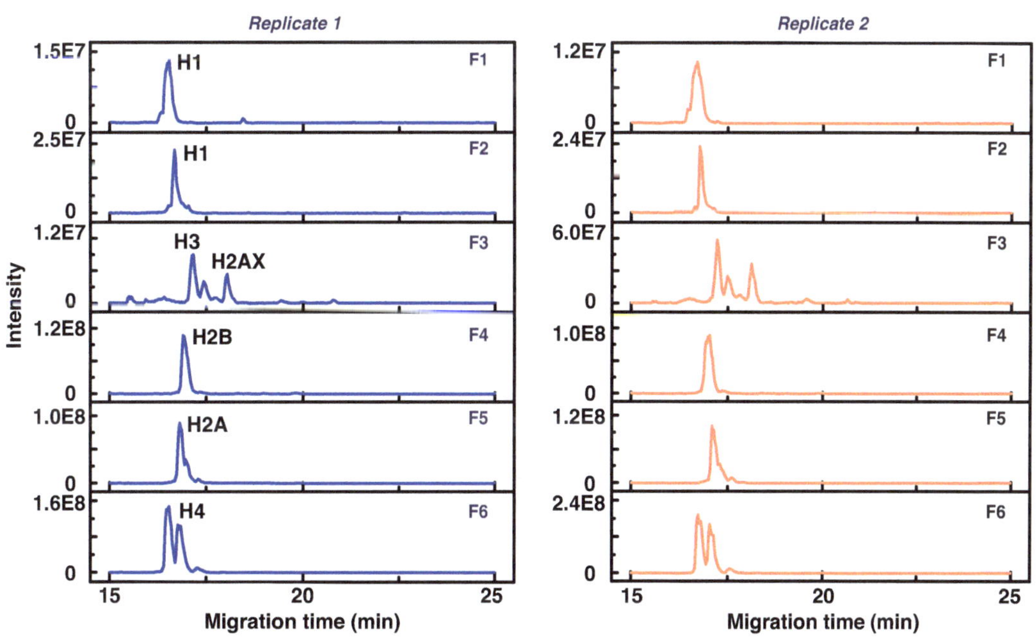

Fig. 2 The electropherograms of the individual histone protein fractions (F1–F6) recovered from SDS-PAGE gel in duplicateThe figure is reproduced from [14], CC BY 4.0

3.6 Data Analysis

1. Apply the MSConvert software to convert the raw files to mzML file. Set the output format to mzML, select 64-bit for binary encoding precision, apply the Peak Picking filter with the Vendor algorithm, and enable both Write Index and TPP compatibility.
2. Perform TopFD (Top-down mass spectrometry feature detection) software (version 1.6.3) for spectral deconvolution. Set the maximum charge to 50, the maximum mass (Da) to

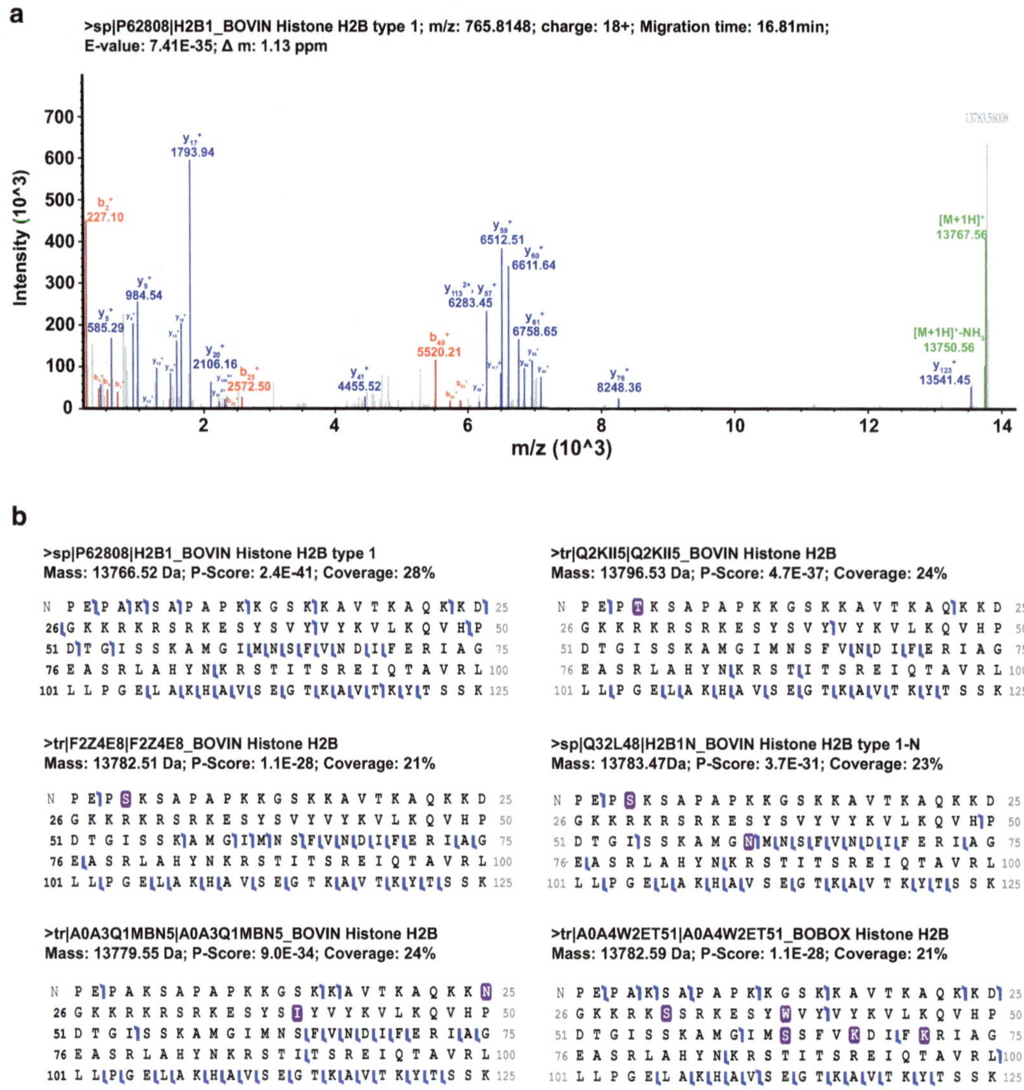

Fig. 3 Fragmentation pattern and sequences of histone H2B variants from the ProSightPD result. (**a**) Fragmentation pattern of histone H2B type 1 (sp|P62808). (**b**) Sequences of histone H2B variants that differ by a few amino acids near both the N- and C-termini. The amino acids marked in white are different from the histone H2B type 1 sequence (sp|P62808)The figure is adapted from [14], CC BY 4.0

30,000, and the MS1 and MS2 S/N ratios to 3 and 1, respectively. Configure the precursor window size (m/z) to 3 and the m/z error to 0.02.

3. Perform TopPIC (Top-down mass spectrometry-based proteoform identification and characterization) software (version 1.6.3) for data analysis. Subject the msalign files generated from TopFD to TopPIC for subsequent database searching. Search the spectra against a *Bos taurus* database downloaded from Swiss-UniProt (October 2023). Set variable modifications, including phosphorylation (S/T/Y), oxidation (M), methylation/dimethylation/trimethylation (K/R), acetylation (K/R), acetylation on protein N-terminus, and citrullination (N/Q/R). Set the maximum number of variable PTMs as 5. The mass error tolerance is 15 ppm for precursor ions and fragment ions. Use a 1.2-Da mass tolerance to identify the Proteoform-Spectrum Match (PrSM) clusters. The maximum mass shift is 500 Da, with the maximum number of mass shifts set to 0. False discovery rate (FDR) is estimated using the target-decoy approach, with a 1% PrSM-level FDR and a 1% proteoform-level FDR employed to filter the identifications.
4. The intact H2B histone variants that differ by a few amino acids near both the N- and C-termini can then be confidently identified (Fig. 3).

4 Notes

1. Having water at the bottom of the cylinder helps to dissolve the Tris relatively easily, allowing the magnetic stir bar to go to work immediately. If using a glass beaker, the Tris can be dissolved faster if the water is warmed to about 37°C. However, the downside is that care should be taken to bring the solution to room temperature before adjusting pH.
2. Concentrated HCl (12 N) can be used at first to narrow the gap from the starting pH to the required pH. From then on, it would be better to use a series of HCl (e.g., 6 and 1 N) with lower ionic strengths to avoid a sudden drop in pH below the required pH.
3. It is best to prepare the APS solution fresh each time.
4. SDS precipitates at −20°C. Therefore, the lysis buffer needs to be warmed before use.
5. Be cautious when handling syringes during use and disposal. Following use, syringes should be promptly disposed of in a sharp container.

6. Wear a mask when weighing acrylamide. To avoid exposing coworkers to acrylamide, cover the weighed boat containing the weighed acrylamide with another weighed boat (similar in size to the original weighed boat containing the weighed acrylamide) when transporting it to the fume hood. Transfer the weighed acrylamide to the cylinder inside the fume hood and mix on a stirrer placed inside the hood. Unpolymerized acrylamide is a neurotoxin, and care should be exercised to avoid skin contact.
7. HF is an extremely dangerous and corrosive liquid that can cause severe tissue damage. Only trained personnel should handle HF, and it is strongly recommended that at least two people be present when working with concentrated HF solutions. Never use glassware to contain HF, as it reacts rapidly and exothermically with glass.
8. Centrifuging the samples before the run helps remove insoluble debris, which could produce streaks in the protein lanes.
9. When dividing sample lanes based on MW, electrophoresis of the MW markers next to the sample lanes facilitates fractionation work; an example of sample and MW marker placement on a gel is shown in Fig. 1a.
10. If the protein extraction is not performed immediately after staining the gel, a fixation treatment before staining is recommended: add 40 mL of 40% (vol/vol) methanol/5% (vol/vol) acetic acid and shake for 30 min before CBB staining.
11. Ensure that the extraction buffer fully covers all gel pieces.
12. Avoid intense homogenization because the presence of SDS will cause excessive foaming.
13. Use the scissors to cutoff the pipette tip at the appropriate position for the convenience of the experiment.
14. The histones precipitate in TCA. The solution will appear milky over time.
15. The dry step should be performed in the hood.
16. The SDS-PAGE image for the histone protein fractions was also obtained after staining with the aqueous CBB (see Fig. 1c).
17. Avoid introducing water in 3-(trimethoxy silyl) propyl methacrylate solution.
18. The incubation time of the 3-(trimethoxysilyl) propyl methacrylate solution should not exceed 1 week.
19. Linear polyacrylamide solution, when properly polymerized, is highly viscous. Make sure you can see the sticky solution pushing out before water.
20. Make sure there is no leaking at the bottom of the tube.

21. An electrospray emitter with an orifice size ranging 25–35 μm was pulled with a Sutter P-1000 flaming/brown micropipette puller.

Acknowledgments

We thank the support from the National Cancer Institute (Grant R01CA247863) and the National Institute of General Medical Sciences (Grant R35GM153479).

References

1. Millán-Zambrano G, Burton A, Bannister AJ, Schneider R (2022) Histone post-translational modifications—cause and consequence of genome function. Nat Rev Genet 23:563–580
2. Ramazi S, Allahverdi A, Zahiri J (2020) Evaluation of post-translational modifications in histone proteins: a review on histone modification defects in developmental and neurological disorders. J Biosci 45:135
3. Rommelfanger SR, Zhou M, Shaghasi H, Tzeng S-C et al (2021) An improved top-down mass spectrometry characterization of Chlamydomonas reinhardtii histones and their post-translational modifications. J Am Soc Mass Spectrom 32:1671–1688
4. Tian Z, Tolić N, Zhao R, Moore RJ et al (2012) Enhanced top-down characterization of histone post-translational modifications. Genome Biol 13: R86
5. Schachner LF, Jooß K, Morgan MA, Piunti A et al (2021) Decoding the protein composition of whole nucleosomes with Nuc-MS. Nat Methods 18:303–308
6. Jeanne Dit Fouque K, Miller SA, Pham K, Bhanu NV et al (2022) Top-"double-down" mass spectrometry of histone H4 proteoforms: tandem ultraviolet-photon and mobility/mass-selected electron capture dissociations. Anal Chem 94:15377–15385
7. Holt MV, Wang T, Young NL (2019) High-throughput quantitative top-down proteomics: histone H4. J Am Soc Mass Spectrom 30:2548–2560
8. Shechter D, Dormann HL, Allis CD, Hake SB (2007) Extraction purification and analysis of histones. Nat Protoc 2:1445–1457
9. Zheng Y, Fornelli L, Compton PD, Sharma S et al (2016) Unabridged analysis of human histone H3 by differential top-down mass spectrometry reveals hypermethylated proteoforms from MMSET/NSD2 overexpression. Mol Cell Proteomics 15:776–790
10. Zhou M, Wu S, Stenoien DL, Zhang Z et al (2017) Profiling changes in histone post-translational modifications by top-down mass spectrometry. Methods Mol Biol 1507:153–168
11. Tvardovskiy A, Wrzesinski K, Sidoli S, Fey SJ et al (2015) Top-down and middle-down protein analysis reveals that intact and clipped human histones differ in post-translational modification patterns. Mol Cell Proteomics 14:3142–3153
12. Chen D, Yang Z, Shen X, Sun L (2021) Capillary zone electrophoresis-tandem mass spectrometry as an alternative to liquid chromatography-tandem mass spectrometry for top-down proteomics of histones. Anal Chem 93:4417–4424
13. Wang Q, Fang F, Wang Q, Sun L (2023) Capillary zone electrophoresis-high field asymmetric ion mobility spectrometry-tandem mass spectrometry for top-down characterization of histone proteoforms. Proteomics 24(3–4):e2200389
14. Fang F, Gao G, Wang Q, Wang Q, Sun L (2024) Combining SDS-PAGE to capillary zone electrophoresis-tandem mass spectrometry for high-resolution top-down proteomics analysis of intact histone proteoforms. Proteomics 24:2300650
15. Fang F, Fries B, Wang Z, Liu X et al (2025) Quantitative top–down proteomics reveals significant differences in histone proteoforms between metastatic and nonmetastatic colorectal cancer cells. Proteomics e202400336
16. Laemmli UK (1970) Cleavage of structural proteins during the assembly of the head of bacteriophage T4. Nature 227:680–685
17. Kline JT, Belford MW, Boeser CL, Huguet R et al (2023) Orbitrap mass spectrometry and high-field asymmetric waveform ion mobility spectrometry (FAIMS) enable the in-depth analysis of human serum proteoforms. J Proteome Res 22:3418–3426
18. Kline JT, Belford MW, Huang J, Greer JB et al (2023) Improved label-free quantification of intact proteoforms using field asymmetric ion mobility spectrometry. Anal Chem 95:9090–9096

19. Takemori A, Kaulich PT, Cassidy L, Takemori N, Tholey A (2022) Size-based proteome fractionation through polyacrylamide gel electrophoresis combined with LC–FAIMS–MS for in-depth top-down proteomics. Anal Chem 94:12815–12821
20. Takemori A, Butcher DS, Harman VM, Brownridge P et al (2020) PEPPI-MS: polyacrylamide-gel-based prefractionation for analysis of intact proteoforms and protein complexes by mass spectrometry. J Proteome Res 19:3779–3791
21. Takemori A, Kaulich PT, Tholey A, Takemori N (2025) PEPPI-MS: gel-based sample pre-fractionation for deep top-down and middle-down proteomics. Nat Protoc 20:1413–1438

Chapter 13

PEPPI-SP3: A Gel-Based High-Resolution Sample Preparation Workflow for In-Depth Top-Down Analysis of Intact Proteoforms by Mass Spectrometry

Ayako Takemori, Jake T. Kline, Luca Fornelli and Nobuaki Takemori

Abstract

Top-down proteomics enables large-scale identification of proteoforms by analyzing intact proteins extracted from biological samples using mass spectrometry systems coupled with liquid chromatography or capillary electrophoresis. In this chapter, we present the PEPPI-SP3 workflow, a high-resolution sample prefractionation method that significantly enhances the analytical depth of top-down proteomics by combining highly efficient passive extraction of in-gel proteins separated by sodium dodecyl sulfate polyacrylamide gel electrophoresis with SP3 bead-based purification of intact proteins.

Keywords Proteoforms, Top-down proteomics, PEPPI-MS, SP3, Sample prefractionation, SDS-PAGE

1 Introduction

Comprehensive analysis of proteoforms, which are diverse molecular forms of proteins derived from a single gene in vivo, remains a major challenge in proteomics [1]. Mass spectrometry (MS) serves as the central analytical platform in proteomics, with bottom-up proteomics (BUP), in which proteins are enzymatically digested into peptides prior to MS analysis, being the current mainstream method [2]. However, it is often difficult to accurately deduce the original intact proteoforms from the peptide-level data obtained [3]. In contrast, top-down proteomics (TDP), which analyzes proteins in their intact form without enzymatic digestion, enables precise identification and characterization of proteoforms [4].

TDP, which enables continuous analysis of intact proteins separated by liquid chromatography (LC) or capillary electrophoresis (CE), has emerged as a powerful strategy for large-scale identification and quantification of proteoforms in biological samples [4]. However, separation by LC or CE alone is often insufficient to resolve the

Xu Zhang et al. (eds.), *Protein Post-Translational Modifications*, Methods in Molecular Biology, vol. 3018, https://doi.org/10.1007/978-1-0716-5166-7_13,

complexity of biological samples, and therefore, a prefractionation step prior to LC/CE is essential for deep proteoform coverage [5].

Sodium dodecyl sulfate polyacrylamide gel electrophoresis (SDS-PAGE) is a classical biochemical technique that separates proteins with high resolution based on molecular weight after denaturation with SDS [6]. Although SDS-PAGE has long been used in protein studies, its application as a prefractionation method for TDP has been limited by the lack of efficient recovery methods for the separated proteins. This challenge was addressed in 2020 by the introduction of PEPPI-MS [7], a passive extraction technique that enables highly efficient recovery of proteins from polyacrylamide gels. PEPPI-MS has since facilitated the development of a simple, cost-effective SDS-PAGE-based prefractionation workflow for TDP, which is now emerging as a standard approach in the field [8–10].

PEPPI-MS improves passive extraction efficiency by employing SDS and Coomassie Brilliant Blue (CBB) as extraction enhancers [7]. However, because both SDS and CBB are incompatible with MS, additional purification steps are required prior to injection. To this end, purification methods such as methanol/chloroform/water (MCW) precipitation and filter aided sample preparation (FASP) with anion-exchange StageTip spin columns have been utilized [10]. While MCW offers a simple workflow, it often results in significant loss of low molecular weight proteoforms [11]. Although FASP, another available option, shows superior recovery of low-molecular-weight proteins compared to MCW [11], it has the disadvantage of requiring repeated solvent exchange steps, making the process labor-intensive and time-consuming.

To overcome the issues described above in the purification process of PEPPI fractionation, we developed a novel PEPPI protocol named PEPPI-SP3 in 2025 [12] that utilizes carboxyl-modified magnetic beads (SP3 beads), originally designed for sample preparation in bottom-up proteomics using the SP3 method [13]. In SP3 for BUP, the proteins are immobilized on the beads, allowing thorough removal of contaminants from the sample, followed by enzymatic digestion directly on the beads and subsequent peptide recovery [14]. Additionally, since the SP3 purification is compatible with commercially available automated handling systems [15, 16], the SP3 protocol is also appropriate for high-throughput sample preparation in large-scale studies. Although SP3 exhibits excellent performance in sample pretreatment for BUP, its application in TDP has been limited due to the difficulty in recovering intact proteins from the beads. As a pioneering work of SP3 in TDP, Webb's group reported a protocol in which proteins were eluted from beads using cold 80% formic acid [17]. However, such harsh acidic conditions pose a risk of protein degradation.

The 2025 iteration of the PEPPI series, the PEPPI-SP3 method, brings SP3 to TDP, enabling the recovery of proteins from beads

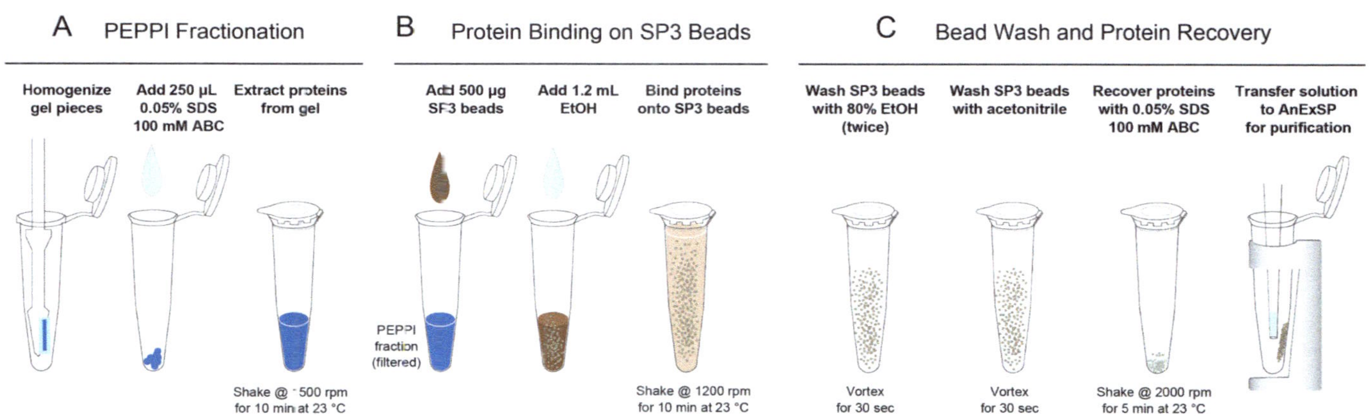

Fig. 1 Schematic illustration of the PEPPI-SP3 protocol. (**A**) PEPPI fractionation. Gel-separated proteins are rapidly recovered into solution via highly efficient passive extraction using CBB and SDS. (**B**) Protein binding on SP3 beads. Proteins in the PEPPI fraction are precipitated with ethanol and adsorbed onto the surface of SP3 beads. (**C**) Bead wash and protein recovery. The beads are washed with 80% (v/v) ethanol and acetonitrile. Proteins are then recovered from the beads using 0.05% (w/v) SDS in 100 mM ammonium bicarbonate (ABC), followed by final purification with an anion-exchange StageTip (AnExSP) (reproduced from [12] with permission from the American Chemical Society)

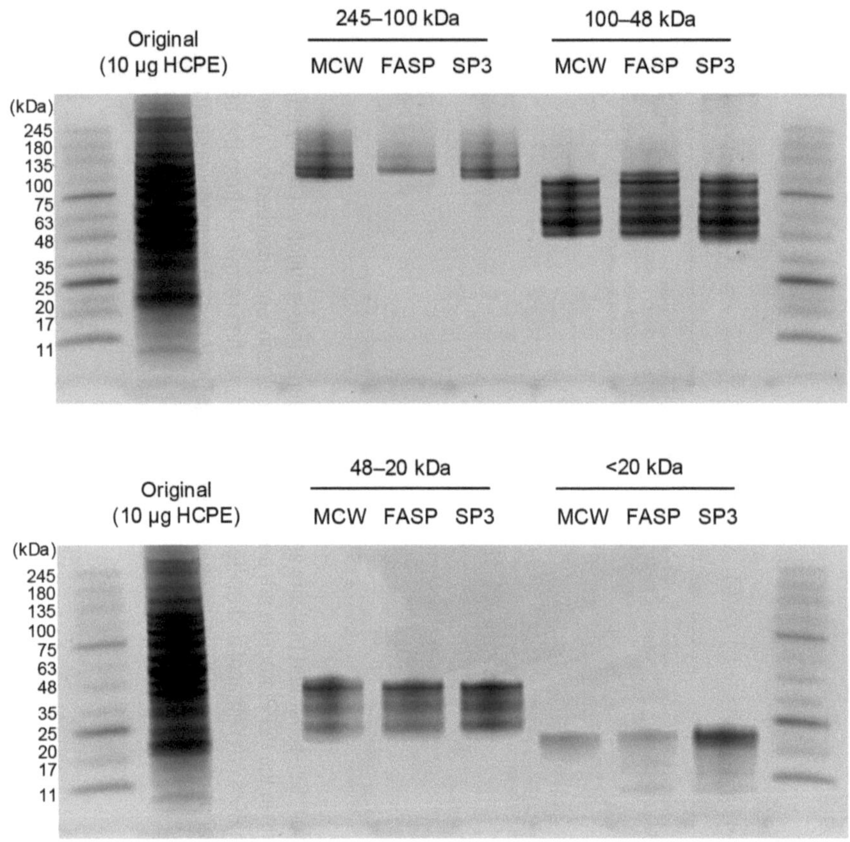

Fig. 2 SDS-PAGE images of fractions obtained using various PEPPI protocols. Human cell protein extract (HCPE, 10 µg) was separated by SDS-PAGE, and the sample lane was divided into four fractions (245–100 kDa; 100–48 kDa; 48–20 kDa; <20 kDa). Each fraction was processed using one of three different PEPPI protocols (MCW, FASP, SP3). The resulting fractions were then reseparated by SDS-PAGE and visualized by CBB staining (reproduced from [12] with permission from the American Chemical Society)

under mild conditions at room temperature [12]. In this protocol, proteins fractionated by PEPPI are first adsorbed onto SP3 beads, thoroughly washed, and then gently eluted at room temperature using a low concentration SDS solution (Fig. 1). The eluate is then rapidly purified using an anion-exchange StageTip spin column. Figure 2 shows the fractionation results of human cell protein extract samples using all three PEPPI methods: the PEPPI-SP3, MCW, and FASP protocols. Notably, PEPPI-SP3 achieved better sample recovery than FASP in the low molecular weight range (below 20 kDa). In the TDP analysis of these fractions, SP3 yielded a significantly higher total number of identified proteoforms compared to the other two methods (Fig. 3A). Furthermore, when examined across different molecular weight ranges, SP3 demonstrated superior performance in proteoform identification in many of these regions (Fig. 3B). However, since each method uniquely identified a subset of proteins

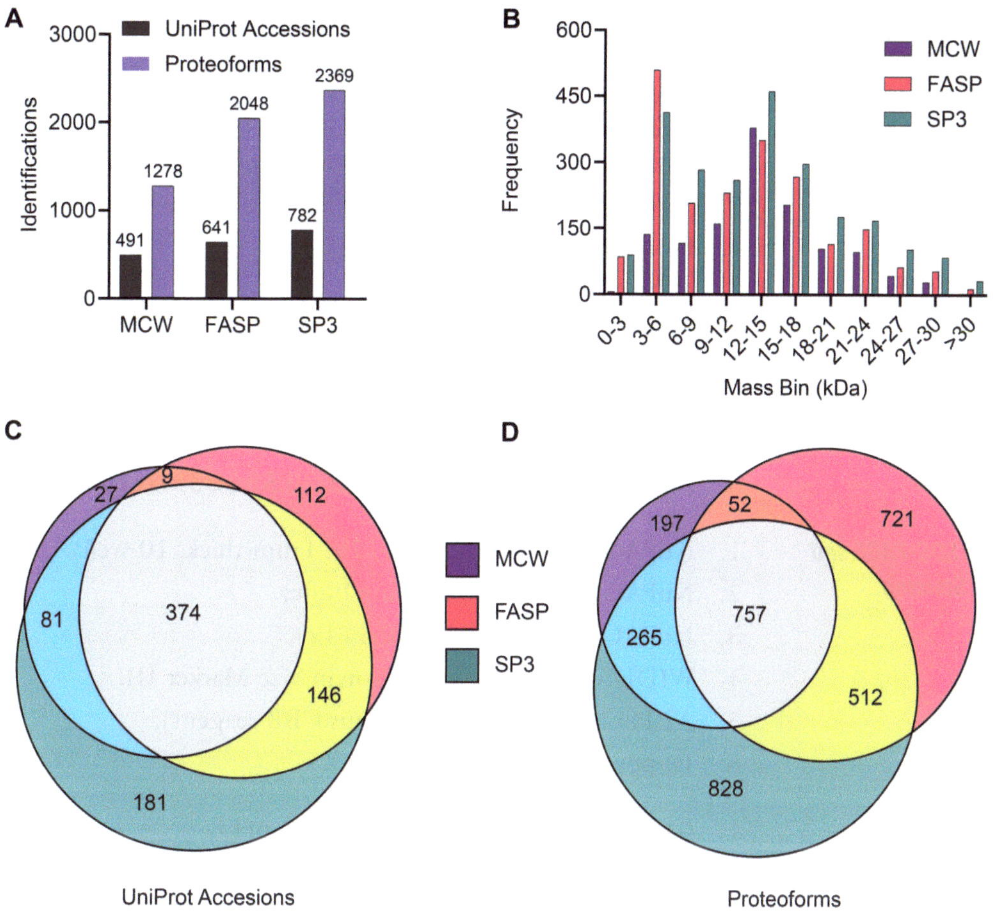

Fig. 3 Results of TDP analysis using various PEPPI fractions. (**A**) Identified UniProt accession numbers and proteoforms. (**B**) Distribution of identified proteoforms across different mass ranges. (**C** and **D**) Venn diagrams of (**C**) UniProt accession numbers (**C** and **D**) proteoforms identified by each PEPPI method (reproduced from [12] with permission from the American Chemical Society)

(Fig. 3C) and proteoforms (Fig. 3D), no single approach is sufficient to achieve comprehensive proteoform detection. Furthermore, a comparison of the characteristics of the identified proteoforms revealed that phosphorylation and acetylation, key posttranslational modifications involved in intracellular signaling, were more frequently detected with SP3 and MCW than with FASP [12]. This is likely due to sample loss caused by protein adsorption to the ultrafiltration membrane used in the FASP method. Each version of PEPPI has its own strengths, and the choice of method should therefore be guided by the specific goals of the study (*see* **Note 1**).

2 Materials

In Sects. 2.1, 2.3, and 2.5, chemical reagents of either LC–MS grade or proteomics grade were used, and LC–MS grade ultrapure water was used for the preparation of aqueous solutions. It is recommended that aqueous solutions be freshly prepared as needed.

2.1 Sample Pretreatment

1. Promega MS Compatible Human Cell Protein Extract (HCPE), intact.
2. 500 mM dithiothreitol.
3. 1 M iodoacetamide.
4. 0.05% (w/v) SDS in 100 mM ammonium bicarbonate (ABC), pH 8.
5. Amicon Ultra Centrifugal Filter (0.5 mL, 3 kDa cutoff).

2.2 SDS-PAGE and CBB Staining

1. NuPAGE Bis-Tris gel (4%–12%, 1 mm-thick, 10-well).
2. NuPAGE MES SDS running buffer.
3. NuPAGE LDS sample buffer (4×).
4. WIDE-VIEW Prestained Protein Size Marker III.
5. ATTO EzStain AQua (aqueous CBB reagent).
6. Disposable plastic tray.

2.3 PEPPI Fractionation

1. 0.05% (w/v) SDS in 100 mM ABC, pH 8.
2. Nippi BioMasher II disposable homogenizer tube.
3. Corning Costar Spin-X centrifugal tube filter (0.45 μm pore CA membrane).

2.4 SP3

1. Sera-Mag SpeedBead Carboxylate-Modified Magnetic Particles, Hydrophylic (SP3 Beads A).
2. Sera-Mag SpeedBead Carboxylate-Modified Magnetic Particles, Hydrophobic (SP3 Beads B).
3. Magnetic rack.

2.5 Anion-Exchange StageTip

1. Empore disc, anion-exchange.
2. Methanol.
3. 100 mM ABC, pH 8.
4. Elution solution #1: 0.5% (v/v) formic acid/50% (v/v) acetonitrile.
5. Elution solution #2: 0.5% (v/v) formic acid/50% (v/v) ethanol.
6. 0.1% (v/v) formic acid/4.9% (v/v) acetonitrile.
7. Plastic straw (inner diameter: 6 mm).

8. Flat-tip plastic rod (for packing membrane disc into a 200 μL micropipette tip).
9. Liquid nitrogen (for flash-freezing samples).

3 Methods

Perform all procedures at room temperature unless otherwise specified.

3.1 Sample Pretreatment

1. In a 1.5 mL microtube, mix 100 μL of HCPE (10 mg/mL) with 1 μL of 500 mM dithiothreitol and incubate at 37°C for 90 min (*see* **Note 2**). After incubation, add 1.5 μL of 1 M iodoacetamide and incubate in the dark at 25°C for 30 min. Adjust the volume to 1 mL with ultrapure water, aliquot, and store frozen until use.
2. Add 300 μL of HCPE stock to a centrifugal ultrafiltration filter unit and centrifuge at 13,500 × *g* for 20 min at 25°C. Add 0.05% (w/v) SDS in 100 mM ABC to bring the volume up to 500 μL and centrifuge again at 13,500 × *g* for 25 min at 25°C. Repeat this washing step once more.
3. Transfer the concentrated sample into a new 1.5 mL tube, add one-third the volume of 4× LDS sample buffer, and mix gently. Perform a protein assay and adjust the final protein concentration to 2 μg/μL using 1× LDS sample buffer. Use the sample for SDS-PAGE analysis.

3.2 SDS-PAGE and CBB Staining

1. Set the NuPAGE gel into an appropriate electrophoresis chamber and add the volume of MES running buffer specified in the manual. Carefully remove the comb from the sample wells, rinse the wells with running buffer, and load the samples. Perform electrophoresis at a constant voltage of 180 V (*see* **Note 3**).
2. After electrophoresis, promptly remove the gel from the gel cassette and transfer it to a disposable tray. Proceed with CBB staining as follows: first, pour 40 mL of EzStain AQua into the tray and shake the gel on a shaker for 8 min (*see* **Note 4**). Next, completely discard the EzStain AQua, add 200 mL of distilled water to the tray, and shake for 30 min (*see* **Note 5**).

3.3 PEPPI Fractionation

1. Cut out the sample lanes from the gel using a craft knife and divide them into segments corresponding to molecular weight regions below 35 kDa, which are suitable for TDP analysis (*see* **Note 6**). Individually transfer each gel band into a separate BioMasher tube and homogenize using the provided pestle (*see* **Note 7**).

2. Add 250 μL of 0.05% (w/v) SDS in 100 mM ABC to each tube and further grind the gel to ensure that no visible gel fragments remain. Once the gel is fully homogenized, remove the pestle, close the cap, and shake the tube on a tube shaker for 10 min.
3. After shaking, place the BioMasher tubes in a benchtop centrifuge and spin at 2000 rpm for 10 s. Open the tube cap and use a 1 mL micropipette to collect the gel suspension, transferring it to a Spin-X filter device. Centrifuge in the Spin-X filter device at 2000 rpm for 3 min (*see* **Note 8**). After centrifugation, remove the filter cup and transfer the filtrate in the collection tube to a new 1.5 mL microcentrifuge tube.

3.4 Pretreatment of SP3 Beads

1. Add 50 μL each of SP3 bead reagents A and B to a new 1.5 mL microcentrifuge tube (*see* **Note 9**). Place the tube on a magnetic rack and let stand for 3 min.
2. Carefully remove the supernatant using a 200 μL micropipette. Add 800 μL of ultrapure water to the tube and gently resuspend the beads using a 1 mL micropipette. Spin the tube briefly (10 s) in a benchtop centrifuge, return it to the magnetic rack, and let stand for 3 min. Remove the supernatant completely using a 1 mL micropipette.
3. Repeat the wash step (Step 2) two more times (total of three washes). Finally, add 250 μL of ultrapure water to resuspend the beads and store at 4°C until use.

3.5 Protein Purification Using SP3 Beads

1. Add the entire PEPPI fraction obtained in Sect. 3.3 to a new 1.5 mL microcentrifuge tube. Sequentially add 25 μL of the SP3 bead suspension prepared in Sect. 3.4 and 1.2 mL of ethanol. Close the cap and immediately place the tube on a tube mixer. Shake at 1200 rpm for 10 min. After shaking, spin the tube for 10 s in a benchtop centrifuge, place it on a magnetic rack, open the cap, and let stand for 3 min. Remove the supernatant completely using a 1 mL micropipette.
2. Add 800 μL of 80% (v/v) ethanol to the tube, close the cap, and vortex vigorously for 30 s. Spin the tube for 10 s, then place it on the magnetic rack, and let stand for 3 min. Open the cap and remove the supernatant as completely as possible. Repeat this wash step once more (total of two ethanol washes).
3. Add 800 μL of acetonitrile to the tube, close the cap, and vortex for 30 s. Spin the tube for 10 s, place it on the magnetic rack and let stand for 3 min. Open the cap and remove the supernatant and air-dry at room temperature (25°C) for 10 min (*see* **Note 10**).
4. Add 20 μL of 0.05% (w/v) SDS in 100 mM ABC to the tube, close the cap, and mix at 2000 rpm for 5 min using a tube

mixer (*see* **Note 11**). After mixing, place the tube on a magnetic rack and let stand for 3 min. Transfer the supernatant to a new 1.5 mL microcentrifuge tube.

3.6 Preparation of StageTip

1. Use a plastic straw (inner diameter: 6 mm) to punch out a disc from an anion-exchange Empore membrane. Insert the cut-out disc into a disposable 200 μL micropipette tip, and pack it firmly using a flat-tip plastic rod.
2. Set the required number of 2 mL collection tubes into the centrifuge rotor and place the StageTips with spin adapters attached onto the tubes.

3.7 StageTip-Based Sample Purification

Perform all the following centrifugation steps at 7000 × *g* for 3 min unless otherwise specified. To each StageTip:

1. Load 40 μL of methanol and centrifuge.
2. Load 40 μL of 100 mM ABC and centrifuge.
3. Load the entire protein solution obtained from Sect. 3.5 and centrifuge.
4. Load 40 μL of 100 mM ABC again and centrifuge.
5. Transfer the StageTip to a new 1.5 mL microcentrifuge tube.
6. Load 40 μL of elution solution #1 and centrifuge.
7. Load 40 μL of elution solution #2 and centrifuge (*see* **Note 12**).
8. Remove the StageTip from the collection tube and immerse the entire tube containing the eluate in liquid nitrogen to flash-freeze the sample.
9. Immediately after freezing, perform solvent evaporation using SpeedVac without heating until the sample volume is reduced to approximately 5 μL (*see* **Note 13**). Following concentration, adjust each sample to a final volume of 20 μL using 0.1% (v/v) formic acid/4.9% (v/v) acetonitrile and transfer to the LC–MS vial.

4 Notes

1. This protocol focuses on PEPPI-SP3; for details on other PEPPI protocols, please refer to our previously published protocols [10].
2. In this protocol, commercially available cellular protein extracts are used as test samples. If preservation of disulfide bonds within the proteins is desired, omit the reduction and alkylation steps during gel electrophoresis.

3. This protocol is designed for application in top-down proteomics. Electrophoresis is to be stopped when the dye front reaches 10 cm from the starting point. Under these conditions, the protocol assumes the preparation of six gel fractions from the sample lanes in the region below 35 kDa.
4. Other commercially available aqueous CBB stains, such as Bio-Rad's BioSafe CBB or Fujifilm Wako's Quick CBB plus, can be used in place of EzStain AQua. If aqueous CBB stains are not available, traditional organic solvent-based CBB stains may also be used, although they require a longer staining time.
5. If stained bands are not visible after 30 min of soaking, additional shaking is recommended. To enhance destaining, place a folded Kimwipe in the tray along with the gel during shaking.
6. The sample lane can be divided according to the specific purpose of the study. However, if molecular weight markers are run alongside the sample, using them as reference points enables more consistent and reproducible gel segmentation.
7. To avoid overloading during homogenization and filtration, large gel bands should be avoided. For 1 mm-thick gels, it is recommended to keep each gel band within 6 mm in width and 10 mm in length per tube.
8. If the volume of gel is large and filtration is not completed within 3 min, centrifuge for an additional 3 min. When filtration is still incomplete after the additional centrifugation, transfer the remaining gel suspension to a new Spin-X filter device and repeat the centrifugation.
9. SP3 beads tend to sediment during storage. Mix thoroughly with a vortex mixer prior to use to ensure complete resuspension.
10. Incomplete drying may reduce protein recovery. Ensure that the beads are fully dried. If 10 min is insufficient, extend air-drying by 5–10 min while checking bead dryness.
11. If only a lower-speed tube mixer is available, extend the shaking time at the maximum speed of the device—for example, shake for 10 min at 1400 rpm.
12. For evaluating SDS levels in small-volume eluates, the use of the ContamSPOT assay [18] is recommended. If purification is insufficient, it is recommended to confirm steps in Sects. 3.5–3.7.
13. The duration of this process may vary depending on the sample, but typically takes 60–120 min.

Acknowledgments

This work was supported by JSPS KAKENHI 22KK0077 and 23K04963 (NT) and NIH R35GM147397 (LF). The authors thank Erika Teraoka for editing and reviewing this manuscript for the use of the English language and helpful discussion.

References

1. Smith LM, Kelleher NL, Consortium for Top Down P (2013) Proteoform: a single term describing protein complexity. Nat Methods 10(3): 186–187. https://doi.org/10.1038/nmeth.2369
2. Aebersold R, Mann M (2003) Mass spectrometry-based proteomics. Nature 422(6928):198–207. https://doi.org/10.1038/nature01511
3. Nesvizhskii AI, Aebersold R (2005) Interpretation of shotgun proteomic data: the protein inference problem. Mol Cell Proteomics 4(10):1419–1440. https://doi.org/10.1074/mcp.R500012-MCP200
4. Roberts DS et al (2024) Top-down proteomics. Nat Rev Methods Primers 4(1):38. https://doi.org/10.1038/s43586-024-00318-2
5. Tran JC et al (2011) Mapping intact protein isoforms in discovery mode using top-down proteomics. Nature 480(7376):254–258. https://doi.org/10.1038/nature10575
6. Laemmli UK (1970) Cleavage of structural proteins during the assembly of the head of bacteriophage T4. Nature 227(5259):680–685. https://doi.org/10.1038/227680a0
7. Takemori A et al (2020) PEPPI-MS: polyacrylamide gel based prefractionation for analysis of intact proteoforms and protein complexes by mass spectrometry. J Proteome Res 19(9):3779–3791. https://doi.org/10.1021/acs.jproteome.0c00303
8. Takemori A, Takemori N (2023) Sample preparation for structural mass spectrometry via polyacrylamide gel electrophoresis. Methods Enzymol 682:187–210. https://doi.org/10.1016/bs.mie.2022.08.051
9. Takemori N, Takemori A (2023) In-depth structural proteomics integrating mass spectrometry and polyacrylamide gel electrophoresis. Front Anal Sci 2:1107183
10. Takemori A, Kaulich PT, Tholey A, Takemori N (2025) PEPPI-MS: gel-based sample pre-fractionation for deep top-down and middle-down proteomics. Nat Protoc 20:1413–1438. https://doi.org/10.1038/s41596-024-01100-0
11. Takemori A et al (2022) Size-based proteome fractionation through polyacrylamide gel electrophoresis combined with LC-FAIMS-MS for In-depth top-down proteomics. Anal Chem 94(37):12815–12821. https://doi.org/10.1021/acs.analchem.2c02777
12. Takemori A et al (2025) Gel-based sample fractionation with SP3-purification for top-down proteomics. J Proteome Res 24(2): 850–860. https://doi.org/10.1021/acs.jproteome.4c00941
13. Hughes CS et al (2014) Ultrasensitive proteome analysis using paramagnetic bead technology. Mol Syst Biol 10:757. https://doi.org/10.15252/msb.20145625
14. Hughes CS et al (2019) Single-pot, solid-phase-enhanced sample preparation for proteomics experiments. Nat Protoc 14(1):68–85. https://doi.org/10.1038/s41596-018-0082-x
15. Muller T et al (2020) Automated sample preparation with SP3 for low-input clinical proteomics. Mol Syst Biol 16(1):e9111. https://doi.org/10.15252/msb.20199111
16. Muller T, Cremonini MA, Kliewer G, Krijgsveld J (2023) Automated sample preparation for mass spectrometry-based clinical proteomics. Methods Mol Biol 2718:181–211. https://doi.org/10.1007/978-1-0716-3457-8_11
17. Dagley LF et al (2019) Universal solid-phase protein preparation (USP(3)) for bottom-up and top-down proteomics. J Proteome Res 18(7):2915–2924. https://doi.org/10.1021/acs.jproteome.9b00217
18. Smeriglio N et al (2024) Contaminant spot check and removal assay (ContamSPOT) for mass spectrometry analysis. Anal Chem 96(6): 2574–2581. https://doi.org/10.1021/acs.analchem.3c05020

Chapter 14

Profiling O-Acetylation in Sialoglycans Using MALDI-TOF and LC–MS/MS with Methylamidation and Permethylation Derivatization

Zhaoguan Wu and Isabelle Sirois

Abstract

O-acetylation of sialoglycans plays a critical role in modulating the structural and functional properties of glycoproteins, with implications in immune response, cell signaling, and disease pathology. Among sialic acids, N-acetylneuraminic acid (Neu5Ac) and its derivatives have garnered attention as biomarkers for various cancers and inflammatory conditions due to their terminal positioning on glycan chains and their susceptibility to biochemical modification. Despite their biological relevance, comprehensive analysis of O-acetylation patterns remains challenging due to their chemical lability and loss during conventional glycomic workflows, such as permethylation. This protocol outlines a robust workflow combining methylamidation and permethylation with porous graphitic carbon (PGC)–based purification and high-resolution MALDI–MS and LC–MS/MS to analyze O-acetylation in serum/plasma–derived N-sialoglycans. We demonstrate sensitive and reproducible profiling of sialylated glycan structures. This method enables deeper insight into glycan modification dynamics and can be broadly applied to biomarker discovery and comparative glycomics.

Keywords Sialoglycans, O-acetylation, N-acetylneuraminic acid, MALDI–MS, LC–MS/MS, Methylamidation, Glycomics, PGC purification, Cancer biomarkers, Posttranslational modification

1 Introduction

Glycosylation is a ubiquitous and essential posttranslational modification of proteins in eukaryotes, contributing to various biological processes including protein folding, cellular recognition, immune response, and cancer progression [1–4]. Among the diverse glycans, sialic acids—typically capping the termini of glycan chains—play especially critical roles. The most prevalent form in humans, N-acetylneuraminic acid (Neu5Ac), is a nine-carbon, carboxylated monosaccharide that modulates numerous biological interactions [5, 6].

In recent years, Neu5Ac has gained considerable attention as a potential biomarker for cancer diagnosis and progression monitoring.

Xu Zhang et al. (eds.), *Protein Post-Translational Modifications*, Methods in Molecular Biology, vol. 3018,
https://doi.org/10.1007/978-1-0716-5166-7_14,

Aberrant sialylation patterns have been associated with multiple tumor types, including breast, colon, pancreatic, prostate, and gynecological cancers [4, 7–9]. Elevated levels of sialylated glycans, especially α2,6-linked sialic acids, have been linked to enhanced metastatic potential and immune evasion in malignant cells [10, 11]. Furthermore, Neu5Ac overexpression or altered modification states have been proposed as hallmarks of malignant transformation and cancer aggressiveness [12–14].

One of such modifications is O-acetylation, a naturally occurring and biologically significant alteration of sialic acids. O-acetylation typically occurs at the 4-, 7-, 8-, or 9-hydroxyl positions of the Neu5Ac molecule [15, 16]. These modifications are known to influence the physicochemical properties of sialoglycans, such as reducing hydrophilicity and altering conformational stability of glycoproteins [17]. More importantly, O-acetylation can affect molecular recognition events, involving lectins, antibodies, and pathogens, thus having significant implications for immunity and disease [18–20]. Importantly, altered O-acetylation patterns of Neu5Ac have been implicated in oncogenesis. For instance, reduced 9-O-acetylation has been observed in colorectal cancer and leukemia, correlating with disease progression [21, 22]. Conversely, overexpression of sialic acid O-acetyltransferases (SOATs) or reduced expression of sialic acid acetylesterases (SIAEs) may contribute to tumor immune evasion and metastasis [23, 24]. Despite its importance, O-acetylation remains poorly characterized in glycomics studies, largely due to the labile nature of acetyl esters under conventional analytical workflows, such as permethylation or acid hydrolysis.

Mass spectrometry (MS), including electrospray ionization (ESI) and matrix-assisted laser desorption/ionization (MALDI), has emerged as a powerful platform for glycan analysis [25–28]. However, most MS-based glycomic workflows rely on derivatization techniques, such as permethylation, to improve glycan stability and ionization efficiency, which unfortunately results in the loss of labile O-acetyl groups [29–31]. To overcome this, alternative derivatization strategies targeting the carboxyl group of sialic acids, such as methylamidation, dimethylamidation, and methyl/ethyl esterification, have been developed to preserve and stabilize O-acetylation patterns during analysis [32–38].

In this protocol, we present an optimized procedure for the detection and characterization of O-acetylated sialic acids in N-glycans using MALDI-TOF MS and LC–MS/MS, combined with selective chemical derivatization. As a proof of concept, we applied our workflow to sera derived from seven species of freshwater carp: crucian carp, grass carp, silver carp, bighead carp, common carp, bream, and black carp, demonstrating interspecies variation in O-acetylation profiles [39]. Our approach enables high-sensitivity, structure-specific profiling of sialoglycan modifications, offering a valuable tool for functional glycomics and biomarker discovery.

2 Materials

2.1 Chemicals and Materials

1. 2,5-Dihydroxybenzoic acid (DHB).
2. Dimethyl sulfoxide (DMSO).
3. Sodium hydroxide (NaOH) beads.
4. N-methylmorpholine (NMM).
5. Acetonitrile (ACN).
6. Methylamine hydrochloride ($CH_3NH_2 \cdot HCl$).
7. (7-azabenzotriazol-1-yloxy) trispyrrolidinophosphonium hexafluorophosphate (PyAOP).
8. Trifluoroacetic acid (TFA).
9. 1-Butanol (BuOH).
10. Ethanol (EtOH).
11. Formic Acid (FA).
12. Porous graphitic carbon (PGC).
13. Microcrystalline cellulose (MCC).
14. N-glycosidase F (PNGase F) Kit, New England Biolabs (Ipswich, MA, USA). The kit includes endoglycosidase buffer sodium phosphate (20 mM, pH 7.5) containing 0.2% SDS and 0.1 M DTT and 10% NP-40 in separate vials.
15. Methyl iodide (CH_3I).
16. Chloroform.
17. Empty cartridges and frits, 1 mL volume size.
18. Water for sample preparation: deionized (DI) water purified by a Milli-Q purification system.
19. Water for mass spectrometry analysis: Fisher Optimal™ LC–MS.

2.2 Buffers and Solutions to-be-Prepared

1. 80% ACN/water (v/v) containing 0.1% TFA.
2. 25% ACN/water (v/v) containing 0.1% TFA.
3. Water containing 0.1% TFA.
4. Water containing 0.1% FA.
5. Mobile phase A for liquid chromatography system: 0.1% FA in water.
6. Mobile phase B for liquid chromatography system: 0.1% FA in 90% ACN/water (v/v).
7. BuOH/EtOH/H_2O (4:4:1, v/v/v).
8. EtOH/H_2O (1:2, v/v).

9. PGC cartridges: PGC powder 100 mg was resuspended in ACN and loaded into empty cartridges with a bottom frit. Flush down the powder from the inner wall with ACN and apply an upper frit. Dry naturally in a clean air environment (*see* **Note 1**).
10. 1 M of CH_3NH_2·HCl and 0.5 M of NMM in DMSO.
11. 50 mM PyAOP in DMSO.

2.3 Devices and Instruments

1. Vortex mixer.
2. Thermomixer, with 1.5 mL tubes plate.
3. Speed-Vac concentrator.
4. SPE Vacuum Manifold, Supelco™.
5. MALDI-TOF/TOF-MS 5800™ System, SCIEX (Framingham, MA, USA).
6. Nano-LC Ultra System, Eksigent (Temecula, CA, USA).
7. Triple-TOF 5600™ System, SCIEX (Concord, ON, Canada).

3 Methods

3.1 N-Glycans Preparation

1. Dissolve serum/plasma (10 μL) in 50 μL of sodium phosphate (20 mM, pH 7.5) containing 0.2% SDS and 0.1 M DTT.
2. Denature in 100°C water bath for 10 min. Loosen the vial cap a little to avoid ejection (*see* **Note 2**).
3. Cool the vial to room temperature.
4. Add 12 μL of 10% NP-40 and 38 μL of DI water, vortex.
5. Add 10 units PNGaseF, vortex.
6. Incubation for 24 h at 37°C on the thermomixer at 600 rpm.
7. Stop reaction by 100°C water bath for 5 min. Loosen the vial cap a little to avoid ejection (*see* **Note 2**).
8. Dilute the sample to 1.0 mL with DI water.
9. Purify the released glycans by using PGC cartridges as follows: wash cartridge three times with 1.0 mL of 80% ACN/water containing 0.1% TFA, and condition cartridge three times with 1.0 mL of water containing 0.1% TFA. Load each sample in ~1.0 mL solution to the cartridge and wash three times with 1.0 mL of distilled water. Elute glycans two times with 600 μL of 25% ACN in 0.1% TFA.
10. Dry the eluted liquid phase by using a Speed-Vac concentrator.

3.2 Permethylation of N-Glycans

1. In a small agate mortar, add 1 mL of DMSO then add some NaOH beads (make sure the DMSO could cover the beads, then grinding the NaOH beads into small particles and homogenize with the DMSO to make a slurry). This DMSO-NaOH slurry should be prepared freshly right before the following steps (*see* **Note 3**).
2. Resuspend the dried oligosaccharide sample (from Sect. 3.1, Step 10) in an Eppendorf tube with 50 μL of DMSO.
3. Add 100 μL of DMSO-NaOH slurry and 50 μL of methyl iodide.
4. Tightly cap the tube and stir for about 10 min at room temperature (*see* **Note 4**).
5. Quench the reaction by adding 500 μL of DI water.
6. Add 250 μL of chloroform to extract the permethylated oligosaccharides.
7. Vortex the mixture and then remove the upper layer water phase.
8. Add 500 μL of DI water, vortex the mixture and remove the upper water phase.
9. Repeat the washing (Step 8) twice.
10. Transfer the lower chloroform phase into a clean micro spin tube.
11. Dry the chloroform phase by using a Speed-Vac concentrator.

3.3 Methylamidation of N-Glycans

1. Resuspend the dried oligosaccharide sample (from Sect. 3.1, Step 10) in an Eppendorf tube with 25 μL of DMSO solution containing 1 M of $CH_3NH_2{\cdot}HCl$ and 0.5 M of NMM, vortex.
2. Add to the mixture 25 μL of 50 mM PyAOP in DMSO solution.
3. Vortex and stood at room temperature for 30 min.
4. Dilute the sample to 0.5 mL with BuOH/EtOH/H_2O (4:4:1, v/v/v).
5. Purifying the derived glycans by using MCC particles according to the following steps.
6. Add 0.5 mL of distilled water to the Eppendorf tube with 5 mg cellulose particles to wash particles, vortex and discard the supernatant, repeat twice.
7. Add 0.5 mL of solvent mixture of BuOH/EtOH/H_2O (4:4:1, v/v/v) to condition particles, vortex and spin down, then discard the supernatant, repeat twice.
8. Mix the sample (in ~0.5 mL solution) with the washed cellulose particles. Gently shake the mixture for 45 min at 400 rpm, spin down then discard the supernatant.

9. Add 1.0 mL of solvent mixture of BuOH/EtOH/H_2O (4:4:1, v/v/v), vortex and spin down, then discard the supernatant, repeat twice.
10. To elute, incubate the cellulose particles with 0.5 mL of a mixture of EtOH/H_2O (1:2, v/v) for 30 min on a shaker at 400 rpm. Spin down and recover the liquid phase.
11. Dry the eluted liquid phase by using a Speed-Vac concentrator.

3.4 MALDI–MS Analysis

1. Set parameters of 5800 MALDI-TOF/TOF as follows: Nd:YAG Laser with 355 nm wavelength of <500 ps pulse and 200 Hz repetition rate. Reflectron mode. Spectra accumulation: 1000 laser shots.
2. Load samples onto MALDI target in 0.5 μL of water and mixed with 0.5 μL of freshly prepared DHB solution (10 mg/mL in 50% ACN) and allowed to dry in a gentle stream of air.
3. Review and process data using Dataexplorer® 4.0, and perform Glycans annotation with GlycoWorkBench [40] (*see* **Note 5**). As an example, the test results of crucian carp plasma samples are presented in Fig. 1 to demonstrate the format in which the experimental data are displayed.

The permethylation process effectively eliminates O-linked modifications, resulting in a markedly simplified MALDI–MS spectrum. This simplification facilitates the determination of N-glycan compositions. Figure 1a presents a representative MALDI–MS profile of permethylated N-glycans derived from the serum of crucian carp (*Carassius carassius*). The N-glycan structures identified in this species exhibit strong similarity to those previously reported in salmon serum [41, 42].

Among the predominant glycoforms, ions at m/z 2431.1, 2792.2, 3241.4, and 3602.5 correspond to mono-, di-, and tri-sialylated oligosaccharides as the following compositions: $Neu5NAc_1Hex_5HexNAc_4$, $Neu5NAc_2Hex_5HexNAc_4$, $Neu5NAc_2Hex_6HexNAc_5$, and $Neu5NAc_3Hex_6HexNAc_5$, respectively. Additionally, two high-mannose-type glycans were identified at m/z 1579.7 and 1783.8, corresponding to $Hex_5HexNAc_2$ and $Hex_6HexNAc_2$, respectively. A nonsialylated biantennary glycan was observed at m/z 2069.9 with a composition of $Hex_5HexNAc_4$. Furthermore, peaks at m/z 4051.7 and 4412.9 were attributed to tetra-antennary glycans with compositions $Neu5NAc_3Hex_7HexNAc_6$ and $Neu5NAc_4Hex_7HexNAc_6$, respectively, indicating the presence of more complex branching structures.

Figure 1b displays the MALDI-TOF MS spectrum of methylamidated N-glycans from the same crucian carp serum sample. As anticipated, methylamidation successfully derivatized all sialic acid residues while preserving labile functional groups, such as O-acetyl moieties. The resulting spectrum exhibited distinct ion clusters

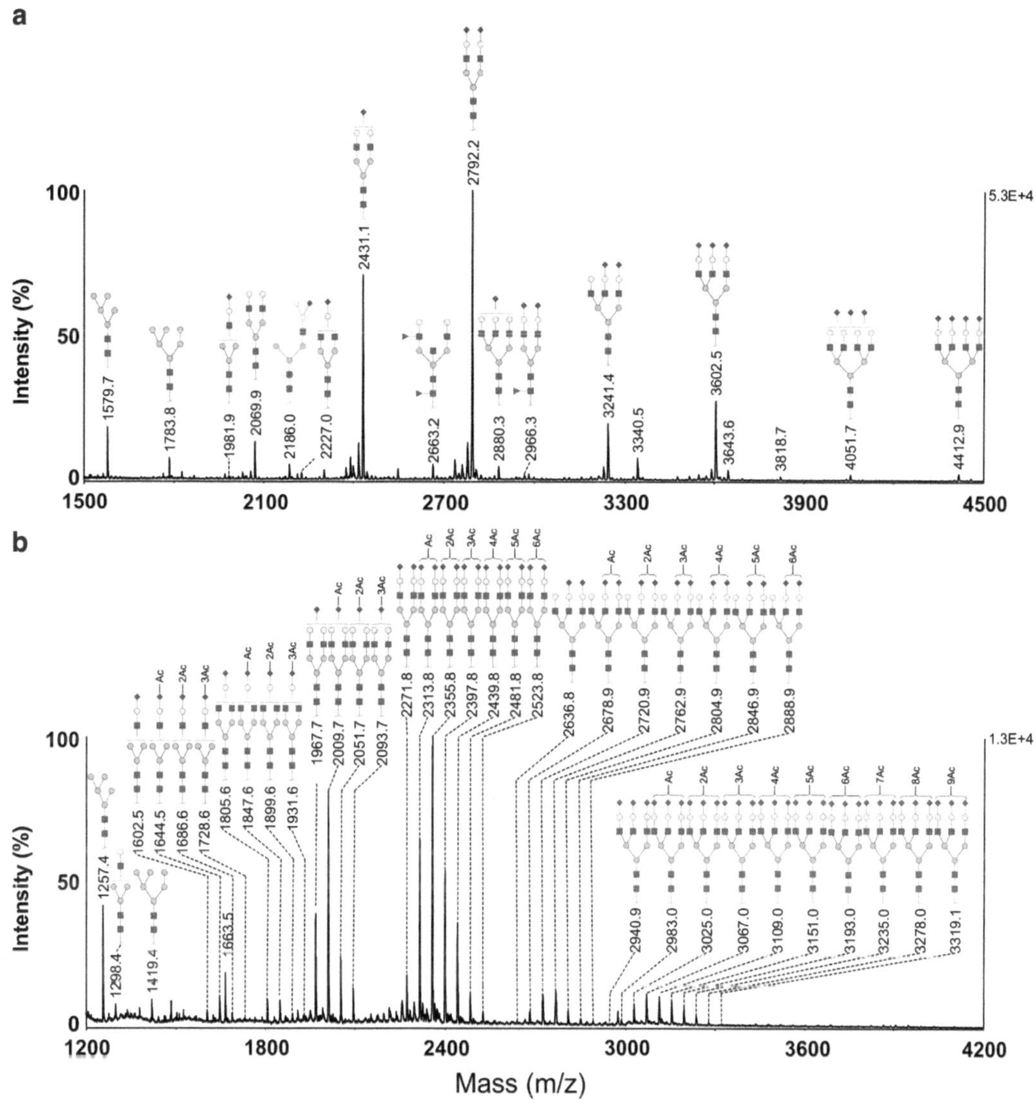

Fig. 1 MALDI–MS analysis of N-glycans from a representative serum sample of crucian carp. Monosaccharides represented as galactose (yellow circle), mannose (green circle), N-acetylglucosamine (blue rectangle), N-acetylneuraminic (pink diamond), and fucose (red triangle). (**a**) MS spectrum of permethylated N-glycans and (**b**) MS spectrum of methylamidated N-glycans(Figure reused with copyright permit from published article [39])

separated by 42 Da intervals, indicative of progressive O-acetylation on a single sialic acid residue—up to three O-acetyl groups per molecule. For instance, the ion series at m/z 1967.7, 2009.7, 2051.7, and 2093.7 were assigned to monosialylated glycans bearing zero to three O-acetyl groups, respectively.

In the case of di-sialylated glycans, ions detected at m/z 2271.8 through 2523.8 represent sequential addition of up to six

O-acetyl groups, yielding the following compositions: $Neu5NAc_2Hex_5HexNAc_4$ to $OAc_6Neu5NAc_2Hex_5HexNAc_4$. This extent of O-acetylation—up to six modifications per glycan—constitutes the highest level of O-acetylation reported for fish N-glycans to date [41, 42].

All identified ions and their corresponding structural compositions are summarized in Table 1.

O-acetylation in sialic acids may occur at multiple hydroxyl positions, specifically the 4-, 7-, 8-, or 9-positions [15, 16]. For di-sialylated glycans carrying six O-acetyl groups, at least two isomeric forms are theoretically possible: either each sialic acid contains three O-acetyl groups (3 + 3), or one residue contains four while the other carries two (4 + 2). Thus, tandem mass spectrometry analysis is essential to resolve the positional isomerism and distribution patterns of O-acetyl groups on these highly modified glycans.

3.5 LC–MS/MS Analysis

1. Use TripleTOF 5600 System (SCIEX, Canada) with NanoLC Ultra System (Eksigent, USA) for LC–MS/MS analysis.
2. Set-up column as follows: trap column, 150 μm I.D. × 1 cm length, PGC, 5 μm; analytical column, 75 μm I.D. × 10 cm length, PGC, 5 μm.
3. Set-up LC gradient as follows: started at 1% B (5 min), increased to 40% B over 30 min, then to 60% B over 15 min, followed by a ramp to 95% B over 10 min (held for 5 min), and re-equilibrated at 1% B for 10 min. Flow rate was 0.3–0.5 μL/min.
4. Set-up mass spectrometer parameters as follows: positive-ion mode, mass range 500–3000 m/z, MS/MS with information dependent acquisition (IDA) mode with a mass range of 100–2000 m/z: Top 20 abundant precursor ions, charge numbers 2 to 5. Duty cycle: MS 0.25 s followed by 20 MS/MS scans for 2 s.
5. Resuspend each sample in 10 μL of 0.1% FA/water (v/v) in sample vial for sample loading and LC–MS/MS measurement.
6. Review and process the MS data using PeakView® 2.1, and perform glycan annotation with GlycoWorkBench [40]. As an example, the test results of crucian carp plasma samples are presented in Fig. 2 to demonstrate the format in which the experimental data are displayed.

Tandem mass spectrometry (LC–MS/MS was conducted on the methylamidated N-glycan samples, with representative MS/MS spectra displayed in Fig. 2. In the case of non-O-acetylated structures, the dominant fragment ions observed at m/z 305.1 and 287.1 were assigned to methylamidated Neu5NAc (incorporating a +13 Da mass shift) and its dehydrated form, respectively (Fig. 2a). Additionally, a characteristic fragment ion at m/z

Table 1
Detected ions and their corresponding compositions of N-glycans from crucian carp serum samples. (Data reused with copyright permit from published article [39])

	Permethylation		Methylamidation	
	Theoretical[a]	Detected	Theoretical	Detected
Proposed composition	$[M+Na]^+$ (*m/z*)	$[M+Na]^+$ (*m/z*)	$[M+Na]^+$ (*m/z*)	$[M+Na]^+$ (*m/z*)
$Hex_5HexNAc_2$	1579.8	1579.7	1257.4	1257.4
$Hex_6HexNAc_2$	1783.9	1783.8	1419.5	1419.4
$Neu5NAc_1Hex_4HexNAc_3$	1982.0	1981.9	1602.6	1602.5
$OAc_1Neu5NAc_1Hex_4HexNAc_3$			1644.6	1644.5
$OAc_2Neu5NAc_1Hex_4HexNAc_3$			1686.6	1686.6
$OAc_3Neu5NAc_1Hex_4HexNAc_3$			1728.6	1782.6
$Hex_5HexNAc_4$	2070.0	2069.9	1663.6	1663.5
$Neu5NAc_1Hex_4HexNAc_4$	2227.1	2227.0	1805.7	1805.6
$OAc_1Neu5NAc_1Hex_4HexNAc_4$			1847.7	1847.6
$OAc_2Neu5NAc_1Hex_4HexNAc_4$			1889.7	1899.6
$OAc_3Neu5NAc_1Hex_4HexNAc_4$			1931.7	1931.6
$Neu5NAc_1Hex_5HexNAc_4$	2431.2	2431.1	1967.7	1967.7
$OAc_1Neu5NAc_1Hex_5HexNAc_4$			2009.7	2009.7
$OAc_2Neu5NAc_1Hex_5HexNAc_4$			2051.7	2051.7
$OAc_3Neu5NAc_1Hex_5HexNAc_4$			2093.7	2093.7
$Neu5NAc_2Hex_5HexNAc_4$	2792.4	2792.2	2271.8	2271.8
$OAc_1Neu5NAc_2Hex_5HexNAc_4$			2313.9	2313.8
$OAc_2Neu5NAc_2Hex_5HexNAc_4$			2355.9	2355.8
$OAc_3Neu5NAc_2Hex_5HexNAc_4$			2397.9	2397.8
$OAc_4Neu5NAc_2Hex_5HexNAc_4$			2439.9	2439.8
$OAc_5Neu5NAc_2Hex_5HexNAc_4$			2481.9	2481.8
$OAc_6Neu5NAc_2Hex_5HexNAc_4$			2523.9	2523.8
$Neu5NAc_1Hex_6HexNAc_5$	2880.4	2880.3	2332.8	2332.7
$Neu5NAc_2Hex_6HexNAc_5$	3241.6	3241.4	2637.0	2336.8
$OAc_1Neu5NAc_2Hex_6HexNAc_5$			2679.0	2678.9
$OAc_2Neu5NAc_2Hex_6HexNAc_5$			2721.0	2720.9
$OAc_3Neu5NAc_2Hex_6HexNAc_5$			2763.0	2762.9
$OAc_4Neu5NAc_2Hex_6HexNAc_5$			2805.0	2804.9
$OAc_5Neu5NAc_2Hex_6HexNAc_5$			2847.0	2864.9
$OAc_6Neu5NAc_2Hex_6HexNAc_5$			2889.0	2888.9

(Continued)

Table 1
(continued)

Proposed composition	Permethylation Theoretical[a] [M+Na]+ (m/z)	Permethylation Detected [M+Na]+ (m/z)	Methylamidation Theoretical [M+Na]+ (m/z)	Methylamidation Detected [M+Na]+ (m/z)
$Neu5NAc_3Hex_6HexNAc_5$	3602.8	3602.5	2941.1	2940.9
$OAc_1Neu5NAc_3Hex_6HexNAc_5$			2983.1	2983.0
$OAc_2Neu5NAc_3Hex_6HexNAc_5$			3025.1	3025.0
$OAc_3Neu5NAc_3Hex_6HexNAc_5$			3067.1	3067.0
$OAc_4Neu5NAc_3Hex_6HexNAc_5$			3109.1	3109.0
$OAc_5Neu5NAc_3Hex_6HexNAc_5$			3151.1	3151.0
$OAc_6Neu5NAc_3Hex_6HexNAc_5$			3193.2	3193.0
$OAc_7Neu5NAc_3Hex_6HexNAc_5$			3235.2	3235.0
$OAc_8Neu5NAc_3Hex_6HexNAc_5$			3277.2	3278.0
$OAc_9Neu5NAc_3Hex_6HexNAc_5$			3319.2	3319.1
$Neu5NAc_3Hex_7HexNAc_6$	4052.0	4051.7	3306.2	ND[b]
$OAc_3Neu5NAc_3Hex_7HexNAc_6$			3432.3	3433.1
$OAc_4Neu5NAc_3Hex_7HexNAc_6$			3747.3	3475.3
$OAc_5Neu5NAc_3Hex_7HexNAc_6$			3616.3	3517.1
$OAc_6Neu5NAc_3Hex_7HexNAc_6$			3558.3	3559.2
$Neu5NAc_4Hex_7HexNAc_6$	4413.2	4412.9	3610.4	ND
$OAc_4Neu5NAc_4Hex_7HexNAc_6$			3778.4	3779.1
$OAc_5Neu5NAc_4Hex_7HexNAc_6$			3820.4	3821.1
$OAc_6Neu5NAc_4Hex_7HexNAc_6$			3862.4	3863.2
$OAc_7Neu5NAc_4Hex_7HexNAc_6$			3904.4	3905.1
$OAc_8Neu5NAc_4Hex_7HexNAc_6$			3946.4	3947.1
$OAc_9Neu5NAc_4Hex_7HexNAc_6$			3988.4	3989.1
$OAc_{10}Neu5NAc_4Hex_7HexNAc_6$			4030.5	4031.3

[a]Monoisotopic mass units were used for the calculation of molecular mass based on the proposed composition
[b]Not detected.

670.1, corresponding to $Neu5NAc_1Hex_1HexNAc_1$, was consistently detected from non-O-acetylated parent ions.

When a single O-acetyl group was present on one of the two sialic acid residues, the MS/MS spectrum revealed a pair of fragment ions at m/z 670.3 and 712.3 with a 42 Da mass difference, indicative of partial O-acetylation (Fig. 2b). Notably, the fragment ions

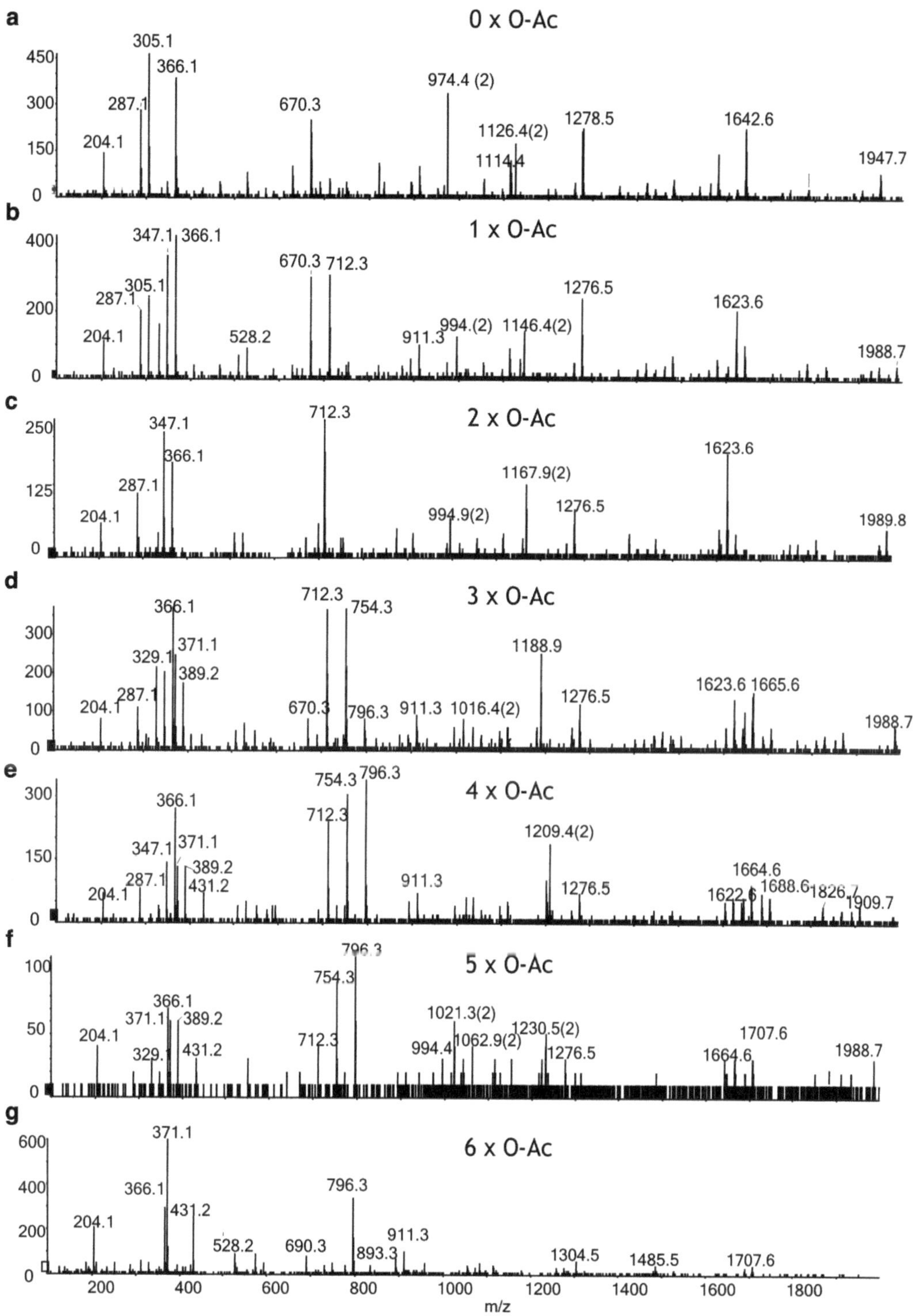

Fig. 2 LC–MS/MS analysis of methylamidated N-glycans isolated from a representative serum sample of crucian carp. (**a**) MS/MS spectrum of [M+2H]2+ at m/z 1125.3 (Neu5NAc2Hex5HexNAc4); (**b**) MS/MS spectrum of [M+2H]2+ at m/z 1146.4 (OAc1Neu5NAc2Hex5HexNAc4); (**c**) MS/MS spectrum

corresponding to the tri-saccharide cores—$Neu5NAc_1Hex_1HexNAc_1$ and $OAc_1Neu5NAc_1Hex_1HexNAc_1$—exhibited comparable signal intensities. However, significant intensity differences were evident in the corresponding monosaccharide fragments: Neu5NAc (m/z 305.1) versus $OAc_1Neu5NAc$ (m/z 347.1), as shown in Fig. 2b.

For glycans bearing two O-acetyl groups, the predominant fragment ion at m/z 712.3 suggested a symmetric distribution (1 + 1) of O-acetyl modifications across the two sialic acid residues, rather than an asymmetric (2 + 0) arrangement. Similarly, for tri-O-acetylated glycans, the most abundant isomeric distribution involved two O-acetyl groups on one sialic acid and one on the other (2 + 1), with corresponding fragment ions at m/z 712.3 and 754.3. In contrast, the alternative (3 + 0) isomer produced less intense signals at m/z 796.3 and 670.3 (Fig. 2c).

In the case of glycans modified with six O-acetyl groups, the MS/MS data indicated that each sialic acid was substituted with three O-acetyl groups (3 + 3), as depicted in Fig. 2f. These findings collectively suggest that the maximum number of O-acetylations per individual sialic acid residue does not exceed three. It is also noteworthy that no diagnostic fragments were detected for Neu5NAc–Neu5NAc disialyl linkages, indicating the absence of di-sialic acid structural motifs in these glycan species.

To evaluate interindividual variability, two additional serum samples from crucian carp were analyzed using the same LC–MS/MS workflow. Although variations in the relative abundances of O-acetylated glycoforms were observed among individuals, the maximal degree of O-acetylation in di-sialylated N-glycans remained consistent across all three crucian carp specimens. Moreover, this analytical approach was extended to include serum samples from six additional freshwater fish species to evaluate interspecies variability: grass carp, silver carp, bighead carp, common carp, bream, and black carp, collected from wild fisheries. Each species was examined with biological replicates to ensure reproducibility. Detailed results and comparative analyses for these species are available in our published article [39] (*see* **Note 6**).

4 Notes

1. Thermofisher Scientific produces similar PGC cartridges commercial product: Product # 57109-U, Supelco, Supelclean™ ENVI-Carb™ SPE Tube.

Fig. 2 (continued) of [M+2H]2+ at m/z 1167.9 (OAc2Neu5NAc2Hex5HexNAc4); (**d**) MS/MS spectrum of [M+2H]2+ at m/z 1188.9 (OAc3Neu5NAc2Hex5HexNAc4); (**e**) MS/MS spectrum of [M+2H]2+ at m/z 1209.4 (OAc4Neu5NAc2Hex5HexNAc4); (**f**) MS/MS spectrum of [M+2H]2+ at m/z 1230.4 (OAc5Neu5NAc 2Hex5HexNAc4); and (**g**) MS/MS spectrum of [M+3H]3+ at m/z 834.6 (OAc6Neu5NAc2Hex5HexNAc4) (Figure reused with copyright permit from published article [39])

2. When using the boiling water bath, loosen the vial cap a little to avoid ejection. Thus, a micro spin tube with a screwing lid would be a better choice.
3. The DMSO-NaOH slurry should be prepared freshly every time right before the following steps to avoid absorbing too much moisture from the air which would reduce the reaction efficiency.
4. Here, we mentioned the room temperature, our experiments attempt taught us that it is better to perform the reaction in a temperature-controlled environment, ensuring the temperature is below 30°C.
5. With recent advances in data science and computational technologies, a range of more intelligent and powerful glycomics data analysis tools have become available compared to those at the time of the original experiments. Notable examples include GlycoWorkbench (Google Code Archive), GlycoPAT (Biomolecular Systems), Byonic (Protein Metrics), PEAKS Glycan by Bioinformatics Solutions Inc. (BSI), GlycoPOST (Glycosmos), GlycoMod (ExPASy), and MS-GlycoMine (Monash University). These platforms enable comprehensive glycan and glycopeptide characterization through advanced algorithms for structure annotation, site localization, and quantification. With increased automation, improved sensitivity, and broader glycan database integration, they have substantially enhanced the throughput and accuracy of modern glycomics studies.
6. In addition to the methods described in this study, recent advances have introduced several novel approaches for characterizing sialic acid O-acetylation patterns. For example, the integration of ion mobility-mass spectrometry (IM-MS) has shown great potential not only in resolving O-acetylation positional isomers but also in determining glycosidic linkage types, thereby providing deeper structural insights into complex glycans [43].

References

1. Varki A, Lowe JB (2009) Biological roles of glycans in essentials of glycobiology. In: Varki A, Cummings RD, Esko JD, Freeze HH, Stanley P, Bertozzi CR, Hart GW, Etzler ME (Eds.)Eds. Cold Spring Harbor Laboratory Press Copyright © 2009, The Consortium of Glycobiology Editors. Cold Spring Harbor (NY), La Jolla, CA
2. Ohtsubo K, Marth JD (2006) Glycosylation in cellular mechanisms of health and disease. Cell 126:855–867
3. Dube DH, Bertozzi CR (2005) Glycans in cancer and inflammation—potential for therapeutics and diagnostics. Nat Rev Drug Discov 4:477–488
4. Pinho SS, Reis CA (2015) Glycosylation in cancer: mechanisms and clinical implications. Nat Rev Cancer 15:540–555
5. Varki A (2008) Sialic acids in human health and disease. Trends Mol Med 14:351–360
6. Schauer R (2009) Sialic acids as regulators of molecular and cellular interactions. Curr Opin Struct Biol 19:507–514

7. Munkley J (2022) Aberrant sialylation in cancer: therapeutic opportunities. Cancers (Basel) 14(17):4248
8. Varki A, Gagneux P (2012) Multifarious roles of sialic acids in immunity. Ann N Y Acad Sci 1253:16–36
9. Berghuis AY, Pijnenborg JFA, Boltje TJ, Pijnenborg JMA (2022) Sialic acids in gynecological cancer development and progression: Impact on diagnosis and treatment. Int J Cancer 150:678–687
10. Munkley J (2016) The role of sialyl-Tn in cancer. Int J Mol Sci 17:275
11. Hedlund M, Padler-Karavani V, Varki NM, Varki A (2008) Evidence for a human-specific mechanism for diet and antibody-mediated inflammation in carcinoma progression. Proc Natl Acad Sci USA 105:18936–18941
12. Büll C, Stoel MA, den Brok MH, Adema GJ (2014) Sialic acids sweeten a tumor's life. Cancer Res 74:3199–3204
13. Pearce OM, Läubli H (2016) Sialic acids in cancer biology and immunity. Glycobiology 26:111–128
14. Läubli H, Varki A (2020) Sialic acid-binding immunoglobulin-like lectins (Siglecs) detect self-associated molecular patterns to regulate immune responses. Cell Mol Life Sci 77:593–605
15. Schauer R, Kamerling JP (2018) Exploration of the sialic acid world. Adv Carbohydr Chem Biochem 75:1–213
16. Klein A, Roussel P (1998) O-acetylation of sialic acids. Biochimie 80:49–57
17. Visser EA, Moons SJ, Timmermans S, de Jong H, Boltje TJ, Büll C (2021) Sialic acid O-acetylation: from biosynthesis to roles in health and disease. J Biol Chem 297:100906
18. Padler-Karavani V, Varki A (2011) Potential impact of the non-human sialic acid N-glycolylneuraminic acid on transplant rejection risk. Xenotransplantation 18:1–5
19. Argüeso P, Sumiyoshi M (2006) Characterization of a carbohydrate epitope defined by the monoclonal antibody H185: sialic acid O-acetylation on epithelial cell-surface mucins. Glycobiology 16:1219–1228
20. Das K, Schulte M, Gerhart M, Bayoumi H, Paulson D, Fink DM, Parrish C, Willand-Charnley R (2024) Early in vitro results indicate that de-O-acetylated sialic acids increase Selectin binding in cancers. Front Oncol 14:1443303
21. Shen Y, Kohla G, Lrhorfi AL, Sipos B, Kalthoff H, Gerwig GJ, Kamerling JP, Schauer R, Tiralongo J (2004) O-acetylation and de-O-acetylation of sialic acids in human colorectal carcinoma. Eur J Biochem 271:281–290
22. Pal S, Ghosh S, Bandyopadhyay S, Mandal C, Bandhyopadhyay S, Kumar Bhattacharya D, Mandal C (2004) Differential expression of 9-O-acetylated sialoglycoconjugates on leukemic blasts: a potential tool for long-term monitoring of children with acute lymphoblastic leukemia. Int J Cancer 111:270–277
23. Corfield AP, Wagner SA, Paraskeva C, Clamp JR, Durdey P, Reuter G, Schauer R (1992) Loss of sialic acid O-acetylation in human colorectal cancer cells. Biochem Soc Trans 20:94S–94S
24. Filipsky F, Läubli H (2024) Regulation of sialic acid metabolism in cancer. Carbohydr Res 539:109123
25. Zaia J (2008) Mass spectrometry and the emerging field of glycomics. Chem Biol 15:881–892
26. Alley JrWR, Novotny MV (2013) Structural glycomic analyses at high sensitivity: a decade of progress. Annu Rev Anal Chem (Palo Alto Calif) 6:237–265
27. Zaia J (2004) Mass spectrometry of oligosaccharides. Mass Spectrom Rev 23:161–227
28. Dell A, Morris HR (2001) Glycoprotein structure determination by mass spectrometry. Science 291:2351–2356
29. Ciucanu I, Kerek F (1984) A simple and rapid method for the permethylation of carbohydrates. Carbohydr Res 131:209–217
30. Kang P, Mechref Y, Novotny MV (2008) High-throughput solid-phase permethylation of glycans prior to mass spectrometry. Rapid Commun Mass Spectrom 22:721–734
31. Shajahan A, Supekar NT, Chapla D, Heiss C, Moremen KW, Azadi P (2020) Simplifying glycan profiling through a high-throughput micropermethylation strategy. SLAS Technol 25:367–379
32. Liu X, Qiu H, Lee RK, Chen W, Li J (2010) Methylamidation for sialoglycomics by MALDI-MS: a facile derivatization strategy for both α2,3- and α2,6-linked sialic acids. Anal Chem 82:8300–8306
33. Nishikaze T, Kawabata S, Tanaka K (2014) In-depth structural characterization of N-linked glycopeptides using complete derivatization for carboxyl groups followed by positive- and negative-ion tandem mass spectrometry. Anal Chem 86:5360–5369
34. Zhou H, Warren PG, Froehlich JW, Lee RS (2014) Dual modifications strategy to quantify neutral and sialylated N-glycans simultaneously by MALDI-MS. Anal Chem 86:6277–6284
35. de Haan N, Reiding KR, Haberger M, Reusch D, Falck D, Wuhrer M (2015) Linkage-specific sialic acid derivatization for MALDI-TOF-MS profiling of IgG glycopeptides. Anal Chem 87:8284–8291

36. Wheeler SF, Domann P, Harvey DJ (2009) Derivatization of sialic acids for stabilization in matrix-assisted laser desorption/ionization mass spectrometry and concomitant differentiation of alpha(2 —> 3)- and alpha(2 —> 6)-isomers. Rapid Commun Mass Spectrom 23:303–312
37. Bladergroen MR, Reiding KR, Hipgrave Ederveen AL, Vreeker GC, Clerc F, Holst S, Bondt A, Wuhrer M, van der Burgt YE (2015) Automation of high-throughput mass spectrometry-based plasma N-glycome analysis with linkage-specific sialic acid esterification. J Proteome Res 14:4080–4086
38. Reiding KR, Blank D, Kuijper DM, Deelder AM, Wuhrer M (2014) High-throughput profiling of protein N-glycosylation by MALDI-TOF-MS employing linkage-specific sialic acid esterification. Anal Chem 86:5784–5793
39. Wu Z, Li H, Zhang Q, Liu X, Zheng Q, Li J (2017) Characterization of O-acetylation in sialoglycans by MALDI-MS using a combination of methylamidation and permethylation. Sci Rep 7:46206
40. Ceroni A, Maass K, Geyer H, Geyer R, Dell A, Haslam SM (2008) GlycoWorkbench: a tool for the computer-assisted annotation of mass spectra of glycans. J Proteome Res 7:1650–1659
41. Liu X, Afonso L, Altman E, Johnson S, Brown L, Li J (2008) O-acetylation of sialic acids in N-glycans of Atlantic salmon (*Salmo salar*) serum is altered by handling stress. Proteomics 8:2849–2857
42. Jayo RG, Li J, Chen DD (2012) Capillary electrophoresis mass spectrometry for the characterization of O-acetylated N-glycans from fish serum. Anal Chem 84:8756–8762
43. Vos GM, Hooijschuur KC, Li Z, Fjeldsted J, Klein C, de Vries RP, Toraño JS, Boons GJ (2023) Sialic acid O-acetylation patterns and glycosidic linkage type determination by ion mobility-mass spectrometry. Nat Commun 14:6795

Chapter 15

Purification of Protein-Free and Protein-Linked Poly(ADP-Ribose) from Biochemical Reactions and from Human Cells

Marie-France Langelier, Manija Mirhasan and John M. Pascal

Abstract

Poly(ADP-ribose) or PAR regulates multiple aspects of cell biology, both as an independent signaling molecule and as a modification on biomolecules. As a posttranslational modification, PAR can modulate the biochemical properties of target proteins. Isolated free PAR molecules function in cellular signaling. This chapter describes two methods to isolate and purify free PAR and protein-linked PAR from biochemical reactions, one using chemical fractionation and another using physical separation. A method to isolate free PAR and protein-linked PAR from human cells is also presented. These methods allow monitoring of free PAR and protein-linked PAR levels under different biochemical conditions or in response to different cellular stimuli.

Keywords Poly(ADP-ribose) (PAR), PARP1, PARP2, Tankyrases (TNKS), Free PAR, Protein-linked PAR, Biochemical assays, Cellular analysis

1 Introduction

Poly(ADP-ribose) or PAR as a posttranslational modification (PTM) in humans is catalyzed by PARP1, PARP2, and Tankyrases (TNKS1 and TNKS2), which are members of the protein family known as PARPs [originally for poly(ADP-ribose) polymerases] [1]. PARP1, PARP2, and TNKS play various roles in diverse cellular processes, including the DNA damage response, gene expression, and cell signaling [2–5]. PARP enzymes are ADP-ribosyltransferases. PAR as a PTM starts with the covalent attachment of an ADP-ribose molecule to a protein side chain, using NAD^+ as a substrate and donor of the ADP-ribose. Several amino acid residues can be modified with ADP-ribose, including glutamates, aspartates, and serines [6]. An NAD^+ molecule binds to the donor site in the ADP-ribosyltransferase domain (ART), providing an ADP-ribose unit that is added to a protein residue, and in the process, nicotinamide is released

Xu Zhang et al. (eds.), *Protein Post-Translational Modifications*, Methods in Molecular Biology, vol. 3018
https://doi.org/10.1007/978-1-0716-5166-7_15,

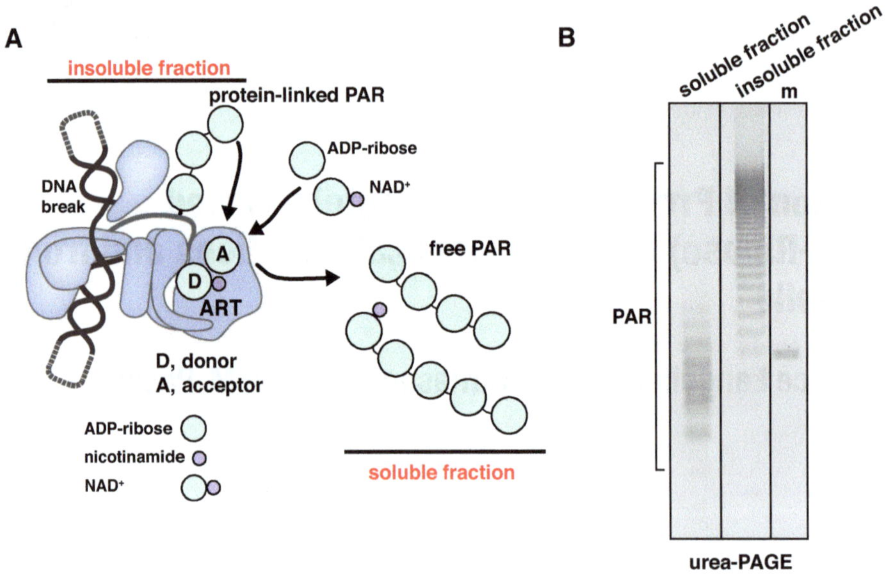

Fig. 1 PARP enzyme reactions yield protein-linked PAR and free PAR. (**A**) A schematic representation of PARP1 multidomain structure bound to a DNA strand break. In this activated conformation, NAD+ molecules have access to the donor site (D) in the ADP-ribosyltransferase (ART) domain. These NAD+ molecules are hydrolyzed into nicotinamide and ADP-ribose. The ADP-ribose is transferred to a protein side chain to initiate protein-linked PAR, which can be extended when the terminal ADP-ribose unit binds to the acceptor site (A), taking on an additional ADP-ribose unit from the donor site. In a different reaction, an individual ADP-ribose molecule or an individual NAD+ molecule binds to the acceptor site to initiate a free PAR molecule that is not protein-linked. During the TCA fractionation protocol, the protein-linked PAR is precipitated into the insoluble fraction. The free PAR predominantly remains in the soluble fraction. (**B**) Urea-PAGE analysis of PAR samples purified from the soluble fraction and the insoluble fraction of a TCA precipitated PARP1 biochemical reaction. The different lengths of PAR form a ladder of bands. Due to initiation on both NAD+ and ADP-ribose molecules, the free PAR is a mixture and exhibits more bands. The size distribution of free PAR from the soluble fraction is smaller than that of the protein-linked PAR isolated from the insoluble fraction. The urea-PAGE is stained with SYBR-Gold. m, PAR size standard with 7 ADP-ribose units

(Fig. 1). The ADP-ribose attached to the protein moves to the acceptor site in the ART, where it will accept an ADP-ribose unit from a new NAD+ molecule bound to the donor site. This step is repeated multiple times to create a PAR protein modification with multiple ADP-ribose units joined through ribose-ribose linkages. PARP1 and PARP2 can create chains of 200 units of ADP-ribose [7], while TNKS synthesizes shorter chains [8]. Other members of the PARP family of enzymes covalently attach a mono(ADP-ribose) modification to protein targets, and lack the capacity to create PAR [9].

Recently, PARP1, PARP2, and TNKS enzymes were discovered to synthesize PAR that is not covalently attached to a protein residue [10]. This novel activity takes place when a free ADP-ribose or a NAD+ molecule binds to the acceptor site while a second molecule of NAD+ binds to the donor site (Fig. 1). This results in the production of free PAR chains that initiate from an individual ADP-ribose or NAD^+ molecule. This finding thus identified a new type of biomolecule by PARP enzymes. The finding was also noteworthy because free PAR was thought to arise solely by the degradation of PAR PTMs through the action of glycohydrolase enzymes. Instead of a degradation pathway, the new biochemical activity pointed to de novo synthesis as a source of free PAR.

The production of free PAR occurred alongside the production of PAR as a PTM. The assessment of both types of PAR product is thus interesting for tracking their capacity to contribute to different biological processes. Here, we describe two methods to isolate free PAR and protein-linked PAR from biochemical reactions. These types of experiments allow testing of the relative output of protein-linked versus free PAR under different conditions and in the presence of regulatory factors. We also describe a protocol to isolate free and protein-linked PAR from human cells. These types of experiments can test the relative production levels of protein-linked and free PAR in response to cellular stimuli. PARP1 and PARP2 are key players in the cellular response to DNA strand break damage. In the described assays, DNA strand breaks are used to stimulate PARP1 and PARP2 activity in biochemical reactions and cell-based analysis. However, the protocols can serve as a general strategy for analyzing PAR as a PTM and signaling molecule in other biochemical compositions or cell biology conditions.

2 Materials

2.1 Boronate Matrix Preparation and Equilibration

1. Bio-Rex 70 resin (Bio-Rad)
2. Aminophenylboronic acid hydrochloride
3. *N*-ethyl-*N*'-(3'-dimethylaminopropyl) carbodiimide
4. 6 M guanidine hydrochloride in 50 mM 2-(*N*-morpholino) ethanesulfonic acid (MES) buffer, pH 6
5. Concentrated HCl
6. AAGE buffer, pH 9: 250 mM ammonium acetate, 6 M guanidine hydrochloride, 10 mM EDTA, adjust pH to 9 using concentrated NaOH
7. pH strips
8. Double distilled water (ddH_2O)
9. Econo-Column chromatography column, 49 mL, Bio-Rad

2.2 In Vitro PARylation Reactions

1. Purified PARP1, PARP2, or TNKS protein
2. PAR buffer 5×: 250 mM Hepes (pH 8), 750 mM NaCl, 5 mM dithiothreitol (DTT), 1% nonidet-P40

3. DNA for PARP1: 18 base-pair duplex with blunt ends (5' GGG TTG CGG CCG CTT GGG annealed to a complementary strand), or a dumbbell DNA containing a central nick (5' GCT GAG CTT CTG GTG AAG CTC AGC TCG CGG CAG CTG GTG CTG CCG CGA) (*see* **Notes 1 and 2**)
4. DNA for PARP2: 28 base-pair duplex with 5'-phosphorylated ends (5' CGA GTC TAC AGC GTT GCG GCC GCT TGG G annealed to a complementary strand) (*see* **Notes 1 and 3**)
5. NAD^+ stock solution (2 mM for PARP1 and PARP2, and 100 mM for TNKS) (*see* **Note 4**)
6. PARP inhibitor Talazoparib (250 μM)
7. $ddH_2 0$

2.3 Trichloro Acetic Acid (TCA) Fractionation of In Vitro and Cellular PAR

1. Benzonase (250 U/μL, New England Biolabs) (*see* **Note 5**)
2. DNAse I (2000 U/mL, Millipore Sigma) (*see* **Note 5**)
3. Bovine serum albumin (BSA) (50 mg/mL) (*see* **Note 5**)
4. 20% ice-cold TCA
5. 70% ice-cold ethanol
6. Refrigerated centrifuge (20,800×*g*)
7. AAGE buffer pH 9: 250 mM ammonium acetate, 6 M guanidine hydrochloride, 10 mM EDTA, adjust pH to 9 using concentrated NaOH
8. Concentrated NaOH
9. pH strips
10. 1 M ammonium acetate, pH 9
11. 1 M KOH/50 mM EDTA, pH 8
12. Filter spin columns (Fisher, F2517-9)
13. Vacuum concentrator
14. Tube rotator

2.4 Urea-PAGE (Polyacrylamide Gel Electrophoresis) Analysis

1. Urea
2. 40% acrylamide:bis-acrylamide, 19:1 ratio
3. TBE buffer (5×): 450 mM Tris, 450 mM boric acid, and 10 mM EDTA
4. 10% ammonium persulfate (APS)
5. Tetramethylethylenediamine (TEMED)
6. SYBR gold stain (10,000×; ThermoFisher Scientific)
7. Gel system that allows migration of gels of roughly 8 inches by 8 inches
8. 95% ethanol
9. Formamide/EDTA (3×): mix 9.8 mL formamide with 200 μL 0.5 M EDTA

10. Formamide/EDTA with bromophenol and xylene cyanol dyes (3×): add a trace of bromophenol and xylene cyanol to formamide/EDTA solution.

2.5 Streptavidin-Agarose Magnetic Beads Fractionation

1. Biotinylated PARP1, PARP2, or TNKS (*see* **Note 6**)
2. Streptavidin-agarose magnetic Dynabeads (ThermoFisher Scientific)
3. Dynabeads magnetic rack
4. PAR buffer (1×, pH 7.4): 50 mM Tris-HCl (pH 7.4), 150 mM NaCl, 1 mM DTT, 0.2% nonidet P-40
5. DNA (same recommendations as described in Sect. 2.2)
6. NAD^+
7. Talazoparib
8. PAR buffer (1×, pH 7.4, high salt): 50 mM Tris-HCl (pH 7.4), 1 M NaCl, 1 mM DTT, 0.2% nonidet P-40
9. 1 M KOH/50 mM EDTA

2.6 Isolation of Free and Protein-Linked PAR from Human Cells

1. Dulbecco's modified Eagle's medium (DMEM) with L-glutamine, supplemented with 10% fetal bovine serum (FBS) and penicillin/streptomycin
2. Cell culture incubator at 37°C, 5% CO_2
3. HEK293 cells or HeLa cells
4. PARG inhibitor PDD00017273 (SelleckChem) resuspended to 15 mM in 100% DMSO
5. Phosphate buffered saline (PBS, 1×)
6. Methyl methane sulfonate (MMS)
7. Trypsin/PARGi/olaparib: trypsin supplemented with 20 μM PDD00017273 and 10 μM olaparib
8. DMEM/PARGi/olaparib: DMEM supplemented with 20 μ M PDD00017273 and 10 μM olaparib
9. PBS/PARGi/olaparib: PBS supplemented with 20 μM PDD00017273 and 10 μM olaparib
10. TCA/PARGi/olaparib: 20% TCA supplemented with 20 μM PDD00017273 and 10 μM olaparib
11. Refrigerated centrifuge that can hold 15 mL conicals

3 Methods

3.1 Isolation of Free and Protein-Linked PAR from In Vitro Reactions

The following methods allow the separation and isolation of free and protein-linked PAR from biochemical PARylation reactions and were developed based on previously published protocols for the purification of protein-linked PAR [11–13]. The reactions can be performed using purified PARP1, PARP2, or TNKS or biotinylated versions of these proteins when using the streptavidin-agarose magnetic beads separation. Reactions can be performed for various amounts of time and are stopped by the addition of a PARP inhibitor. The PAR is then fractionated, purified, and analyzed by urea-PAGE (Fig. 1).

3.1.1 Boronate Matrix Preparation and Equilibration

1. Add 50 g of Bio-Rex70 powder to 500 mL ddH_2O (in a 1 L bottle). Let the mixture swell overnight at room temperature (RT).
2. Remove fines by decantation. Add ddH_2O again, resuspend and transfer to a 250 mL beaker. Let the resin settle and remove fines as much as possible by decantation and with a pipette.
3. Dissolve 5 g of aminophenylboronic acid hydrochloride in 50 mL ddH_2O. Add it to the Bio-Rex70 slurry and mix with a stir bar.
4. Adjust the pH to 5 with concentrated HCl.
5. Add 5 g of *N*-ethyl-*N*'-(3'-dimethylaminopropyl) carbodiimide to the mix.
6. Place the mixture (250 mL bottle) on the rotator for 18 h at RT.
7. Transfer mixture to gravity flow column (e.g., Econo-Column from Bio-Rad) and wash extensively with ddH_2O.
8. Wash with 6 M guanidine hydrochloride in 50 mM MES buffer (pH 6). Prepared matrix can then be transferred to a 250 mL bottle and stored in this same buffer at 4°C for several months.
9. A portion of the prepared matrix is equilibrated for use. From the stored matrix stock in Step 8, transfer desired volume of Boronate matrix (e.g., 10 mL slurry) to a gravity flow column.
10. Allow storage buffer to flow through.
11. To equilibrate the Boronate matrix, pass through 10 column volumes (CV) of AAGE buffer, 20 CV of ddH_2O, and 20 CV of AAGE buffer sequentially. Measure the pH of flow-through (with pH strip) and keep applying AAGE buffer until the pH is higher than 8.5.
12. Once pH of flow-through is higher than 8.5, add AAGE buffer to the matrix (AAGE volume should be 2× that of the boronate resin). Using a pipette, resuspend resin and transfer to a bottle.
13. The equilibrated boronate matrix can be stored at 4°C for several months.

3.1.2 In Vitro PARylation Reactions

Assemble PARylation reactions in a 0.5 or 1.5 mL tube. The final volume of the reaction will be 40 μL after the addition of NAD^+. If several time points are planned, a master mix can be prepared.

1. Add PARP1, PARP2, or TNKS at 5 or 10 μM final concentration in a reaction tube containing 8 μL of 5× PAR buffer and ddH_2O (*see* **Note 7**).
2. Add appropriate DNA at a final concentration of 5 or 10 μM (matching the protein concentration) (*see* **Note 8**).
3. Incubate 10 min at RT.
4. Start reaction by adding 10 μL of 2 mM NAD^+ for a final concentration of 500 μM.
5. After the appropriate time point, stop the reaction by adding 10 μL 250 μM Talazoparib.
6. Add 1 μL of DNAse I (2000 U/mL) and 1 μL of benzonase (250 U/μL) to the reaction and incubate for 30 min at 37°C. This step digests the activating DNA.
7. Centrifuge briefly at 20,800x*g*.

3.1.3 TCA Fractionation of In Vitro PARylation Reactions and Boronate Purification of PAR

The TCA fractionation protocol is performed on ice for a limited amount of time to avoid inadvertent detachment of the protein-linked PAR [14]. Note that a small amount of free PAR is precipitated by the TCA treatment and can contaminate the protein-linked PAR.

1. Add 5 μL of 50 mg/mL BSA to the PARylation reaction mixture.
2. Add 57 μL of 20% ice-cold TCA (should be 1:1 ratio with reaction volume).
3. Place on ice for 15 min.
4. Spin at 20,800×*g* for 20 min at 4°C.
5. Transfer the supernatant, which contains free PAR, to a new tube. Leave on ice.
6. Wash pellets (containing protein-linked PAR) with 200 μL of 70% ice-cold ethanol.
7. Remove supernatant and leave pellets to air dry.
8. Add 100 μL 1 M KOH/50 mM EDTA to pellets.
9. Incubate 1 h at 60°C, vortex every 10 min.
10. Once the KOH treatment is done, add 400 μL AAGE buffer and store all samples at −20°C or proceed directly with boronate purification.
11. Add 400 μL AAGE buffer to supernatant samples. Adjust pH to 9 by adding NaOH (it usually takes 10 μL of concentrated NaOH, but pH should be verified with pH strips).

12. Resuspend/mix the pre-equilibrated boronate resin by swirling the bottle.
13. Add 120 µL of boronate resin to all samples (supernatants and pellets).
14. Incubate samples on a rotator for 1 h at RT.
15. Centrifuge briefly at 20,800×*g*.
16. Transfer samples to 750 µL filter spin columns
17. Centrifuge 1 min at 9000×*g*. Discard flow-through.
18. Add 700 µL AAGE buffer. Spin 1 min at 9000×*g*. Discard flow-through.
19. Add 700 µL 1 M ammonium acetate buffer (pH 9). Centrifuge 4 min at 9000×*g*. Discard flow-through. Repeat this step one more time.
20. Transfer columns to new 1.5 mL tubes.
21. Add 100 µL ddH_2O (37°C) to elute. Let column sit for 1 min, then centrifuge at 9000×*g* for 1 min. Repeat the elution two more times (3 × 100 µL eluates; collect the eluates in the same tube).
22. Dry down eluates overnight using a vacuum concentrator at 30°C.
23. Resuspend dried samples in 50 µL ddH_2O. Vortex and incubate at 37°C for 30 min. Briefly centrifuge and store at −20°C or proceed with visualization on urea-PAGE.

3.1.4 PAR Visualization on Urea-PAGE Denaturating Gel

1. Prepare the gel on the same day that it will be used.
2. Dissolve 13.2 g of urea in a mixture of 27.5 mL 40% acrylamide:bis acrylamide (19:1) and 11 mL of 5× TBE. Complete volume to 55 mL with ddH_2O and filter using a 0.2-micron filter to remove particles.
3. Wash glass plates thoroughly with water and 95% ethanol. Air dry and make sure no dust is present on glass surface since these can interfere with the imaging of the gel.
4. Add 550 µL 10% APS and 77 µL TEMED.
5. Mix briefly and pour immediately.
6. Insert comb and allow to polymerize for 5 min.
7. Wash the wells quickly with ddH_2O.
8. Assemble gel on apparatus, rinse wells with migration buffer and premigrate for 40 min at 600 V.
9. When samples are ready, add 25 µL 3× formamide dye so you have a 1× final concentration.
10. Heat samples for 5–10 min at 90°C.

11. Load samples and a lane with formamide dye/bromophenol/xylene cyanol to monitor migration.
12. Run gel for about 50 min at 600 V.
13. Stain gel in 200 mL 1× TBE with 40 µL 10,000× SYBR gold in a clean glass container for 20 min, and proceed with imaging immediately.

3.1.5 Streptavidin-Agarose Magnetic Bead Fractionation

This type of separation of free PAR and protein-linked PAR can be used as an alternative to TCA fractionation to avoid contamination of free PAR chains in the protein-linked PAR fraction.

1. Transfer 120 µL of 50% streptavidin-agarose magnetic bead slurry to tubes.
2. Wash the beads twice with 1 mL of PAR buffer (pH 7.4), place on magnetic rack for 1 min or until the beads are fully pelleted, and then discard buffer.
3. Add 60 µL of biotinylated protein at 20 µM and incubate for 1 h at RT on a rotator.
4. Put beads on magnetic rack and discard supernatant.
5. Wash beads with 1× 1 mL and 2× 500 mL of PAR buffer (pH 7.4). Place beads on magnetic rack and discard buffer each time.
6. Add 60 µL 20 µM DNA and then add 50 mM NAD^+ to have a final concentration of 500 µM.
7. Incubated for 30 min (*see* **Note 9**) at RT and stop the reaction by adding Talazoparib to 250 µM final concentration.
8. Put the beads on the magnetic rack, incubate for 1 min and transfer the supernatant, which contains free PAR, to a new tube and keep it on ice.
9. Wash the beads twice with 500 µL PAR buffer (pH 7.4, high salt) and once with 500 µL PAR buffer (pH 7.4).
10. To release PAR from the beads, add 100 µL 1 M KOH/50 mM EDTA and incubate 1 h at 60°C with vortexing every 10 min.
11. Follow Sect. 3.1.3 Steps 11–24 for boronate purification and Sect. 3.1.4 for urea-PAGE visualization.

3.2 Isolation of Free and Protein-Linked PAR from Human Cells

Free and protein-linked PAR can be isolated from cultured human cells. The amounts of PAR purified are higher when using PARG inhibitor and MMS, but lower amounts can also be purified without these two components. Starting from the trypsin steps for detaching the cells, a PARG inhibitor is added to all solutions to preserve PAR and the PARP inhibitor olaparib is added to all solutions to prevent new PAR formation due to DNA damage that could arise during the cell harvesting and lysis.

1. Cultivate a 150 mm petri dish with HEK293 or HeLa cells to 80%–90% confluency in 30 mL DMEM with L-glutamine, supplemented with 10% FBS and penicillin/streptomycin at 37°C in 5% CO_2.
2. Once 80%–90% confluency is reached, remove 6 mL of media to reduce the total volume from 30 to 24 mL.
3. Add 32 μL of PARGi at 15 mM to the petri dish and swirl gently to mix.
4. Incubate 1 h at 37°C in cell incubator.
5. Add 1 mL of 0.25% MMS diluted in cell media.
6. Incubate 30 min at 37°C in cell incubator.
7. Remove media and wash with 15 mL of PBS.
8. Add 3 mL of trypsin/PARGi/olaparib.
9. Incubate for 2 min at 37°C in cell incubator.
10. Detach cells gently and transfer them to a 15 mL conical tube.
11. Add 6 mL of DMEM/PARGi/olaparib. Cells can be counted at this step to normalize between petri dishes, if multiple conditions are tested.
12. Centrifuge cells at 250×*g* for 3 min at RT, discard supernatant.
13. Add 2 mL of PBS/PARGi/olaparib to resuspend the cells.
14. Centrifuge cells at 250×*g* for 3 min at RT, discard supernatant.
15. Resuspend pellet in 900 μL ice-cold TCA/PARGi/olaparib.
16. Place tubes on ice for 15 min.
17. Centrifuge for 20 min at 4°C at 3000×*g* .
18. Transfer supernatant to a new 15 mL tube.
19. Wash the pellet with 1 mL of 70% ice-cold ethanol
20. Remove supernatant and leave pellets to air dry.
21. Add 600 μL 1 M KOH/50 mM EDTA to pellets.
22. Incubate 3 h at 60°C, vortex every 15 min and pipet up and down to mix.
23. Once the KOH treatment is done, add 11.4 mL of AAGE buffer to both the pellet sample and supernatant sample.
24. Adjust pH of the supernatant to 9 by adding NaOH (it usually takes 200 μL of concentrated NaOH, but pH should be verified with pH strips).
25. Resuspend/mix the pre-equilibrated boronate resin (Sect. 3.1.1) by swirling the bottle.
26. Add 150 μL of boronate resin to all samples (supernatants and pellets).
27. Incubate samples on a rotator for 1 h at RT.

28. Centrifuge briefly at 3000×*g*.
29. Discard most of the volume but leave a total of about 1 mL of resin plus liquid.
30. Transfer sample gradually to 750 µL filter spin column, spin and discard fluid until the resin is all loaded on the column.
31. Centrifuge for 1 min at 9000×*g*. Discard flow-through.
32. Add 700 µL AAGE buffer. Spin for 1 min at 9000×*g*. Discard flow-through.
33. Add 700 µL 1 M ammonium acetate buffer (pH 9). Centrifuge for 4 min at 9000×*g*. Discard flow-through. Repeat this step one more time.
34. Transfer filter spin columns to new 1.5 mL tubes.
35. Add 200 µL 37°C ddH_2O to elute. Let filter spin column sit for 1 min, then centrifuge at 9000×*g* for 1 min. Repeat the elution for two more times (3× 200 µL eluates, collect all eluates in the same tubes).
36. Dry down eluates overnight using a vacuum concentrator at 30°C.
37. Resuspend dried samples in 50 µL ddH_2O. Vortex and incubate at 37°C for 30 min. Briefly centrifuge and store at −20°C.
38. Run 25 to 50 µL of the reaction on a 20% acrylamide:bis acrylamide 19:1, 4 M urea gel in 1× TBE. Stain gel with SYBR gold and proceed with imaging immediately.

4 Notes

1. Anneal DNA following published protocol [15].
2. PARP1 activation is sequence independent, therefore, a variety of DNA fragments carrying blunt ends, nicks, or gaps can be used.
3. Like PARP1, PARP2 activation is sequence independent, but PARP2 is preferentially activated by DNA ends carrying a 5'phosphate.
4. If using TNKS, the NAD^+ concentration needs to be increased to a final concentration of 20 mM NAD^+, as TNKS has a higher K_M for NAD^+ than PARP1 and PARP2.
5. Benzonase, DNAse I, and BSA are only needed for in vitro reactions.
6. An Avi-tag can be used to create a biotinylated version of a protein. The purified Avi-tagged protein is incubated with D-biotin, the enzyme BirA, ATP, and $MgCl_2$ to add the biotin group. The modified protein is then repurified using size-exclusion chromatography as described [10].

7. Calculate the ddH_2O volume by taking into account how much protein, DNA, and 5× buffer is added so that the final volume of the reaction after adding NAD^+ is 40 μL.
8. DNA is not needed when TNKS reactions are performed.
9. Various time-points can be performed to optimize the reaction time as the reaction rate can vary depending on which protein is being used.

References

1. Lüscher B, Ahel I, Altmeyer M et al (2021) ADP-ribosyltransferases, an update on function and nomenclature. FEBS J. https://doi.org/10.1111/febs.16142
2. Rouleau-Turcotte É, Pascal JM (2023) ADP-ribose contributions to genome stability and PARP enzyme trapping on sites of DNA damage; paradigm shifts for a coming-of-age modification. J Biol Chem 299:105397. https://doi.org/10.1016/J.JBC.2023.105397
3. Jessop M, Broadway BJ, Miller K, Guettler S (2024) Regulation of PARP1/2 and the tankyrases: emerging parallels. Biochem J 481:1097–1123. https://doi.org/10.1042/BCJ20230230
4. Lüscher B, Bütepage M, Eckei L et al (2018) ADP-ribosylation, a multifaceted posttranslational modification involved in the control of cell physiology in health and disease. Chem Rev 118:1092–1136. https://doi.org/10.1021/acs.chemrev.7b00122
5. Suskiewicz MJ, Prokhorova E, Rack JGM, Ahel I (2023) ADP-ribosylation from molecular mechanisms to therapeutic implications. Cell 186:4475–4495. https://doi.org/10.1016/J.CELL.2023.08.030
6. Palazzo L, Leidecker O, Prokhorova E et al (2018) Serine is the major residue for ADP-ribosylation upon DNA damage. Elife 7. https://doi.org/10.7554/eLife.34334
7. Reber JM, Mangerich A (2021) Why structure and chain length matter: on the biological significance underlying the structural heterogeneity of poly(ADP-ribose). Nucleic Acids Res 49:8432–8448. https://doi.org/10.1093/NAR/GKAB618
8. Rippmann JF, Damm K, Schnapp A (2002) Functional characterization of the poly(ADP-ribose) polymerase activity of tankyrase 1, a potential regulator of telomere length. J Mol Biol 323:217–224
9. Richard IA, Burgess JT, O'Byrne KJ, Bolderson E (2022) Beyond PARP1: the potential of other members of the poly (ADP-ribose) polymerase family in DNA repair and cancer therapeutics. Front Cell Dev Biol 9:801200
10. Langelier MF, Mirhasan M, Gilbert K et al (2024) PARP enzyme de novo synthesis of protein-free poly(ADP-ribose). Mol Cell 84. https://doi.org/10.1016/J.MOLCEL.2024.10.024
11. Tan ES, Krukenberg KA, Mitchison TJ (2012) Large-scale preparation and characterization of poly(ADP-ribose) and defined length polymers. Anal Biochem 428:126–136. https://doi.org/10.1016/j.ab.2012.06.015
12. Wielckens K, Bredehorst R, Adamietz P, Hilz H (1981) Protein-bound polymeric and monomeric ADP-ribose residues in hepatic tissues. Comparative analyses using a new procedure for the quantification of poly(ADP-ribose). Eur J Biochem 117:69–74. https://doi.org/10.1111/J.1432-1033.1981.TB06303.X
13. Juarez-Salinas H, Sims JL, Jacobson MK (1979) Poly(ADP-ribose) levels in carcinogen-treated cells. Nature 282(5740):740–741. https://doi.org/10.1038/282740a0
14. Longarini EJ, Matić I (2024) Preserving ester-linked modifications reveals glutamate and aspartate mono-ADP-ribosylation by PARP1 and its reversal by PARG. Nat Commun 15. https://doi.org/10.1038/S41467-024-48314-0
15. Langelier M-F, Planck JL, Servent KM, Pascal JM (2011) Purification of human PARP-1 and PARP-1 domains from *Escherichia coli* for structural and biochemical analysis. Methods Mol Biol 780:209–226. https://doi.org/10.1007/978-1-61779-270-0_13

Check for updates

Chapter 16

An NMR-Based Approach for Global Arginine Methylation Analysis

Tobias Huberts, Hansjörg Habisch and Tobias Madl

Abstract

Protein arginine methylation (ArgMet) plays a crucial role in the regulation of cellular processes, including transcription, RNA processing, signal transduction, and DNA damage response. However, the mechanisms linking protein ArgMet dynamics to (patho)physiology remain unclear due to the lack of global analysis methods. In this chapter, we present a robust protocol for the quantification of global protein ArgMet, including asymmetric dimethylarginine, symmetric dimethylarginine, and monomethylarginine. After isolation of proteins from biological fluids, tissues, or cell lysates, a straightforward method for homogenization, precipitation, and hydrolysis of proteins is outlined. As the hydrolysates contain a high variability of components, nuclear magnetic resonance (NMR)-based detection was used, providing a robust tool to study arginine methylome with high specificity.

Keywords Arginine methylation, Asymmetric dimethylarginine, Protein hydrolysis, Solid-phase extraction, NMR spectroscopy

1 Introduction

Posttranslational modifications (PTMs) are key regulatory mechanisms that significantly affect protein function, stability, and interactions within cells. These modifications can alter protein activity, localization, and their ability to interact with other cellular components, playing key roles in various cellular processes [1, 2]. Therefore, quantitative information on the levels and dynamics of PTMs is essential for understanding cellular functions. Probably the most well-known PTM is (de-)phosphorylation. Another example is ArgMet, which plays a key role in regulating processes like transcription, RNA processing, signal transduction, DNA damage response, liquid–liquid phase separation (LLPS), and apoptosis [3–7]. This modification involves the transfer of methyl groups to the guanidine group of arginine side chains in proteins. Arginine methylations are catalyzed by protein arginine methyltransferases (PRMTs) and have a major impact on the structure, stability, function, and physicochemical

Xu Zhang et al. (eds.), *Protein Post-Translational Modifications*, Methods in Molecular Biology, vol. 3018,
https://doi.org/10.1007/978-1-0716-5166-7_16,

properties of proteins, affecting their interactions with other molecules, including proteins and nucleic acids [3, 4].

The number and types of PRMTs vary between species, from unicellular eukaryotes to humans. While yeast has at least one or two primary PRMTs (HMT1/RMT1 and potentially Hsl7), mammals have a larger family of nine PRMTs, which can be classified into four main types based on the type of methylated arginine they form (Fig. 1) [4, 8]. Type I PRMTs, including PRMT1, PRMT2, PRMT3, PRMT4 (also known as CARM1), PRMT6, and PRMT8, are responsible for the formation of ω-monomethylarginine and ω,ω-asymmetric dimethylarginine. Type II PRMTs, such as PRMT5, PRMT9, or

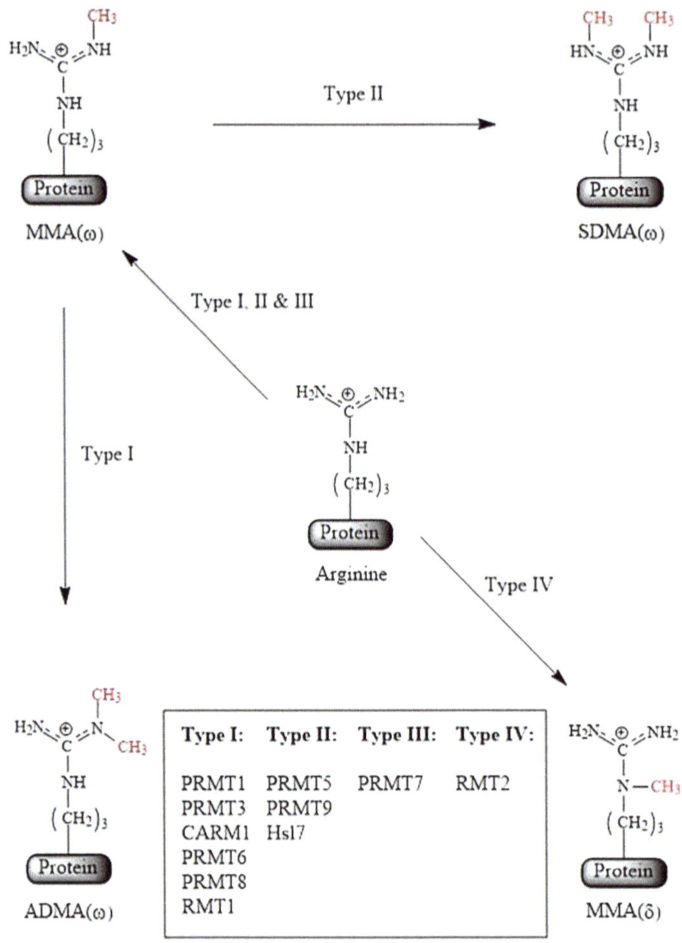

Fig. 1 Four types of PRMTs and the known enzymes involved in these processes. Types I, II, and III PRMTs produce MMA(ω) by adding a methyl group to one of the two terminal (ω) nitrogen atoms of the guanidino group. These nitrogens are chemically identical. Type I PRMTs can further convert MMA(ω) to ADMA(ω), while type II PRMTs produce SDMA(ω). In contrast, type III PRMTs only catalyze the formation of MMA(ω) and do not proceed to dimethylation. A distinct type IV activity, involving monomethylation of the internal nitrogen of the guanidino group, has been observed exclusively in yeast

Hsl7, catalyze the formation of MMA and ω,ω-symmetric dimethylarginine [9]. Type III PRMTs, such as PRMT7, catalyze only the methylation to MMA [10]. In yeast, only one type IV PRMT, RMT2, has been identified and is known to methylate the delta nitrogen of arginine residues [11].

Most PRMTs target protein regions rich in glycine and arginine, commonly referred to as RG or RGG motifs [12–14]. These motifs are found in over a thousand human proteins, particularly those involved in RNA binding [15, 16]. However, glycine is not necessarily required to be adjacent to arginine for methylation; PRMT1, for instance, can modify RXG motifs, where "X" is typically lysine, phenylalanine, threonine, or leucine [17–19]. While PRMT7 has shown to methylate substrates with motifs containing two arginine residues separated by another amino acid (RXR motifs), other sequences such as RPAAPR and APR have also been identified as methylation sites [20, 21]. PRMT4/CARM1 tends to methylate arginine residues in regions rich in proline, glycine, and methionine [22, 23]. PRMT5, based on mass spectrometry data, prefers glycine-flanked arginine (GRG) motifs [14, 24], while PRMT6 targets arginine residues within positively charged domains and avoids regions with nearby acidic residues [25]. Although intrinsically disordered protein regions often undergo ArgMet, it is not their only site of action [26].

The resulting derivatives monomethylarginine (MMA), asymmetric dimethylarginine (ADMA), and symmetric dimethylarginine (SDMA), are linked to a broad spectrum of cellular processes [3, 4]. Dysregulated PRMT activity is related to the development of multiple diseases, including cardiovascular, metabolic, and neurodegenerative disorders, as well as viral infections, and various forms of cancer [27]. Due to the relevance of ArgMet, a comprehensive investigation of these methylation dynamics and their associated metabolomic changes is crucial.

Current methods predominantly use antibodies to detect ArgMet. Nevertheless, these antibodies exhibit context-dependent binding specificity, potentially limiting their ability to detect diverse sequence motifs surrounding methylated arginine residues [28, 29]. Another possibility for the quantitative measurement of ArgMet levels is genetically encoded fluorescence resonance energy transfer (FRET)-based biosensors [30, 31]. These advanced genetically encoded biosensors are specifically designed to monitor the real-time dynamics of intracellular ArgMet in living cells, enabling sensitive detection of spatial and temporal variations in ArgMet levels. However, its sensitivity remains insufficient to enable precise absolute quantification of global ArgMet with high accuracy. In addition to immunoaffinity-based techniques, several chromatography-based methods have been used to enrich methylated proteins. Strong cation exchange (SCX) chromatography shows potential for enrichment of lysine and ArgMet [32–36]. However, this method is significantly affected by histidine-containing peptides, which also carry additional positive charges under SCX separation conditions, leading to reduced specificity [35]. Hydrophilic interaction chromatography (HILIC) is another enrichment approach, although the identification of methylated peptides is limited [37]. Steric effect-based chemical enrichment (SECEM), which uses 1,2-cyclohexanedione to block unmethylated arginine, improves specificity for dimethylated peptides but also blocks monomethylated residues, preventing their enrichment [38]. Despite these advances, comprehensive and selective enrichment methods are still required for sensitive detection of all forms of ArgMet.

Recent advances in analytical techniques, such as nuclear magnetic resonance (NMR) spectroscopy, have provided powerful tools for studying the quantitative aspects of protein ArgMet. NMR covers metabolites in the high nanomolar—to micromolar concentration range with outstanding reproducibility, making it an ideal technique for the quantification of methylated arginine derivatives in complex biological samples. The ability to detect protein ArgMet levels in a variety of matrices, including cells, tissues, and biofluids, offers a promising approach for investigating the role of ArgMet in health and disease [39–41].

Given the broad impact of protein ArgMet on cellular processes and its implications in ubiquitous diseases such as cancer, cardiovascular and neurodegenerative diseases, the development of reliable and scalable methods to investigate protein ArgMet is essential. This understanding could lead to better therapeutic strategies, including the development of PRMT inhibitors, and enhance our knowledge of how ArgMet modulates cellular functions.

Therefore, this chapter describes an analytical technique to quantify the total content of the methylated arginine derivatives ADMA, SDMA, and MMA using NMR spectroscopy. This technique includes an acidic hydrolysis of the proteins as a pretreatment to generate free amino acids [42]. Solid-phase extraction (SPE) allows selective enrichment of target compounds [43] and subsequent ^{1}H-NMR spectroscopy, as a quantitative, highly reproducible, and robust technique, allowing the simultaneous detection of arginine and its methylated derivatives [44, 45].

The method outlined in this protocol has been successfully applied to a variety of research topics to investigate ArgMet profiles. These include the differentiation of ArgMet phenotypes in frontotemporal dementia and Alzheimer's disease [46], the impact of elevated plasma homocysteine and vitamin/choline-deficient diets on ArgMet levels in blood and liver tissues [47], and the support of cell line characterization and assessment of PRMT1 inhibitor antitumor activity [48]. Furthermore, this method has been employed to analyze homocysteine-related methylation patterns in Drosophila and yeast models, demonstrating its versatility in both clinical and model organism studies [49].

2 Materials

All solutions are prepared using ultrapure water, with a resistance of ≥18 MΩ·cm^{-1} at 25°C, corresponding to double-distilled grade, unless specified otherwise. Waste materials must be disposed of in strict accordance with applicable waste disposal regulations. Pipettes with a working range of 10–1000 μL should be prepared in advance, along with their corresponding sterile tips. All procedures involving concentrated hydrochloric acid (HCl) were conducted within a certified chemical fume hood. An overview of the whole protocol is shown in Fig. 2.

2.1 Solid Sample and Cell Preparation

1. Phosphate-buffered saline (PBS).
2. 1.5 mL Eppendorf tubes.
3. Microcentrifuge.

2.2 Initial Sample Preparation

1. Methanol (MeOH) ≥99% p.a.
2. 1.5 mL Eppendorf tubes.
3. Binary solvent mixture: combining two parts of ≥99% methanol with one part of deionized water (H2O/MeOH) (*see* **Note 1**).
4. Precellys 2 mL reinforced pulping tubes filled with Zirconium oxide beads (1.4 mm diameter) (*see* **Note 2**).
5. Precellys instrument (Bertin) or equivalent tissue homogenizer.

2.3 Acidic Hydrolysis

1. 9 M Hydrochloric acid.
2. 1.5 mL Eppendorf tubes.
3. Beaker filled with sand.
4. Duran® Glass tubes, diameter 18 mm.
5. Appropriate screw caps with gray polytetrafluoroethylene (PTFE) sealing foil.
6. Sea sand.
7. Drying cabinet
8. Chloroform ≥99% p.a.
9. Refrigerated Cooling Trap.
10. SpeedVac vacuum concentrator.
11. Deep vacuum oil pump.
12. Sodium hydroxide (NaOH).

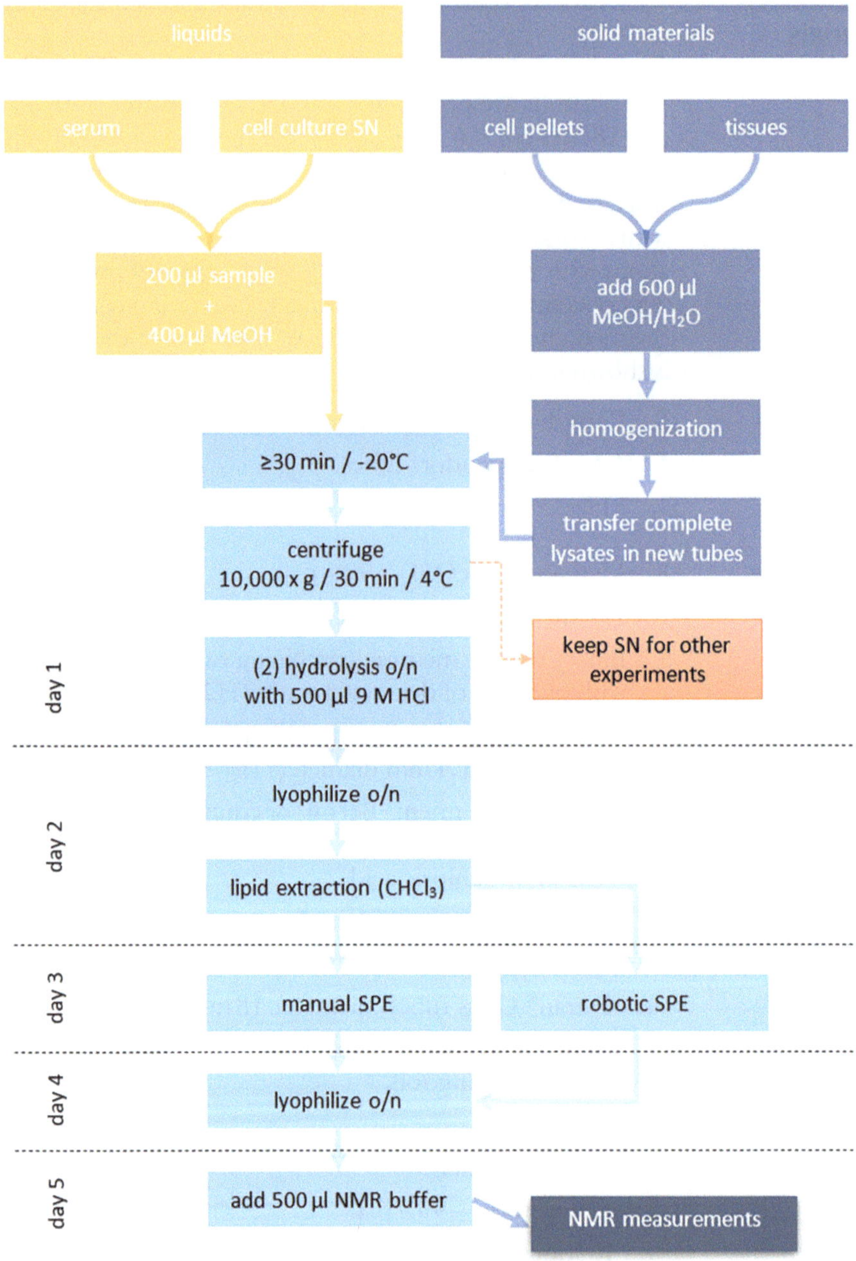

Fig. 2 Overview of the sample preparation outlining the key steps involved from initial sample collection through to NMR analysis. Estimated durations for each phase are indicated, with these durations being subject to variation depending on the number of samples processed. If liquid samples are not homogeneous, a preliminary centrifugation step can be performed to remove debris—this step is not illustrated. The supernatant of the resultant mixture, obtained subsequent to the initial centrifugation process, which may contain intracellular compounds or unbound arginine metabolites, can be stored at −20°C for subsequent use. In most cases, the pellets of methanol precipitate from step 2 are then subjected to further analysis. Abbreviations: *$CHCl_3$* chloroform, *HCl* hydrochloric acid, *MeOH* methanol, *NMR* nuclear magnetic resonance, *o/n* overnight, *SN* supernatant, *SPE* solid-phase extraction

2.4 Solid-Phase Extraction

1. 0.1 M Hydrochloric acid.
2. Replacement solution for SPE: Mix 10% saturated aqueous ammonia solution (prepared from a 30% w/w stock), 50% methanol, and 40% water.
3. Phosphate-buffered saline (PBS).
4. 15 mL tubes.
5. Gilson pipetting robot GX-241 Aspec or equivalent liquid handling robot.
6. Oasis MCX 1cc/30 mg, 1 mL cartridges.

2.5 NMR Measurements

1. NMR buffer: Dissolve 5.56 g of disodium hydrogen phosphate (Na_2HPO_4, final concentration 0.08 M), 0.4 g of 3-(trimethylsilyl)propionic acid-2,2,3,3-d4 sodium salt (TSP, 5 mM), and 0.04% (w/v) sodium azide (NaN_3) in 500 mL of deuterium oxide (D_2O). Adjust the pH to 7.4 using HCl (12 M or 37%). Evaluate the freshly prepared buffer batch by ^{1}H-NMR spectroscopy to confirm purity (*see* **Note 3**).
2. Borosilicate glass NMR tube (outer diameter: 5 mm, wall thickness: 0.38 mm, length: 103.5 mm).
3. Bruker 600 MHz NMR spectrometer, equipped with a TXI probe head.
4. Dimethyl sulfoxide-d_6 (d_6-DMSO).

3 Methods

3.1 Cell Preparation

1. Harvest 3–5 million cells from culture. For adherent cells, wash them with cold PBS, scrape, and centrifuge at 10,000×*g* for 10 min at 4°C. For suspension cultures, simply centrifuge the cells.
2. Separate the supernatant from the solid cells by pipetting into a separate tube (*see* **Note 4**).
3. Flash-freeze the resulting cell pellets and supernatants in liquid nitrogen and store them at −80°C.

3.2 Solid Sample Preparation

1. Collect and store the solid sample, e.g., tissue or cell pellets, according to the experimental requirements. Use a sample amount of 30–70 mg for the following sample preparation protocol.
2. For uniform tissues, such as liver or muscle, any section can be used. For more complex tissues, such as mouse brain, either analyze the whole brain or divide it into sections to assess global ArgMet (*see* **Note 5**).

3. Store the solid samples (tissues or cell pellets) at −80°C until the commencement of the sample preparation.

3.3 Initial Sample Preparation

1. For liquid samples, thaw the frozen samples and mix 200 μ L with 400 μL of ice-cold methanol (−20°C) in a 1.5 mL Eppendorf tube and continue with Step 5 (*see* **Note 6**). If the sample volume is lower than 200 μL, adjust the methanol proportionally in a ratio of 1:2.
2. For cell pellets, thoroughly resuspend in 600 μL of a binary solvent mixture (H_2O/MeOH, 1:2), and transfer the suspension to pulping tubes. For tissue samples, place them directly into pulping tubes (weighing them if necessary), and add 600 μL of binary solvent mixture per 30–60 mg of tissue. Adjust the volume if the tissue weight varies significantly. Always store the samples on ice.
3. Place tubes in a tissue homogenizer and homogenize as follows. For soft tissues or cell pellets, shake vigorously for 20 s once. For stiffer tissues, homogenize for 20 s twice, with a 5-min cooling interval on ice between sessions. If the sample is still not homogenized, repeat the procedure (*see* **Note 7**).
4. After homogenization, put the tubes on ice immediately and transfer the lysates to new 1.5 mL Eppendorf tubes.
5. Store the prepared samples at −20°C for at least 30 min (*see* **Note 8**).
6. Centrifuge the samples at 10,000×*g* for 30 min at 4°C.
7. Separate the supernatant from the cell pellet by pipetting the liquid phase into another tube and use the supernatant for further analysis or discard it, as appropriate (*see* **Note 9**).
8. Store the pellet at −20°C or immediately continue with the acidic hydrolysis (*see* **Note 10**).

3.4 Acidic Hydrolysis

1. Thaw the sample to room temperature and add 500 μL of 9 M HCl to each sample. Using scissors, carefully cut the caps off the Eppendorf tubes containing the sample and place them in the glass culture tubes. Securely close the tubes with the red screw caps (*see* **Note 11**).
2. Once all samples are sealed, place the tubes in a beaker partially filled with sand to improve heat distribution. Place the beaker in a drying chamber and incubate the samples at 110°C for approximately 16 h to complete hydrolysis (Fig. 3A).
3. After hydrolysis, allow the samples to cool down to room temperature for approximately 1 h.
4. Take out the 1.5 mL tubes from the glass tubes and lyophilize the samples overnight using a vacuum centrifuge capable of attaining pressures below 100 Pa (*see* **Note 12**).

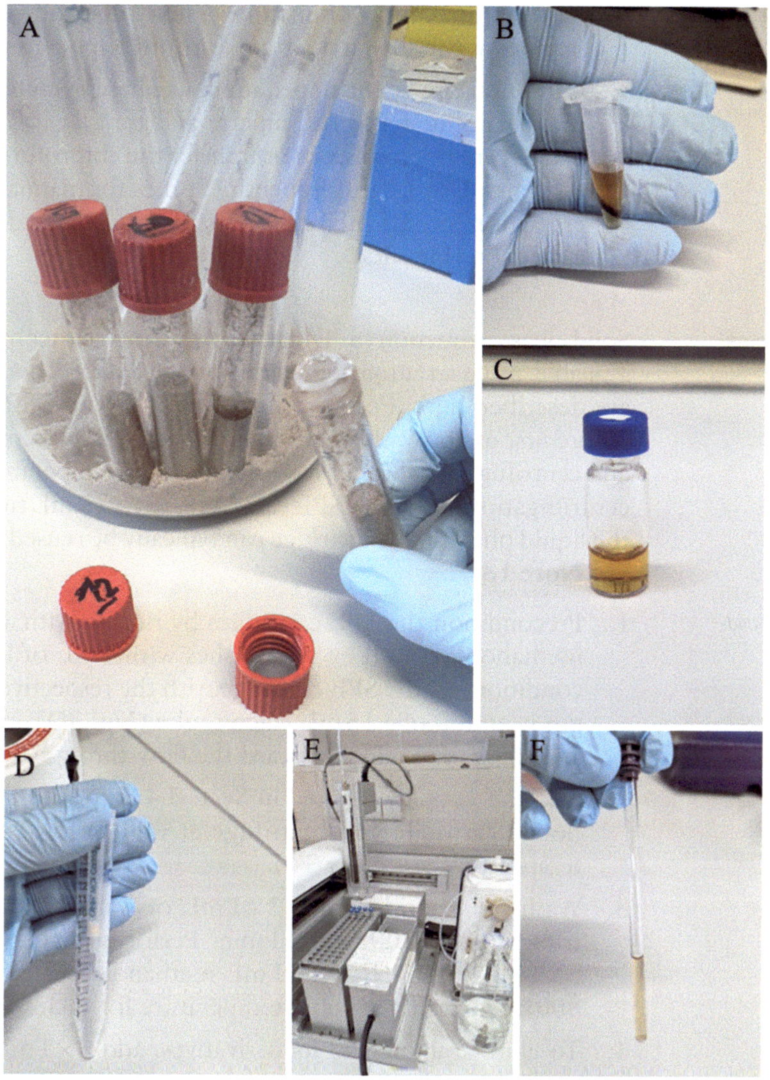

Fig. 3 Preparation and treatment of protein hydrolysates for solid phase extraction: **(A)** A setup showing a sand-filled beaker, glass tubes with screw caps lined with grey polytetrafluoroethylene (PTFE) sealing foil, and sample tubes (1.5 mL) with caps removed. **(B)** A sample tube after hydrolysis and centrifugation. The dark residue is composed of insoluble charred organic matter. **(C)** Aqueous phase after transfer in a glass vial for automated SPE. In some cases, samples may retain colored compounds. **(D)** Setup for the manual SPE by sequential centrifugation steps showing a SPE cartridge in a 15 mL falcon. **(E)** Pipetting robot for automated SPE. **(F)** Final sample dissolved in D_2O buffer in a 5 mm NMR tube

5. After 16 h ensure the pellets are fully desiccated, otherwise continue lyophilization.
6. Dissolve the pellets with 1000 μL of 0.1 M HCl.
7. Add 50 μL of chloroform ($CHCl_3$) to each tube and use a 1000 μL pipette to thoroughly mix and dissolve the pellet completely.

8. Transfer the entire volume into a new tube and centrifuge the samples at 8000×*g* for 10 min at room temperature (Fig. 3B).
9. Carefully remove the upper aqueous phase from the two-phase mixture using a 1000 μL pipette and transfer it to new tubes, taking care not to disturb or transfer the chloroform layer or its contents. Use 1.5 mL Eppendorf tubes for manual solid-phase extraction or 1.5 mL glass vials for automated solid-phase extraction (Fig. 3C).

3.5 Solid-Phase Extraction

Solid-phase extraction (SPE) can be performed either manually or, if available, using an automated SPE robot. Therefore, both procedures are described here. For both methods, clean new SPE cartridges before first use by washing twice with 1 mL of replacement solution and centrifuging at 800×*g* for 1 min at room temperature. For each centrifugation step of an SPE cartridge, use a 15 mL tube to collect the liquid phase. Each cartridge can typically be reused 10–20 times (*see* **Note 13**).

3.5.1 Manual Solid-Phase Extraction

1. Precondition the SPE cartridges by rinsing with 1 mL of pure methanol followed by two washes with 1 mL of PBS. To precondition, fill the SPE cartridge with the respective solvent and put it on top of a 15 mL Falcon tube (Fig. 3D) and centrifuge at 800×*g* for 1 min to discard the flow-through.
2. Add the sample collected in Sect. 3.4 Step 9 to the preconditioned cartridge and centrifuge at 600×*g* for 2 min at room temperature (*see* **Note 14**).
3. Wash the cartridge with 3×1 mL of water followed by centrifugation at 800×*g* for 1 min. Perform further washes with 5×1 mL 0.1 M HCl, 2×1 mL methanol, and centrifuging at 800×*g* for 1 min at room temperature for each wash.
4. To elute arginine and its derivatives, add 3×1 mL of replacement solution to the cartridge, centrifuge at 800×*g* for 1 min after each addition. Collect the eluate in a new 15 mL tube.
5. Lyophilize the purified samples overnight using a vacuum centrifuge capable of attaining pressures below 100 Pa to remove any remaining ammonia and water.

3.5.2 Automated Solid-Phase Extraction

1. Transfer the samples collected in Sect. 3.4 Step 9 to glass vials with PTFE-sealed caps and place them in the pipetting robot (Fig. 3E) (*see* **Note 15**).
2. Prepare sufficient volumes (approximately 100 mL each) of all required reagents, including replacement solution, 99% methanol, PBS, water, and 0.1 M HCl. Also, prepare 5 mL collection tubes for eluates and water to wash the pipetting needle.
3. Program the robot using a method that includes the following steps, in accordance with the manual SPE (*see* **Note 16**).

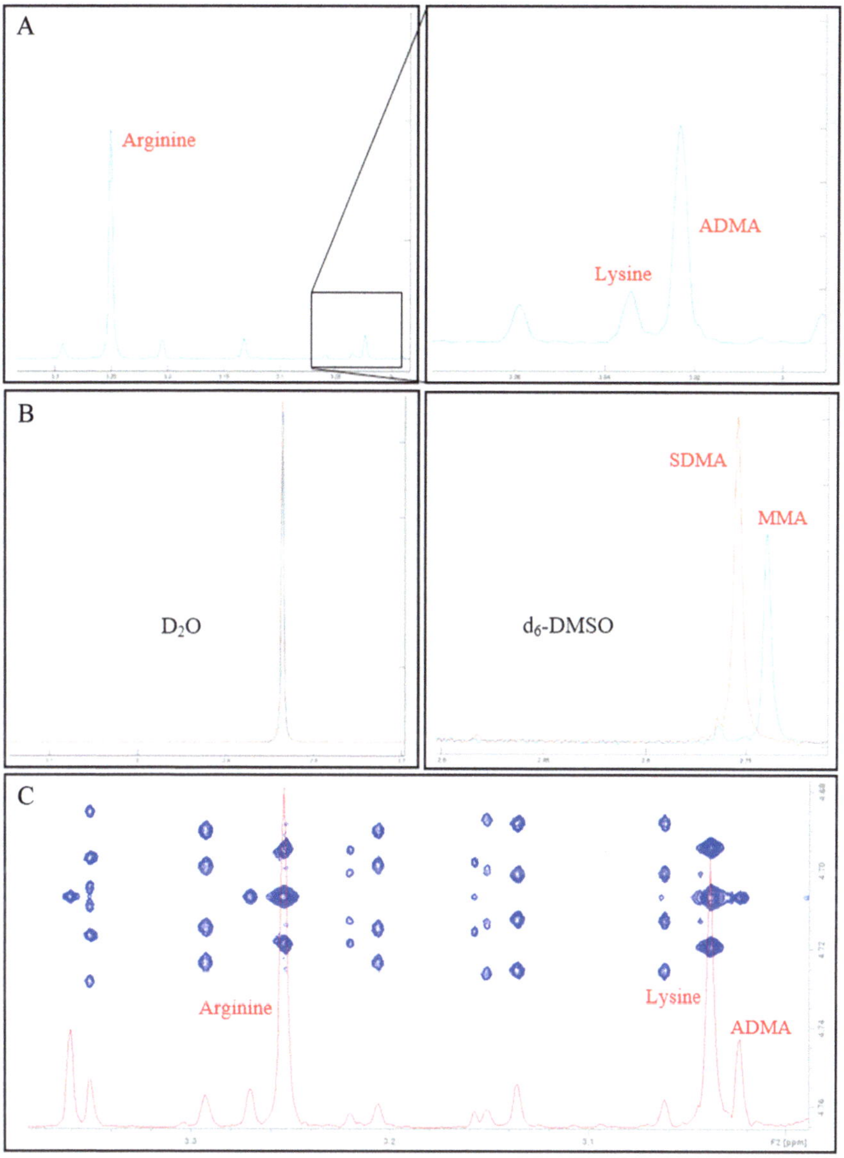

Fig. 4 Representative NMR spectra of arginine and its methylated derivatives: One-dimensional ^{1}H J-resolved NMR spectra allow a clear distinction between the δ-(CH_2) protons of arginine (at 3.25 ppm) and the ω-N-CH_3 protons of ADMA, at 3.02 ppm. **(A)** Yeast pellet sample can be quantitatively analyzed by integrating the signal intensities, even when arginine is present in large amounts. **(B)** Because SDMA and MMA are difficult to distinguish in D_2O buffer, the samples are instead dissolved in d_6-DMSO, where their signals (shown here for 100 μmol/L of each) appear clearly separated at 2.75 and 2.76 ppm, allowing accurate quantification. All spectra shown were exported directly from the NMR spectrometer's control software, providing the initial data that must be verified before more detailed, large-scale statistical analysis can be performed. **(C)** Overlap of a 2D-JRES spectrum and the resulting 1H-decoupled 1D spectrum after processing of an *E. coli* pellet hydrolysate without SPE, spiked with ADMA (100 μmol/L). As the lysine peak is normally obtained as a triplet due to the two hydrogen atoms at the binding carbon atom, there may be some overlap with the ADMA peak. Therefore, a 1 H decoupled proton spectrum is used for analysis, which allows a clear distinction between lysine and ADMA

The SPE cartridges should be preconditioned with 1 mL of pure methanol followed by two washes with 1 mL of PBS. Afterward, the cartridge should be loaded with the sample, followed by washing steps with 3 × 1 mL water, 5 × 1 mL 0.1 M HCl, and 2 × 1 mL methanol. Finally, the arginine and its derivatives should be eluted and collected using 3 × 1 mL replacement solution.

4. Lyophilize the purified samples overnight using a vacuum centrifuge capable of attaining pressures below 100 Pa to remove any remaining ammonia and water.

3.6 NMR Measurements

1. Reconstitute each sample in 500 μL of NMR buffer and transfer the solution into NMR tubes, ensuring the volume is consistent across all samples (Fig. 3F). The solution must be homogeneous, without any visible residue or lipid aggregates that may interfere with the spectra.
2. Acquire a 1D ^{1}H NMR spectrum using the CPMG (Carr–Purcell–Meiboom–Gill) pulse sequence (cpmgpr1d, 128 scans, size of fid 73728, 11904.76 Hz spectral width, recycle delay 4 s) on a 600 MHz NMR and apply presaturation to suppress the water signal. In order to eliminate ^{1}H scalar couplings in these experiments, we use decoupling by JRES (jresgpprqf, 16 scans, size of fid 16384 and 64, 10000.00 Hz and 78.042 spectral width recycle delay 2 s), which helps to minimize signal overlap and improve spectral clarity (Fig. 4A) (*see* **Note 17**).
3. For absolute quantification by integration of the respective peak areas, use standards of known concentration (e.g., 100 μ M arginine) (*see* **Note 18**).
4. To differentiate between SDMA and MMA, remove any D_2O from the previous NMR buffer by lyophilizing the samples overnight using a vacuum centrifuge (*see* **Note 19**). Reconstitute the dried samples in 500 μL of deuterated dimethyl sulfoxide (d6-DMSO) and repeat the previous NMR procedures (Fig. 4B).

4 Notes

1. Store both pure MeOH and the H_2O/MeOH mixture at −20°C to minimize methanol volatilization and to ensure sample stability.
2. Fill the tubes for tissue or sample homogenization with approximately 10–20 Zirconium oxide beads (1.4 mm diameter). Alternatively, prefilled tubes can be used when available.

3. Take special care to minimize contamination in NMR buffer from volatile compounds, such as ethanol, as they may interfere with spectral analysis.
4. The supernatant can be used for further analysis, either measured directly by NMR spectroscopy or used after soluble protein precipitation.
5. For whole brain analysis, for example, freeze and pulverize the entire brain and take aliquots. Alternatively, dissected regions, such as the frontal lobe, cerebellum, occipital lobe, or an entire hemisphere, can be used.
6. Following the collection of liquid samples, such as serum, plasma, or cell culture supernatants, immediately snap-freeze the samples using liquid nitrogen to preserve their integrity, and store them at −80°C until they are ready to be processed or analyzed.
7. This procedure should not be repeated excessively, even if some insoluble material remains. To avoid possible sample degradation, it is recommended to centrifuge the samples for 10 min at 10,000×*g* and separate the suspension from the remaining tissue.
8. After the homogenization process, the samples can be stored at −20°C for several days without any risk of damage or changes in composition.
9. The corresponding supernatant can be discarded, lyophilized, processed directly for NMR analysis, or used for other types of analysis depending on the approach. It can be used for the quantification of metabolites in complex matrices by ^{1}H-NMR spectroscopy, as NMR spectroscopy represents a reliable and rapid analytical method with robustness. However, this protocol focuses on the use of the methanol-insoluble pellet containing components, including nucleic acids, lipids, and proteins.
10. In this state, the cell pellets can be stored at −20°C for several days without risk of damage or change in composition. However, prolonged storage in this state may impede resuspension in the next step.
11. Always inspect the red screw caps, especially the gray PTFE lining, to ensure they are properly sealed and nothing of the sample can leak out. Clean them regularly with water to maintain a tight closure.
12. In order to protect the vacuum pump from the gaseous hydrochloric acid, it is highly recommended to place a cooling trap between the SpeedVac and the vacuum pump. Furthermore, prefilling the cooling trap with 6 M NaOH solution will provide additional protection by neutralizing the hydrochloric

acid immediately after it has been collected in the cooling trap.

13. The replacement solution has a dual function. It elutes the target analytes and also acts as a cleaning agent due to its basic pH-value. Although cartridges are reusable, prolonged reuse should be avoided to minimize the risk of contamination.
14. The limit of detection for ADMA is approximately 100 nmol/L, as previously determined by serial dilution experiments using the established NMR acquisition settings [39]. However, at very high arginine concentrations above 3 mmol/L, the binding capacity of the SPE column may become saturated, potentially leading to an overestimation of the ADMA/arginine ratio.
15. A pipetting robot setup typically includes a pipetting needle, washing station, reagent reservoirs, and designated positions for sample and eluate collection. If a different system is used, ensure that it is equipped with equivalent components. Robotic SPE systems typically apply air pressure, rather than centrifugation, to move liquids through the cartridges. When setting up the protocol for the first time, include quality controls (e.g., with known concentrations of arginine) to validate the methods performance.
16. As protocols may differ depending on the robot manufacturer, refer to the instrument software manual for specific programming instructions.
17. If peaks of interest overlap with signals from other constituents of the hydrolysates or more detailed investigations for complex matrices are necessary, then further NMR-experiments can be performed. Detailed NMR procedures are beyond the scope of this protocol and can be found in the references provided [50–52]. This section provides only the essential steps and requirements for sample preparation. Quantification of arginine and its derivatives is typically done using a 600 MHz NMR spectrometer, but other field strengths can also be used. Any probe head with z-axis gradients that supports ^{1}H spectroscopy is suitable, assuming appropriate pulse sequences are used. For complex matrices or more detailed investigations, consider recording 2D spectra, such as ^{1}H-^{13}C HSQC (hsqcetgpsisp2). If the methylated arginine peaks overlap with signals from other metabolites (e.g., lysine), acquire J-resolved (JRES) ^{1}H spectra [52] for better peak resolution.
18. In general, ADMA, SDMA, and MMA are quantified relative to arginine, which is used as an internal standard. This method eliminates the need for normalization based on cell

number, protein content, or tissue mass, making data interpretation straightforward.

19. As demonstrated, arginine and ADMA can be clearly distinguished by their unique ^{1}H NMR chemical shifts at 3.25 ppm and 3.02 ppm, respectively. In contrast, SDMA and MMA show overlapping methyl proton signals around 2.85 ppm when measured in D2O-based NMR buffer. The two compounds can be resolved by redissolving the samples in d6-DMSO. In this solvent, SDMA and MMA show distinct and quantifiable signals at 2.76 ppm (SDMA) and 2.74 ppm (MMA), respectively.

References

1. Mann M, Jensen ON (2003) Proteomic analysis of post-translational modifications. Nat Biotechnol 21(3):255–261. https://doi.org/10.1038/nbt0303-255
2. Ramazi S, Zahiri J (2021) Post-translational modifications in proteins: resources, tools and prediction methods. Database 2021:baab012. https://doi.org/10.1093/database/baab012
3. Bachand F (2007) Protein arginine methyltransferases: from unicellular eukaryotes to humans. Eukaryot Cell 6(6):889–898. https://doi.org/10.1128/ec.00099-07
4. Bedford MT, Clarke SG (2009) Protein arginine methylation in mammals: who, what, and why. Mol Cell 33(1):1–13. https://doi.org/10.1016/j.molcel.2008.12.013
5. Bedford MT, Richard S (2005) Arginine methylation: an emerging regulator of protein function. Mol Cell 18(3):263–272. https://doi.org/10.1016/j.molcel.2005.04.003
6. Lee J, Sayegh J, Daniel J et al (2005) PRMT8, a new membrane bound tissue-specific member of the protein arginine methyltransferase family. J Biol Chem 280(38):32890–32896. https://doi.org/10.1074/jbc.M506944200
7. Pahlich S, Zakaryan RP, Gehring H (2006) Protein arginine methylation: cellular functions and methods of analysis. Biochim Biophys Acta (BBA) Proteins Proteom 1764(12):1890–1903. https://doi.org/10.1016/j.bbapap.2006.08.008
8. Guccione E, Richard S (2019) The regulation, functions and clinical relevance of arginine methylation. Nat Rev Mol Cell Biol 20(10):642–657. https://doi.org/10.1038/s41580-019-0155-x
9. Sayegh J, Clarke SG (2008) Hsl7 is a substrate-specific type II protein arginine methyltransferase in yeast. Biochem Biophys Res Commun 372(4):811–815. https://doi.org/10.1016/j.bbrc.2008.05.121
10. Zurita-Lopez CI, Sandberg T, Kelly R et al (2012) Human protein arginine methyltransferase 7 (PRMT7) is a type III enzyme forming ω-NG-monomethylated arginine residues*. J Biol Chem 287(11):7859–7870. https://doi.org/10.1074/jbc.M111.336271
11. Chern M-K, Chang K-N, Liu L-F et al (2002) Yeast ribosomal protein L12 is a substrate of protein-arginine methyltransferase 2. J Biol Chem 277(18):15345–15353. https://doi.org/10.1074/jbc.M111379200
12. Feng Y, Maity R, Whitelegge JP et al (2013) Mammalian protein arginine methyltransferase 7 (PRMT7) specifically targets RXR sites in lysine- and arginine-rich regions. J Biol Chem 288(52):37010–37025. https://doi.org/10.1074/jbc.M113.525345
13. Guo A, Gu H, Zhou J et al (2014) Immunoaffinity enrichment and mass spectrometry analysis of protein methylation. Mol Cell Proteomics 13(1):372–387. https://doi.org/10.1074/mcp.O113.027870
14. Tewary SK, Zheng YG, Ho M-C (2019) Protein arginine methyltransferases: insights into the enzyme structure and mechanism at the atomic level. Cell Mol Life Sci 76(15):2917–2932. https://doi.org/10.1007/s00018-019-03145-x
15. Chong PA, Vernon RM, Forman-Kay JD (2018) RGG/RG motif regions in RNA binding and phase separation. J Mol Biol 430(23):4650–4665. https://doi.org/10.1016/j.jmb.2018.06.014
16. Thandapani P, O'Connor TR, Bailey Timothy L et al (2013) Defining the RGG/RG Motif. Mol Cell 50(5):613–623. https://doi.org/10.1016/j.molcel.2013.05.021
17. Uhlmann T, Geoghegan VL, Thomas B et al (2012) A method for large-scale identification of protein arginine methylation. Mol Cell Proteomics 11(11):1489–1499. https://doi.org/10.1074/mcp.M112.020743

18. Wooderchak WL, Zang T, Zhou ZS et al (2008) Substrate profiling of PRMT1 reveals amino acid sequences that extend beyond the "RGG" paradigm. Biochemistry 47(36):9456–9466. https://doi.org/10.1021/bi800984s
19. Chang K, Gao D, Yan J et al (2023) Critical roles of protein arginine methylation in the central nervous system. Mol Neurobiol 60(10):6060–6091. https://doi.org/10.1007/s12035-023-03465-x
20. Lee J, Bedford MT (2002) PABP1 identified as an arginine methyltransferase substrate using high-density protein arrays. EMBO Rep 3(3):268–273. https://doi.org/10.1093/embo-reports/kvf052
21. Bondoc TJ, Lowe TL, Clarke SG (2023) The exquisite specificity of human protein arginine methyltransferase 7 (PRMT7) toward Arg-X-Arg sites. PLoS One 18(5):e0285812. https://doi.org/10.1371/journal.pone.0285812
22. Cheng D, Côté J, Shaaban S et al (2007) The arginine methyltransferase CARM1 regulates the coupling of transcription and mRNA processing. Mol Cell 25(1):71–83. https://doi.org/10.1016/j.molcel.2006.11.019
23. Shishkova E, Zeng H, Liu F et al (2017) Global mapping of CARM1 substrates defines enzyme specificity and substrate recognition. Nat Commun 8(1):15571. https://doi.org/10.1038/ncomms15571
24. Musiani D, Giambruno R, Massignani E et al (2020) PRMT1 Is recruited via DNA-PK to chromatin where it sustains the senescence-associated secretory phenotype in response to cisplatin. Cell Rep 30(4):1208–1222.e9. https://doi.org/10.1016/j.celrep.2019.12.061
25. Hamey JJ, Rakow S, Bouchard C et al (2021) Systematic investigation of PRMT6 substrate recognition reveals broad specificity with a preference for an RG motif or basic and bulky residues. FEBS J 288(19):5668–5691. https://doi.org/10.1111/febs.15837
26. Low JKK, Im H, Erce MA et al (2016) Protein substrates of the arginine methyltransferase Hmt1 identified by proteome arrays. Proteomics 16(3):465–476. https://doi.org/10.1002/pmic.201400564
27. Blanc RS, Richard S (2017) Arginine methylation: the coming of age. Mol Cell 65(1):8–24. https://doi.org/10.1016/j.molcel.2016.11.003
28. Lee Y-H, Stallcup MR (2009) Minireview: protein arginine methylation of nonhistone proteins in transcriptional regulation. Mol Endocrinol 23(4):425–433. https://doi.org/10.1210/me.2008-0380
29. Bhatter N, Roy R, Shah S et al (2019) Arginine methylation augments Sbp1 function in translation repression and decapping. FEBS J 286(23):4693–4708. https://doi.org/10.1111/febs.15057
30. Zhang F, Bischof H, Burgstaller S et al (2024) Genetically encoded fluorescent sensor to monitor intracellular arginine methylation. J Photochem Photobiol B 252:112867. https://doi.org/10.1016/j.jphotobiol.2024.112867
31. Sun X, Chen F, Zhang L et al (2023) A gene-encoded FRET fluorescent sensor designed for detecting asymmetric dimethylation levels in vitro and in living cells. Anal Bioanal Chem 415(8):1411–1420. https://doi.org/10.1007/s00216-023-04541-w
32. Wang K, Dong M, Mao J et al (2016) Antibody-free approach for the global analysis of protein methylation. Anal Chem 88(23):11319–11327. https://doi.org/10.1021/acs.analchem.6b02872
33. Li Z, Wang Q, Wang Y et al (2021) An efficient approach based on basic strong cation exchange chromatography for enriching methylated peptides with high specificity for methylproteomics analysis. Anal Chim Acta 1161:338467. https://doi.org/10.1016/j.aca.2021.338467
34. Wang Q, Li Z, Zhou J et al (2021) Chemical depletion of histidine-containing peptides allows identification of more low-abundance methylation sites from proteome samples. J Proteome Res 20(5):2497–2505. https://doi.org/10.1021/acs.jproteome.0c00976
35. Deng Z, Dong M, Wang Y et al (2017) Biphasic affinity chromatographic approach for deep tyrosine phosphoproteome analysis. Anal Chem 89(4):2405–2410. https://doi.org/10.1021/acs.analchem.6b04288
36. Liu Z, Fang Z, Wang K et al (2023) Hydrophobic derivatization strategy facilitates comprehensive profiling of protein methylation. J Proteome Res 22(10):3275–3281. https://doi.org/10.1021/acs.jproteome.3c00318
37. Ma M, Zhao X, Chen S et al (2017) Strategy based on deglycosylation, multiprotease, and hydrophilic interaction chromatography for large-scale profiling of protein methylation. Anal Chem 89(23):12909–12917. https://doi.org/10.1021/acs.analchem.7b03673
38. Wang Q, Li Z, Zhang S et al (2022) Global profiling of arginine dimethylation in regulating protein phase separation by a steric effect–based chemical-enrichment method. Proc Natl Acad Sci U S A 119(43):e2205255119. https://doi.org/10.1073/pnas.2205255119
39. Zhang F, Kerbl-Knapp J, Rodriguez Colman MJ et al (2021) Global analysis of protein arginine methylation. Cell Rep Methods 1(2):100016. https://doi.org/10.1016/j.crmeth.2021.100016
40. Zinellu A, Sotgia S, Scanu B et al (2008) Determination of protein-incorporated

methylated arginine reference values in healthy subjects whole blood and evaluation of factors affecting protein methylation. Clin Biochem 41(14):1218–1223. https://doi.org/10.1016/j.clinbiochem.2008.07.011
41. Habisch H, Zhang F, Zhou Q et al (2021) Exploring the arginine methylome by nuclear magnetic resonance spectroscopy. JoVE 178:e63245. https://doi.org/10.3791/63245
42. Weiss M, Manneberg M, Juranville J-F et al (1998) Effect of the hydrolysis method on the determination of the amino acid composition of proteins. J Chromatogr A 795(2):263–275. https://doi.org/10.1016/S0021-9673(97)00983-7
43. Davids M, Swieringa E, Palm F et al (2012) Simultaneous determination of asymmetric and symmetric dimethylarginine, l-monomethylarginine, l-arginine, and l-homoarginine in biological samples using stable isotope dilution liquid chromatography tandem mass spectrometry. J Chromatogr B 900:38–47. https://doi.org/10.1016/j.jchromb.2012.05.025
44. Stryeck S, Birner-Gruenberger R, Madl T (2017) Integrative metabolomics as emerging tool to study autophagy regulation. Microbial cell (Graz, Austria) 4(8):240–258. https://doi.org/10.15698/mic2017.08.584
45. Vignoli A, Ghini V, Meoni G et al (2019) High-throughput metabolomics by 1D NMR. Angew Chem Int Ed Engl 58(4):968–994. https://doi.org/10.1002/anie.201804736
46. Zhang F, Rakhimbekova A, Lashley T et al (2023) Brain regions show different metabolic and protein arginine methylation phenotypes in frontotemporal dementias and Alzheimer's disease. Prog Neurobiol 221:102400. https://doi.org/10.1016/j.pneurobio.2022.102400
47. Tehlivets O, Almer G, Brunner MS et al (2024) Homocysteine contributes to atherogenic transformation of the aorta in rabbits in the absence of hypercholesterolemia. Biomed Pharmacother 178:117244. https://doi.org/10.1016/j.biopha.2024.117244
48. Karner C, Anders I, Vejzovic D et al (2023) Targeting epigenetic features in clear cell sarcomas based on patient-derived cell lines. J Transl Med 21(1):54. https://doi.org/10.1186/s12967-022-03843-4
49. Brunner MS, Habisch H, Hafner J et al (2024) Accumulation of S-adenosyl-L-homocysteine impairs methylation and development in *Saccharomyces cerevisiae* and *Drosophila melanogaster*. https://doi.org/10.21203/rs.3.rs-5144569/v1
50. Carr HY, Purcell EM (1954) Effects of diffusion on free precession in nuclear magnetic resonance experiments. Phys Rev 94(3):630–638. https://doi.org/10.1103/PhysRev.94.630
51. Meiboom S, Gill D (1958) Modified spin-echo method for measuring nuclear relaxation times. Rev Sci Instrum 29(8):688–691
52. Nagayama K, Wüthrich K, Bachmann P et al (1977) Two-dimensional J-resolved 1H n.m.r. spectroscopy for studies of biological macromolecules. Biochem Biophys Res Commun 78(1):99–105. https://doi.org/10.1016/0006-291X(77)91226-8

Chapter 17

Nanopore-Based Detection of Protein Posttranslational Modifications

Meng-Yin Li, Yan Gao, Jie Jiang and Yi-Tao Long

Abstract

Posttranslational modifications (PTMs) are crucial regulators of protein functions, localization, and interactions, contributing to nearly all aspects of cellular physiology. Conventional analytical approaches, such as mass spectrometry (MS), have achieved high sensitivity and coverage, yet they often require complex sample preparation, enrichment, and high-end instrumentation. Nanopore has emerged as a promising single-molecule approach for PTM detection, offering label-free, high-throughput, and potentially portable analysis capabilities. Despite remarkable progress in nucleic acid sequencing with nanopores, its application to PTM detection in proteins remains at an early stage, hindered by challenges in protein capture, controlled unfolding, the interplay of sequence and structural heterogeneity, and diverse PTM chemistries. In this chapter, we summarize the current landscape of nanopore-based PTM identification and describe two experimental strategies that represent the most feasible near-term workflows for translating nanopore technology into practical protein PTM analysis.

Keywords Nanopore, Single-molecule, Aerolysin, Electrochemistry, Sulfation

1 Introduction

Protein posttranslational modifications (PTMs), such as phosphorylation, sulfation, glycosylation, and ubiquitination, are crucial for cellular signaling and proteoform diversity [1, 2]. Conventional analytical platforms, particularly mass spectrometry (MS), have enabled high-sensitivity and large-scale PTM mapping. However, MS-based approaches face several challenges: they require substantial sample amounts and complex pretreatment, may lose labile modifications during ionization or fragmentation, and pinpointing modification sites can be ambiguous, especially when multiple residues are potential targets. Additionally, low-abundance PTMs suffer from poor signal-to-noise ratios, and certain PTMs degrade or change under pretreatment or analysis, leading to incomplete or inaccurate detection.

Xu Zhang et al. (eds.), *Protein Post-Translational Modifications*, Methods in Molecular Biology, vol. 3018,
https://doi.org/10.1007/978-1-0716-5166-7_17,

Nanopore electrochemistry offers an emerging alternative for PTM characterization, enabling label-free, real-time analysis of single molecules [3–5]. As proteins or peptides enter nanometer-scale pores, changes in ionic current reflect variations in their size, charge, conformation, and chemical modifications, producing distinctive blockade events [6–8]. Until now, there are two main strategies to study protein PTMs. One is to acquire the sequence-related PTMs by controlling the movement of the whole peptides or proteins, which, however, normally requires complicated modifications/label or specific structure to target sequence [9, 10]. Thus, due to protein heterogeneity in charge, folding states, and modification patterns, such a strategy can be hardly employed in practical application in PTM detection.

To address this challenge, another strategy that could be applied in real-life system [11, 12] has emerged, including: (i) targeted detection, which often relies on established PTM databases and known modification sites, and (ii) identification of PTM in unknown proteins from natural protein digestion. Taking aerolysin nanopore-based detection of PTMs as a model, this chapter describes the detailed procedures for both methods.

2 Materials

2.1 Nanopore Detection System

1. 1.0 M Tris–KCl buffer: weigh the appropriate amounts of potassium chloride (KCl), tris(hydroxymethyl)aminomethane (Tris), and ethylenediaminetetraacetic acid (EDTA), dissolve them in a specified volume of ultrapure water, and adjust the pH to 8.0 using a 10% (v/v) hydrochloric acid (HCl) solution. Prepare a 1.0 M Tris–KCl solution containing 1.0 mol/L KCl, 10 mmol/L Tris, and 1.0 mol/L EDTA.
2. Tris–HCl buffer: weigh the appropriate amounts of Tris, dissolve them in a specified volume of ultrapure water, and adjust the pH to 7.5 using a 10% (v/v) HCl solution. Prepare a Tris–HCl solution containing 10 mmol/L Tris.
3. Ag/AgCl electrodes: take two pieces of silver wire with a diameter of 1 mm and a length of approximately 3–5 cm. First, use sandpaper to grind off the silver oxide from the surface, then clean the surface with ultrapure water. Next, immerse the silver wire in a 10% HCl solution and prepare the Ag/AgCl electrodes via electroplating (*see* **Note 1**).
4. Trypsin-EDTA solution: weigh the appropriate amounts of trypsin and EDTA, dissolve them in a specified volume of ultrapure water, and adjust the pH to 8.0 using a 10% (v/v) hydrochloric acid (HCl) solution. Prepare a trypsin-EDTA

solution containing 0.5 g/L trypsin and 0.2 g/L EDTA•4Na at pH 8.0.

5. Aerolysin protein solution: take 50 μL of the 1 mg/mL proaerolysin stock solution from the −80°C freezer and add 3 μL of trypsin-EDTA solution. Incubate at room temperature for 6 h. Then, dilute the above solution with Tris–HCl solution to prepare a 10 μg/mL aerolysin protein solution (*see* **Note 2**).
6. Phospholipid *n*-decane solution: retrieve 500 μL of the phospholipid chloroform solution (40 mg/mL, purchased from Avanti Research, Product No. 850356C) from the −20°C freezer. Evaporate the chloroform using a vacuum pump, then add an appropriate volume of *n*-decane to prepare a 30 mg/mL phospholipid *n*-decane solution (*see* **Note 3**).
7. Ultra-low current recording instrument: the current traces were amplified and recorded by using Clampex 10.7 software with a patch-clamp amplifier (Axon 200B) equipped with a Digidata 1440A A/D converter (Molecular Devices), or by using SmartNano 2.0 software with Cube-D2 (Yi-Tao Long's Lab).
8. Polytetrafluoroethylene (PTFE) membrane with a hole: each Teflon sheet contains a hole (~50 μm in diameter) fabricated using an electric spark generator and visually checked under a microscope.

2.2 Tyrosine Desulfation

1. 0.1% TFA solution: take an appropriate amount of trifluoroacetic acid (TFA) and add a sufficient amount of ultrapure water to prepare a 0.1% TFA (v/v) solution.
2. 5% Acetonitrile solution: take an appropriate amount of acetonitrile and add a sufficient amount of ultrapure water to prepare a 75% acetonitrile (v/v) solution.
3. 1.0 M NaOH solution: take an appropriate amount of NaOH solid and add a sufficient amount of ultrapure water to prepare a 1.0 M NaOH solution.

2.3 Digestion of HIRV1 Protein

1. Hirudin variant-1 (HIRV1) protein solution: dissolve 45 μg of HIRV1 protein lyophilized powder in 50 μL of 10 mM Tris–HCl buffer (pH 7.5).
2. 10% HCl solution: take an appropriate volume of concentrated HCl and dilute it to prepare a 10% (v/v) HCl solution.
3. Asp-N enzyme solution: dissolve 2 μg lyophilized Asp-N endopeptidase in 100 μL of 10 mM Tris–HCl buffer (pH 7.5) to achieve a final concentration of 0.02 μg/μL.

3 Methods

3.1 Identification of PTMs in Unknown Peptide Fragments from Natural Protein Digestion: A Case Study on Tyrosine Sulfation

1. Place a polytetrafluoroethylene (PTFE) membrane with a ~50 μm diameter hole between the *cis* and *trans* chambers and secure with screws. Add equal volumes of 1.0 M Tris–KCl solution (containing EDTA) to both chambers, and connect two Ag/AgCl electrodes to the instrument, immersing them in the chambers to apply voltage (*see* **Note 4**).
2. Add phospholipid *n*-decane solution to both chambers, allowing the phospholipids to spread evenly. Pull the solution from one side to form a phospholipid bilayer at the hole. Apply voltage to test if the bilayer thickness is suitable for nanopore insertion.
3. If the bilayer breaks at +300 to +400 mV, it is suitable for nanopore insertion. If the breakdown voltage is higher, destroy the bilayer with bubbles and repeat the process to adjust the thickness. If the breakdown voltage is lower, the bilayer is too thin, and additional phospholipid solution can be added to adjust the thickness.
4. After obtaining the desired phospholipid bilayer thickness, add 0.5–2.0 μL of the prepared aerolysin protein solution to the *cis* chamber. Once aerolysin proteins insert into the bilayer and form heptamers, a step-like change in ionic current will occur. At +100 mV, the ionic current produced by a single aerolysin nanopore inserted into the bilayer is approximately 50 pA (24 ± 2°C). A stable single aerolysin nanopore is then obtained.
5. Add peptides with different molecular weights and tyrosine sulfation at various sites to the *cis* side for nanopore detection, respectively. This allows the correlation between the peptide molecular weight and the average increase in current blockade caused by tyrosine sulfation at different sites to be determined (Fig. 1) (*see* **Note 5**).
6. To digest HIRV1 protein, add 10 μL of Asp-N endoproteinase solution (enzyme to protein mass ratio of 1:225) to 50 μL HIRV1 protein solution and incubate the sample overnight at 37°C. Store the sample at −20°C before use.
7. Add TFA to the enzymatically digested peptide solution to a final concentration of 0.5% TFA (v/v). Incubate the peptide-TFA solution at 50°C for 8 h. Finally, neutralize the incubated mixture with 1.0 M NaOH solution to adjust the pH to 8 before nanopore detection.
8. Nanopore detection is performed on the protein before enzymatic digestion and on the peptides after digestion, both with and without desulfation (Fig. 2). The blockade current (I/I_0) vs. blockade time scatter plots and the blockade current (I/I_0)

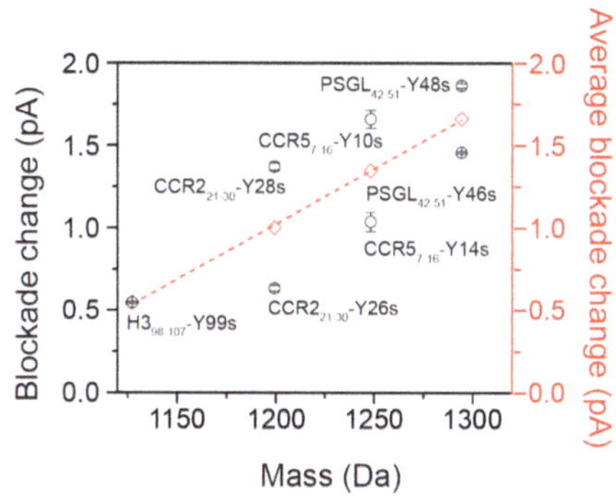

Fig. 1 Correlation between the peptide mass and the blockade enhancement induced by sulfation, with a linear fitting shown in red. The fitting was performed based on the mass of nonsulfated peptides and the average blockade enhancement observed for sulfation at different sites. All means and error bars were from three independent experiments. All data were obtained at the applied voltage of +50 mV and in 1.0 M KCl, 10.0 mM Tris, pH 8.0, 24 ± 2°C. Reproduced from Ref. [11] with permission. (Niu H, Li M-Y, Gao Y, Li J-G, Jiang J, Ying Y-L, Long Y-T (2025) Direct mapping of tyros ine sulfation states in native peptides by nanopore. Nat Chem Biol 21(5):716–726.)

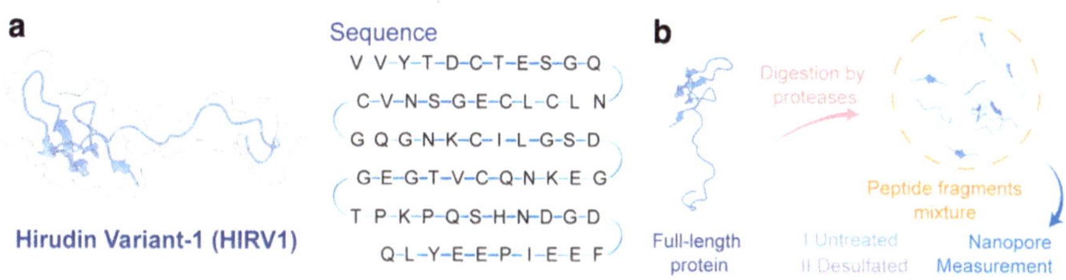

Fig. 2 Illustration of discovering tyrosine sulfation in the native protein HIRV1. (**a**) The structure and sequence of HIRV1 protein. (**b**) First, HIRV1 was digested by Endoproteinase Asp-N enzyme overnight. Then the peptide fragments were directly detected by T232K mutant aerolysin nanopore. Next, the peptide fragments were treated by TFA to desulfate tyrosine sulfation prior to measurement. Reproduced from Ref. [11] with permission. (Niu H, Li M-Y, Gao Y, Li J-G, Jiang J, Ying Y-L, Long Y-T (2025) Direct mapping of tyros ine sulfation statesin native peptides by nanopore. Nat Chem Biol 21(5):716–726.)

histograms are obtained. Based on the previously established linear relationship between peptide molecular weight and the current enhancement caused by sulfation, the molecular weight of the sulfated peptide fragments can be inferred. This allows for the identification of tyrosine sulfation modifications in specific peptide fragments produced after HIRV1 digestion (Fig. 3).

3.2 Targeted Detection Based on Established PTM Databases and Known Modification Sites

1. Select proteolytic fragments that encompass known PTM sites based on existing databases and generate synthetic reference peptides library.
2. Evaluate multiple types of nanopores or mutants to identify the optimal nanopore for detecting PTMs (refer to Sect. 3.1).

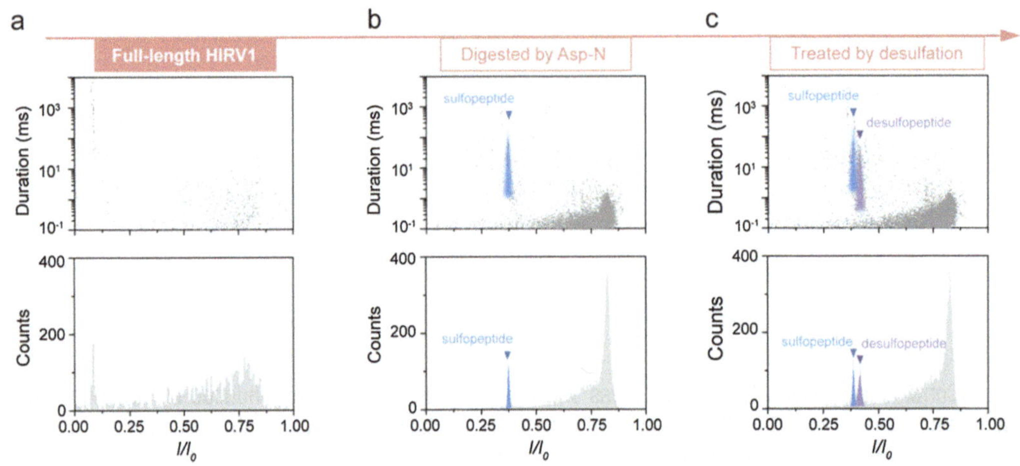

Fig. 3 Scatter plots and histograms of I/I_0 for HIRV1 protein analysis. (**a**) Full-length HIRV1 protein. (**b**) HIRV1 peptide fragments after digestion. (**c**) Digestion solution after desulfation. All measurements were performed at an applied voltage of +80 mV. All experiments were conducted in 1.0 M KCl, 10.0 mM Tris, pH 8.0, at 24 ± 2°C. Reproduced from Ref. [11] with permission. (Niu H, Li M-Y, Gao Y, Li J-G, Jiang J, Ying Y-L, Long Y-T (2025) Direct mapping of tyros ine sulfation states in native peptides by nanopore. Nat Chem Biol 21(5):716–726.)

3. Acquire blockade and dwell time signatures from synthetic reference peptides under controlled conditions, and use machine learning algorithms to process the data. This allows for the development of a peptide library capable of recognizing various PTMs.
4. Use protease digestion to produce target peptides from protein samples, ensuring the preservation of PTMs during the process.
5. For nanopore detection, translocate sample-derived peptides through the selected nanopore and record blockade events to capture PTM-related signals.
6. Match detected events to the reference library for signal classification and compute the frequency of PTMs relative to their unmodified counterparts for quantification.

4 Notes

1. Both electrodes should be plated simultaneously under identical conditions. The process should be stopped once a dense AgCl film has formed on the electrode surface. The plating time should not be excessive to avoid detachment of the AgCl film.

2. Both the wild-type and mutant-type of aerolysin are prepared according to the published protocol [4]. The concentration of the protein that forms a single channel on lipid membranes can be adjusted based on the experimental observations.
3. During the preparation of phospholipid solutions, efforts should be made to minimize exposure to oxygen and moisture. Once prepared, the solution should be sealed promptly and stored at 4°C in a refrigerator.
4. The hole diameter on the PTFE membrane generally ranges from 10 to 100 μm, providing a favorable lipid formation environment and minimized current noise.
5. All peptide sequences for detection are as follows: $CCR_{221\text{-}30}$: TTFFDYDYGA; $CCR2_{21\text{-}30}$-Y26s: TTFFDs(Y)DYGA; $CCR2_{21\text{-}30}$ -Y28s: TTFFDYDs(Y)GA; $PSGL_{42\text{-}5}$: QATEYEYLDY; $PSGL_{42\text{-}51}$-Y46s: QATEs(Y)EYLDY; $PSGL_{42\text{-}51}$-Y48s: QATEYEs(Y)LDY; $CCR5_{7\text{-}16}$: SPIYDINYYT; $CCR_{57\text{-}16}$-Y10s: SPIs(Y)DINYYT; $CCR_{57\text{-}16}$-Y14s: SPIYDINs(Y)YT; $H3_{98\text{-}107}$: AYLVGLFEDT; and $H3_{98\text{-}107}$-Y99s: As(Y) LVGLFEDT.

References

1. Aebersold R, Agar JN, Amster IJ, Baker MS, Bertozzi CR, Boja ES, Costello CE, Cravatt BF, Fenselau C, Garcia BA, Ge Y, Gunawardena J, Hendrickson RC, Hergenrother PJ, Huber CG, Ivanov AR, Jensen ON, Jewett MC, Kelleher NL, Zhang B (2018) How many human proteoforms are there?. Nat Chem Biol 14(3):206–214
2. Zecha J, Bayer FP, Wiechmann S, Woortman J, Berner N, Müller J, Schneider A, Kramer K, Abril-Gil M, Hopf T, Reichart L, Chen L, Hansen FM, Lechner S, Samaras P, Eckert S, Lautenbacher L, Reinecke M, Hamood F, Kuster B (2023) Decrypting drug actions and protein modifications by dose- and time-resolved proteomics. Science 380(6640):93–101
3. Li M-Y, Jiang J, Li J-G, Niu H, Ying Y L, Tian R, Long Y-T (2025) Nanopore approaches for single-molecule temporal omics: promises and challenges. Nat Methods 22(2):241–253
4. Ying Y-L, Hu Z-L, Zhang S, Qing Y, Fragasso A, Maglia G, Meller A, Bayley H, Dekker C, Long Y-T (2022) Nanopore-based technologies beyond DNA sequencing. Nat Nanotechnol 17(11):1136–1146
5. Reiner JE, Balijepalli A, Robertson JWF, Campbell J, Suehle J, Kasianowicz JJ (2012) Disease detection and management via single nanopore-based sensors. Chem Rev 112(12):6431–6451
6. Kasianowicz JJ, Brandin E, Branton D, Deamer DW (1996) Characterization of individual polynucleotide molecules using a membrane channel. Proc Natl Acad Sci U S A 93(24):13770–13773
7. Jiang J, Li M-Y, Wu X-Y, Ying Y-L, Han H-X, Long Y-T (2023) Protein nanopore reveals the renin–angiotensin system crosstalk with single-amino-acid resolution. Nat Chem 15(4):578–586
8. Li M-Y, Ying Y-L, Yu J, Liu S-C, Wang Y-Q, Li S, Long Y-T (2021) Revisiting the origin of nanopore current blockage for volume difference sensing at the atomic level. JACS Au 1(7):967–976
9. Nova IC, Ritmejeris J, Brinkerhoff H, Koenig TJR, Gundlach JH, Dekker C (2024) Detection of phosphorylation post-translational modifications along single peptides with nanopores. Nat Biotechnol 42(5):710–714
10. Martin-Baniandres P, Lan W-H, Board S, Romero-Ruiz M, Garcia-Manyes S, Qing Y, Bayley H (2023) Enzyme-less nanopore detection of post-translational modifications within long polypeptides. Nat Nanotechnol 18(11):1335–1340
11. Niu H, Li M-Y, Gao Y, Li J-G, Jiang J, Ying Y-L, Long Y-T (2025) Direct mapping of tyrosine sulfation states in native peptides by nanopore. Nat Chem Biol 21(5):716–726
12. Restrepo-Pérez L, Wong CH, Maglia G, Dekker C, Joo C (2019) Label-free detection of post-translational modifications with a nanopore. Nano Lett 19(11):7957–7964

Part IV

PTMs of Therapeutic Proteins

Check for updates

Chapter 18

Mass Spectrometry Analysis of Biotherapeutic Monoclonal Antibody Glycosylation

Min Fang, Timothy R. Rudd and Gail C. Whiting

Abstract

Glycosylation plays a critical role in the efficacy of therapeutic monoclonal antibodies (mAbs), the major category of glycoprotein-based therapeutic medicines. It is claimed that glycosylation of therapeutic mAbs has implications on their half-life, immunogenicity, and pharmacokinetics. Comprehensive characterization of glycosylation enables a deeper understanding of the complexity of therapeutic mAbs. The method presented here describes an integrated approach that incorporates filtered-aided sample preparation, mass spectrometry with different ion activation techniques, and a software that enables intact glycopeptide database search and de novo sequencing, to provide both N- and O-linked glycan profiling with structural resolution, without performing pre-glycoprotein/glycopeptide enrichment.

Keywords Therapeutic monoclonal antibody, Glycosylation, Unenriched intact glycopeptide, De-glycosylation, PEAKS glycan search, FASP, Mass spectrometry

1 Introduction

Monoclonal antibodies (mAbs) are important targeted therapeutics, which are involved in the treatment of a number of diseases, e.g., autoimmune disorders (Infliximab) and cancer (Rituximab) through targeting a specific antigen, with infliximab targeting tumor necrosis factor-alpha (TNF-α) and Rituximab targeting B-lymphocyte antigen CD20. Traditionally, mAbs are expressed in mammalian cell lines, such as Chinese hamster ovary (CHO) cells, hence an inherent degree of structural heterogeneity occurs in proteins due to the biosynthetic processes used by living organisms to produce them. mAbs can undergo several posttranslational modifications (PTMs), e.g., oxidation, acetylation, phosphorylation, methylation, and glycosylation [1]. The resulting product can be a mixture of posttranslationally modified forms, which dramatically increases the molecular complexity of these mAbs.

Xu Zhang et al. (eds.), *Protein Post-Translational Modifications*, Methods in Molecular Biology, vol. 3018,
https://doi.org/10.1007/978-1-0716-5166-7_18,

Glycosylation is one of the most commonly observed PTMs found in biotherapeutic mAbs and considered to be one of the critical quality attributes (CQAs) in biologic manufacturing [2]. Glycosylation of mAb is a diverse, enzyme-mediated, and nontemplate driven process by which an oligosaccharide, a carbohydrate consisting of several sugar molecules, is covalently attached to protein molecules, and therefore, modulates their functions, such as biological activity, solubility, immunogenicity, serum half-life, and safety of a given biotherapeutic mAb [3–5]. There are two major types of glycosylation: N- and O-linked glycosylation. N-linked glycosylation is the attachment of an oligosaccharide, e.g., *N*-acetylglucosamine (GlcNAc), to the amide nitrogen of an asparagine (Asn/N) located within the sequence motif Asn-X-Serine (Ser) or Threonine (Thr), i.e., N-X-S/T sequon, where X is any amino acid except proline, by an *N*-glycosidic bond. The process is facilitated by glycosyltransferases [6]. O-linked glycosylation is the attachment of an oligosaccharide to the hydroxyl oxygen of either serine (S) or threonine (T) residues, and, to a lesser extent, to hydroxyproline and hydroxylysine, via an *O*-glycosidic bond. The two most common oligosaccharides linked to the hydroxyl group (-OH) of Ser/Thr are O-linked-β-D-*N*-acetylglucosamine (O-GlcNAc) and *O*-acetylgalactosamine (O-GalNAc) [7].

Most glycoproteins emerge with characteristic glycosylation patterns, however, the glycans at each glycosylation site are highly heterogeneous [8]. The structural complexity of glycans (~10^{12} different branched glycan structures may exist [9]) and sialylation, where sialic acid (*N*-acetyl neuraminic acid) is covalently attached to the terminal residues of oligosaccharides and glycoproteins, makes in-depth characterisation of glycosylated therapeutic mAbs extremely challenging. Recent advancements in mass spectrometry (MS), computational tools and search engines make MS-based methods the gold standard in therapeutic mAb analysis [10–13]. Glycosylation analysis of therapeutic mAbs can be performed variously by analysis of released glycans (chemically or enzymatically) [14], intact glycopeptide analysis (through trypsin digestion) [15], mAb subunit analysis (e.g., IdeS cleaves at a single amino acid site below the hinge to yield homogenous F(ab')2 and Fc fragments) [16], or intact mAb analysis [17]. Intact glycopeptide analysis has received increasing attention in recent years; not only is it useful in characterizing glycan heterogeneity, but also providing the site-specific glycan information (also known as site-specific glycan analysis). Tandem mass spectrometry (MS/MS) coupled with liquid chromatography (LC) is becoming the method of choice in site-specific intact glycopeptide analysis.

The MS dissociation energies required for glycan and peptide dissociation are vastly different, rendering it impossible to acquire glycopeptide spectra using a single MS fragmentation technique. A quintessential-MS/MS acquisition should generate the most comprehensive and complementary fragmentation in a single spectrum for each intact glycopeptide. Advanced MS instruments that integrate different dissociation techniques, such as collision-induced dissociation (CID), higher-energy collisional dissociation (HCD), electron-transfer dissociation (ETD) [18, 19], electron-activated dissociation (EDA) [20], or hybrid dissociation techniques, such as those that couple ETD with HCD (EThcD) [21, 22] or ETD with CID (ETciD) [23], or EDA with CID [24, 25], have the capacity to generate complicated and informative fragment ions, such as *b/y*, *c/z*, or glycan B/Y ions, which can be used to elucidate glycan structure and peptide sequence. HCD can generate glycopeptide Y ions but few B ions and a limited number of peptide *b/y* ions, thus limiting its usefulness in identifying specific glycosylation sites on a peptide sequence. ETD is a relatively soft dissociation technique, breaking N-Cα bonds of the peptide backbone to generate *c/z* ions while maintaining the integrity of the glycans. HCD produces abundant diagnostic oxonium ions, such as m/z 138.055 (HexNac fragment), 204.087 (HexNac), and 366.140 (HexNac-Hex fragment) [26, 27]. With the development of the ion-trigger technique in MS, e.g., a feature of the Orbitrap Eclipse mass spectrometer, the diagnostic oxonium ions produced by HCD can trigger next-stage EThcD fragmentation to yield B- and Y-ions for site-specific glycan identification and partial *b/y* ion and *c/z* ion series for peptide identification. The diagnostic oxonium ions can also trigger a CID fragmentation to generate glycan fragments to further support glycan and peptide identification. This could be carried out in parallel with EThcD fragmentation.

In comparison with other PTMs, spectra generated by LC–MS/MS-based glyco-proteomics present an extreme challenge due to the complexity of the structures present in a sample, including all potential peptide sequences, glycan compositions, and structures. The search engines used in general proteomics are not suitable for the analysis of the enormous number of variable modifications. To overcome the challenge posed by glycoproteomics, software tools have been developed for automated glycopeptide identification [28–32]. PEAKS Studio 12 (Bioinformatics Solution Inc.) [33], Byonic (Protein Metrics) [34], and GlycoScape (Bruker) [35] are three commercial software packages commonly used for glycopeptide analysis. Many academically developed computational tools for glycopeptide identification are also available and can be downloaded freely, e.g., StrucGP [36] https://github.com/Sun-GlycoLab/StrucGP/releases, pGlyco3 [37] https://github.com/pFindStudio/pGlyco3.

By harnessing the advances in MS hardware and software, in this chapter, we describe an LC–MS/MS-based method for intact glycopeptides analysis, using the originator Remicade (infliximab), and the biosimilars Remsima and Flixabi, and the NIBSC infliximab standard, as examples. The integrated method includes filter-aided sample preparation method (FASP), LC–MS/MS using dissociation techniques CID, HCD, and EThcD, and glycoprotein analysis using the PEAKS Glycan search engine in PEAKS Studio 12, which integrates peptide- and glycan-based search strategies to address the challenge of complex fragmentation of glycopeptides, achieving in-depth unenriched glycopeptide analysis. De-glycosylation was also employed to aid the identification of glycosylation sites on the protein core.

2 Materials

2.1 Chemicals and Consumables

All chemicals should be of high purity. We use Optima LC/MS Grade for acetonitrile (ACN), formic acid (FA), and water.

1. Vivacon 500, 10 kDa FASP tubes.
2. Eppendorf Lo-Bind tubes, 0.5 and 1.5 mL.
3. 0.3 mL glass screw top microvials with fixed insert.
4. 9 mm pre-slit screw caps.

2.2 Solutions and Solvents

1. 200 mM TEAB buffer, pH 8.5: prepare a 1 in 5 dilution of 1.0 M triethylammonium bicarbonate (TEAB) in water.
2. 20 mM TCEP in 8 M urea. Add 143.3 mg of tris(2-carboxyethyl)phosphine hydrochloride (TCEP) to 25 mL of 8 M urea solution. This can be aliquoted and stored at −20°C.
3. 50 mM IAA solution: add 9.25 mg of iodoacetamide (IAA) to 1 mL of 200 mM TEAB buffer.
4. 0.5 μg/μL trypsin working solution: add 20 μL of 200 mM TEAB buffer to the vial containing 20 μg of lyophilized trypsin.
5. PNGase-F working solution: prepare immediately before use by adding 2 μL of supplied glycobuffer and 2 μL of PNGase-F to 18 μL of water (New England BioLabs) (*see* **Note 1**).
6. 40 units/μL OglyZOR working solution: reconstitute vial in 50 μL water.
7. 40 units/μL SialEXO working solution: reconstitute vial in 50 μL water.

2.3 Equipment

1. LC–MS system: Vanquish Neo ultra-high performance liquid chromatography (UHPLC) system coupled with Orbitrap Eclipse Tribrid mass spectrometer, all from Thermo Scientific.
2. Analytical column: EASY-Spray PepMap Neo UHPLC Columns, 150 mm, 2 μm, 75 μm, 100 Å.
3. Trap column: PepMap Neo Trap Cartridge, C18, 5 μm, 300 μm × 5 mm.

4. Eppendorf concentrator plus.
5. Eppendorf mini-spin centrifuge.
6. Eppendorf thermomixer.

2.4 Software

1. Xcalibur (version 4.7.69.37), Thermo Scientific.
2. PEAKS Studio 12, Bioinformatics Solutions Inc.

2.5 Samples

1. Remicade (Originator) Lot # 6RMA68807.
2. Remsima (Biosimilar) Lot # 16B1C11BHA1.
3. Flixabi (Biosimilar) Lot # EC0585.
4. NIBSC internal reference (SS-717).

3 Methods

3.1 Filter-Aided Sample Preparation

A FASP protocol to generate enzymatic peptides is described here. The method combines the use of cartridge with a defined molecular weight cut-off (MWCO) membrane as a "reactor," on which proteins are denatured, reduced, alkylated, de-glycosylated, and then digested (Fig. 1).

1. Use Vivacon 500, 10 kDa FASP cartridges.
2. Set centrifuge at 13,500 rpm.
3. Add 200 μL of 200 mM TEAB to condition the cartridge.
4. Centrifuge to remove the buffer but leave a thin layer of solution on surface by adjusting spin time (say starting with 5 min) (*see* **Note 2**).
5. Add 200 μL of 200 mM TEAB.
6. Add sample (50 μg) (*see* **Note 3**).

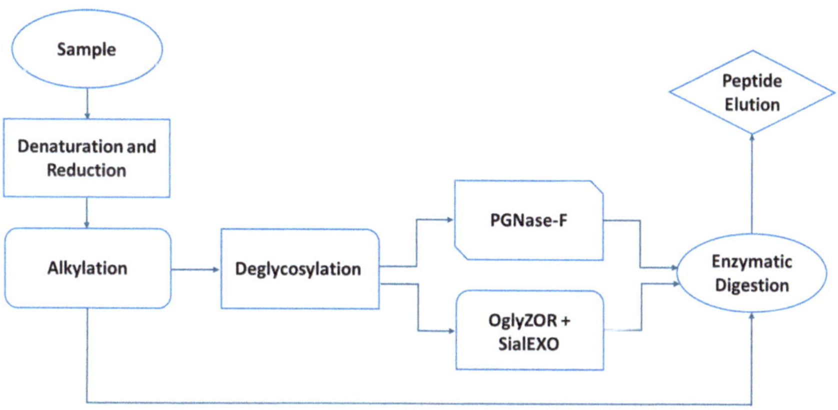

Fig. 1 Sample preparation workflow

7. Agitate by pipetting up and down using 200 μL pipettor for ten times. Be careful not to touch the membrane surface.
8. Centrifuge to remove the solution as described in Step 5.
9. Repeat Steps 3 and 4 twice.
10. Add 200 μL of 20 mM TCEP/8 M urea solution to cartridge for denaturation and reduction.
11. Agitate by pipetting up and down for ten times. Be careful not to touch the membrane surface.
12. Incubate for 1 h at 25°C and shake at 350 rpm in the Thermomixer.
13. Centrifuge to remove the solution.
14. Add 200 μL of 50 mM IAA solution for alkylation.
15. Incubate for 1 h at 25°C and shake at 350 rpm in the Thermomixer.
16. Add 200 μL of 200 mM TEAB buffer.
17. Centrifuge to remove the solution.
18. Repeat Steps 3 and 4 twice.
19. Change to a new cartridge collection tube.
20. Leave the cartridge at 4°C until around 4 p.m. of the day to continue the next step.
21. Add 50 μL of 200 mM TEAB buffer.
22. Add 4 μL of freshly prepared trypsin working solution (optimized for 50 μg protein with an enzyme:protein ratio of 1:25).
23. Incubate at 37°C overnight and shake at 350 rpm in the Thermomixer.
24. Add 200 μL of 0.1% TFA in water to stop enzymatic digestion and leave it on bench for 10 min.
25. Centrifuge at 13,500 rpm for 10 min to collect the solution.
26. Add 200 μL of 0.1% TFA in 50% ACN solution and leave it on bench for 10 min.
27. Centrifuge at 13,500 rpm for 10 min to collect the solution into the same collection tube.
28. Add 200 μL of 0.1% TFA in ACN and leave it on bench for 10 min.
29. Centrifuge at 13,500 rpm for 15 min to collect the solution into the same collection tube.
30. Discard the cartridge.
31. Eluates can go to next step or stored at 4°C.
32. Vacuum centrifuge the sample to remove solvent using Eppendorf Concentrator plus. Temperature is set at 45°C.

33. Check regularly to avoid over-drying.
34. Reconstitute the peptides using 100 μL of 0.1% TFA in water, to make a concentration of 500 ng/μL and store at 4°C until LC–MS/MS analysis (*see* **Note 4**).

3.2 De-Glycosylation

De-glycosylation of the glycopeptides before tryptic digestion increases protein sequence coverage and improves protein identification, confirms glycosylation sites on the protein core, confirms presence of glycans, and reduces sample heterogeneity. De-glycosylation can also be used to study the functional role of the glycans.

3.2.1 Removal of N-Linked Glycans Using PNGase-F

The following instructions are for one sample. The working solution volume can be scaled up proportionally for multiple samples.

1. Add 10 μg (*see* **Note 5**) of glycoprotein sample to a Vivacon 500, 10 kDa FASP cartridge, perform protein denaturation, reduction, and alkylation as described in Sect. 3.1 (Steps 1–19).
2. Add 20 μL of TEAB buffer.
3. Add 22 μL of PNGase-F working solution (*see* **Notes 6 and 7**).
4. Agitate by pipetting up and down for ten times. Be careful not to touch the membrane surface.
5. Incubate at 37°C for 16 h and shake at 350 rpm.
6. Add 200 μL of 200 mM TEAB buffer.
7. Centrifuge to remove the solution.
8. Repeat Steps 6 and 7 twice.
9. Perform enzymatic digestion as described in Sect. 3.1 (Steps 22–34).

3.2.2 Removal of O-Glycans Using OglyZOR and SialEXO

1. Add 10 μg (*see* **Note 5**) of glycoprotein sample to a Vivacon 500, 10 kDa FASP cartridge, perform protein denaturation, reduction, and alkylation as described in Sect. 3.1 (Steps 1–19).
2. Add 1 unit of OglyZOR and 1 unit of SialEXO working solutions per μg glycoprotein to the sample (*see* **Notes 8–10**).
3. Incubate overnight (16–18 h) at 37°C and shake at 350 rpm.
4. Add 200 μL of 200 mM TEAB buffer.
5. Centrifuge to remove the solution.
6. Repeat Steps 4 and 5 twice.
7. Perform enzymatic digestion as described in Sect. 3.1 (Steps 22–34).

3.3 LC–MS/MS Method

This section describes the parameters to be used to acquire data on a Thermo Orbitrap Eclipse mass spectrometer paired with a Thermo Vanquish Neo UHPLC System (Fig. 2).

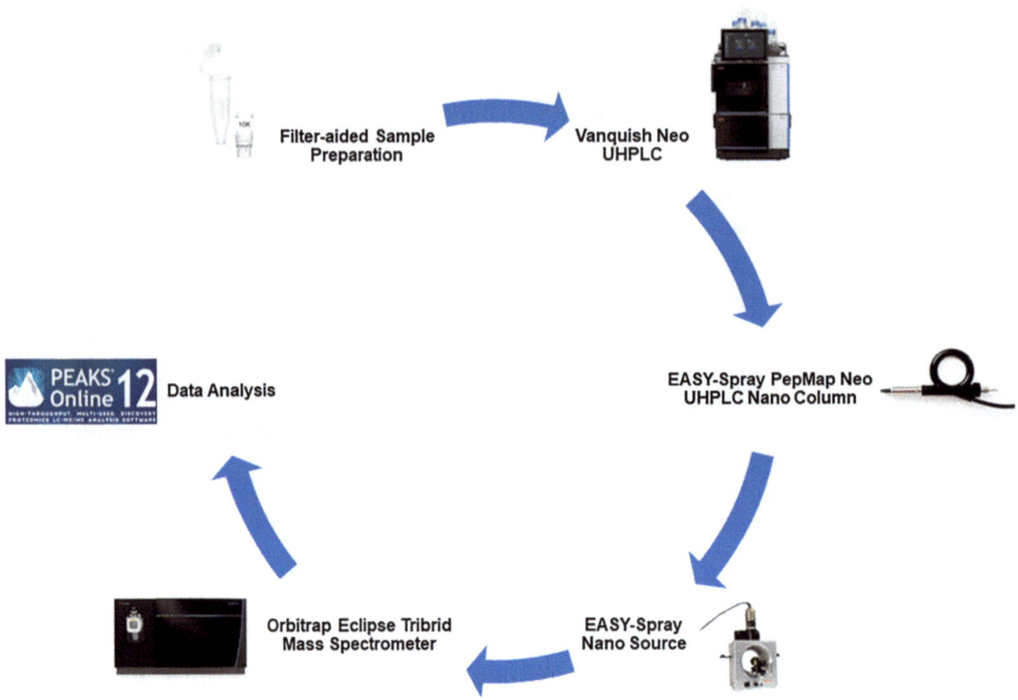

Fig. 2 Data-dependent acquisition (DDA)-based unenriched glycopeptide and glycan analysis workflow

Table 1
Specifications of chromatographic columns used in this protocol

Analytical column specifications		
Inner diameter	75 μm	
Length	15 cm	
Maximum pressure	1450 bar	
Maximum temperature	60°C	Set to 40°C
Max. flow	5 μL/min	Set to 0.3 μL/min
Max. pressure change up/down	300 bar/min	
Trap column specifications		
Inner diameter	300 μm	
Cartridge length	0.5 cm	
Maximum pressure	800 bar	
Maximum temperature	50°C	
Max. flow	100 μL/min	

3.3.1 Liquid Chromatographic Method

For Vanquish Neo system using trap-and-elute injection settings. See Table 1 for specifications of trap and analytical columns.

Mobile phases A and B consist of water with 0.1% formic acid and 80% acetonitrile in water with 0.1% formic acid, respectively. See Table 2 for the mobile phase gradient; 71-min method duration with 60-min gradient.

3.3.2 Mass Spectrometric Method—Glycopeptide and Glycan Method

The method used CID, HCD, and hybrid EThcD to generate different types of fragment ions, such as *b/y*, *c/z*, or glycan *B/Y* ions (Fig. 3).

Mass spectrometric settings are detailed in Table 3. ddMS² OT EThcD and ddMS² OT CID are performed in parallel.

Table 2
Mobile phase gradient

Time (min)	Duration (min)	Flow rate (μL/min)	%mobile phase B
0	0	0.3	5
40	40	0.3	35
60	20	0.3	55
61	1	0.3	99
71	10	0.3	99
Column equilibration	Built in instrument		

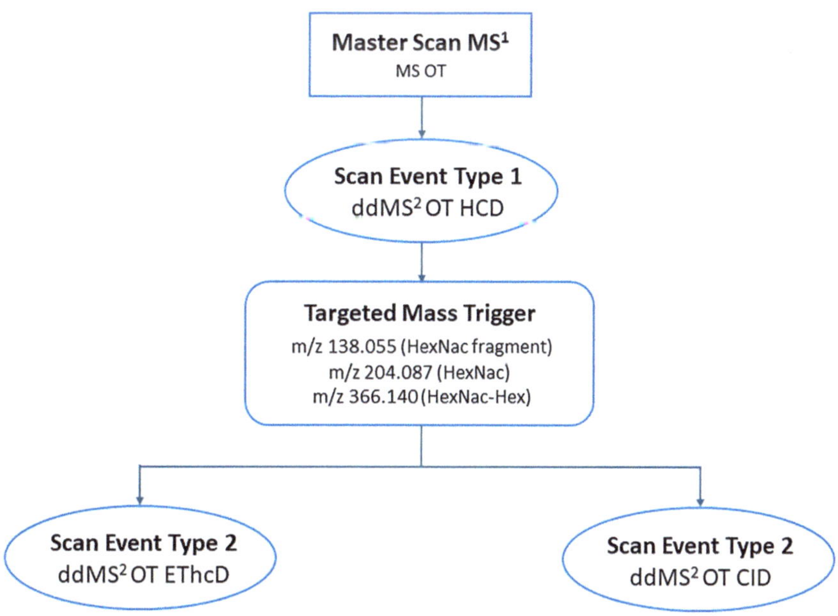

Fig. 3 MS/MS method workflow. Scan Event Type 2 is trigged by diagnostic oxonium ions, m/z 138.055 (HexNac fragment), 204.087 (HexNac), and 366.140 (HexNac-Hex). ddMS² OT EThcD and ddMS² OT CID are carried out in parallel. *OT*, Orbitrap

Table 3
Glycopeptide and glycan unenriched MS method parameters

Global parameters		
	Expected peak width (s)	30
	Default charge state	2
Master scan (MS^1)		
	Detector type	Orbitrap
	Scan range (m/z) MS^1	350–1800
	Resolution	120,000
	RF lens (%)	30
	Data type	Profile
Scan event type 1		
$ddMS^2$ OT HCD	Isolation mode	Quadrupole
	Isolation window (m/z)	2
	Activation type	HCD
	HCD collision energy (%)	28
	Detector type	Orbitrap
	Resolution MS^2	30,000
	Data type	Profile
Targeted mass trigger	Compound	m/z
	HexNac	204.0867
	HexNac fragment	138.0545
	HexNac-Hex	366.1396
Scan event type 2		
$ddMS^2$ OT EThcD	Isolation mode	Quadrupole
	Activation type	ETD
	SA collision energy type	EThcD
	SA collision energy (%)	25
	Detector type	Orbitrap
	Orbitrap resolution	30,000
	Data type	Profile
Scan event type 2		
$ddMS^2$ OT CID	Isolation mode	Ion trap
	Activation type	CID
	CID collision energy %	30
	Detector type	Orbitrap
	Orbitrap resolution	30,000
	Data type	Profile

Table 4
Data analysis workflow parameter settings

Create a project		
	Define instrument type	Orbitrap tribrid
	Define fragment type	HCD
	Define acquisition type	DDA
	Define enzyme type	Trypsin
Select a workflow		
	Glycan search	
Data refine options		
	Mass correction	Yes
	Associate feature with chimera	Yes
Glycan search		
Error tolerance	Precursor mass	10.0 ppm
	Fragment ion	0.02 Da
	Glycan fragment ion	20.0 ppm
Specify enzyme	Specified by each sample (when creating a project)	
	Digestion mode	Semi-specific
	Missed cleavage	1
PTM		
	Maximum allowed variable PTM per peptide	5
Fixed	Carbamidomethylation	Cysteine (C)
Variables	Deamidation	Asparagine (N), glutamine (Q)
	Oxidation	Methionine (M)
	Acetylation	N-term
	Pyro-glu	Glutamine (Q), glutamic acid (E)
Database	Target database	
	Taxonomy	All species
	Peptide length	6–45
Glycan option		
	N-linked database	N-linked
	O-linked database	O-linked
	Maximum adduct per peptide	NH4
	Maximum fucose	2

(Continued)

Table 4
continued

Create a project		
Report filter		
Glycopeptides	N-linked peptide score FDR (%)	1
	N-linked glycan score FDR (%)	1
	O-linked peptide score FDR (%)	1
	O-linked glycan score FDR (%)	1
Proteins	Proteins-10LgP	≥15
	Unique peptides	≥1

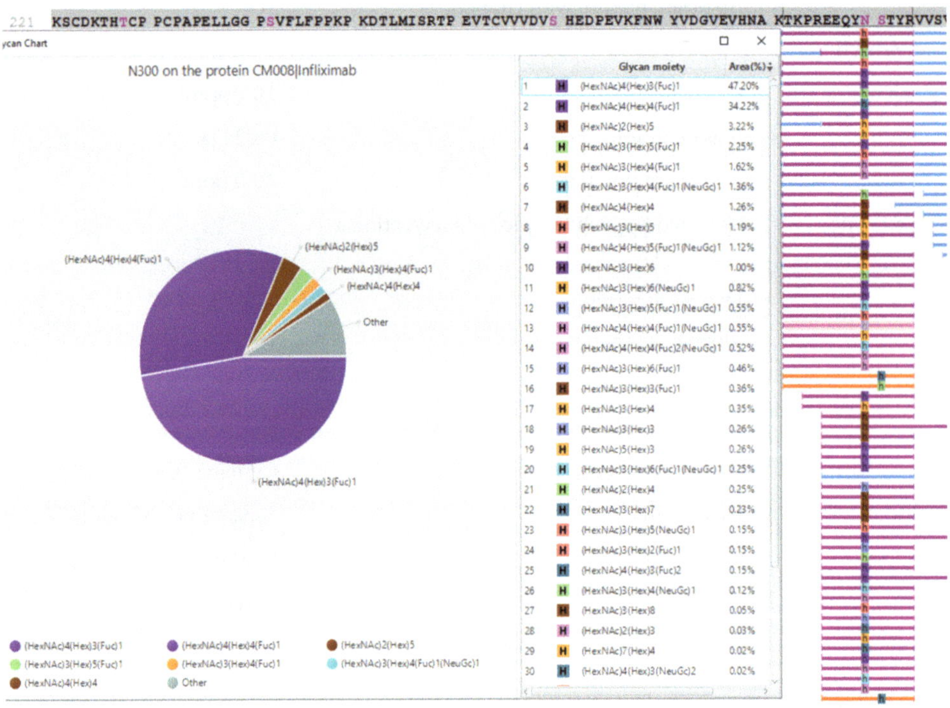

Fig. 4 N-linked glycan profiling of Remicade. N300 on infliximab heavy chain

3.4 Data Analysis

Data analysis is performed using PEAKS Glycan Search module in PEAKS Studio 12. This integrates peptide- and glycan-based search strategies to address the challenge of complex fragmentation of glycopeptides and to achieve an in-depth intact glycopeptide analysis from mass spectrometry data, to provide both N- and O-linked glycan profiling with structural resolution. There are 1828 basic N-linked glycans and 236 basic O-linked glycans in PEAKS Studio 12 glycan database, and the software can also perform de novo sequencing to

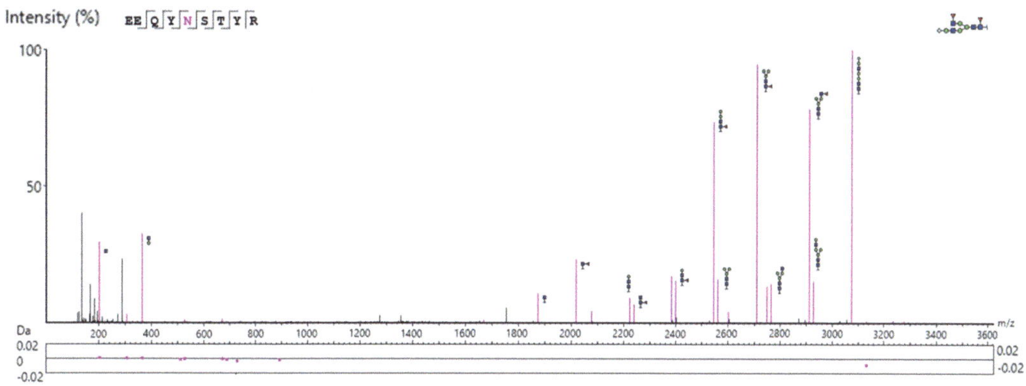

Fig. 5 Annotated spectrum of glycan (HexNAc)4(Hex)4(Fuc)2(NeuGc)1

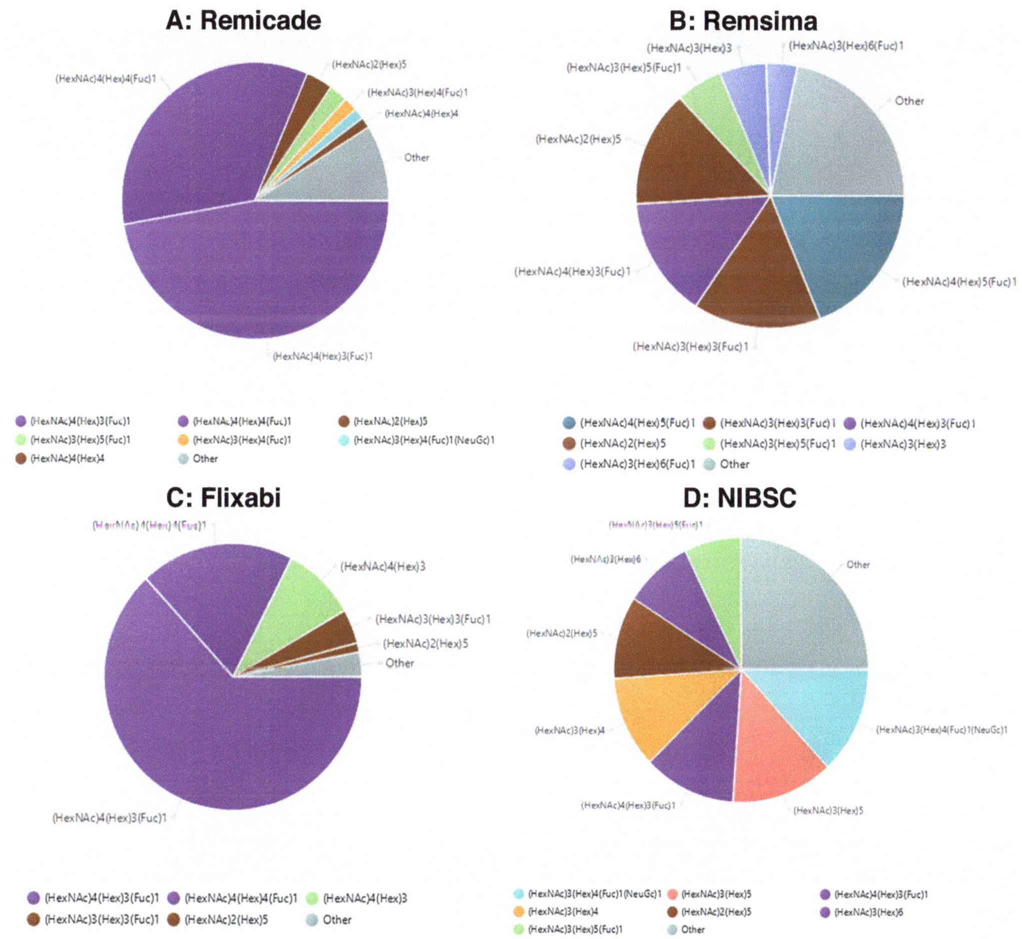

Fig. 6 Glycosylation profiles of Remicade **(A)**, biosimilars Remsima **(B)** and Flixabi **(C)**, NIBSC internal reference **(D)**

identify glycans not included in the database. Using PEAKS, we can identify intact glycopeptides and perform label-free glycopeptide quantification (*see* **Note 11**). The parameters used in our study are listed in Table 4. Examples of the results generated from PEAKS are shown in Figs. 4–8 and Tables 5 and 6 (*see* **Note 12** for infliximab amino acid sequence).

To summarize, the method described here can be used to identify intact glycopeptides without performing glycopeptide enrichment; to generate glycan profiling of biosimilars; to provide glycosylation site information and relative glycan abundance at a given glycosylation

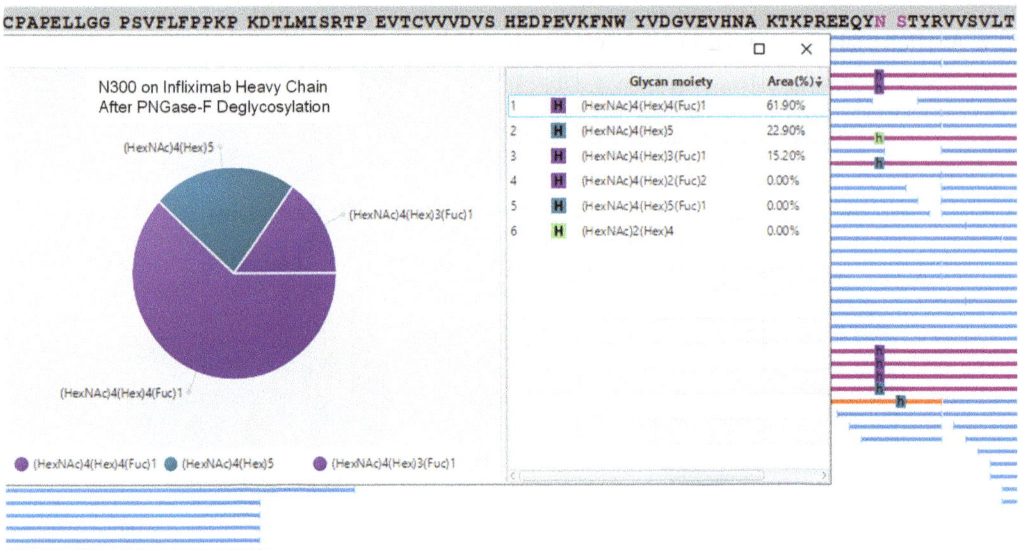

Fig. 7 N-linked glycan profiling of Remicade, after removing *N*-glycans using PNGase-F

Fig. 8 O-linked glycan profiling of Remicade, before **(A)** and after **(B)** removing *O*-glycans using OglyZOR

Table 5
Ion matches for glycan (HexNAc)4(Hex)4(Fuc)2(NeuGc)1

Observed m/z	Theoretical m/z	Moiety
204.09	204.09	2993_HexNAc1
308.10	308.10	2993_NeuGc1
366.14	366.14	2993_HexNAc1Hex1
512.20	512.20	2993_HexNAc1Hex1Fuc1
528.19	528.19	2993_HexNAc1Hex2
673.23	673.23	2993_HexNAc1Hex1NeuGc1
690.25	690.24	2993_HexNAc1Hex3
731.28	731.27	2993_HexNAc2Hex2
893.33	893.32	2993_HexNAc2Hex3
1671.81	1672.79	pep
1874.89	1875.87	pep_2993_HexNAc1
2020.95	2021.93	pep_2993_HexNAc1Fuc1
2077.97	2078.95	pep_2993_HexNAc2
2224.03	2225.01	pep_2993_HexNAc2Fuc2
2240.02	2241.00	pep_2993_HexNAc2Hex1
2386.08	2387.06	pep_2993_HexNAc2Hex1Fuc1
2402.07	2403.06	pep_2993_HexNAc2Hex2
2548.13	2549.11	pep_2993_HexNAc2Hex2Fuc1
2564.13	2565.11	pep_2993_HexNAc2Hex3
2605.15	2606.14	pep_2993_HexNAc3Hex2
2710.18	2711.17	pep_2993_HexNAc2Hex3Fuc1
2751.21	2752.19	pep_2993_HexNAc3Hex2Fuc1
2767.20	2768.19	pep_2993_HexNAc3Hex3
2913.26	2914.25	pep_2993_HexNAc3Hex3Fuc1
2929.26	2930.24	pep_2993_HexNAc3Hex4
3075.31	3076.30	pep_2993_HexNAc3Hex4Fuc1
3075.31	3075.28	pep_2993_HexNAc3Hex3NeuGc1
3133.33	3133.32	pep_2993_HexNAc4Hex4
3237.37	3237.33	pep_2993_HexNAc3Hex4NeuGc1
3278.39	3279.38	pep_2993_HexNAc4Hex4Fuc1

Table 6
O-linked glycan profiling of Remicade, biosimilars Remsima and Flixabi, and NIBSC infliximab standard

		Glycan	Glycosylation position in protein	Peak area
Remicade				
	HC	(HexNAc)1(Fuc)1	270	4.15E+04
	HC	(HexNAc)1	301	9.64E+04
	HC	(HexNAc)1	301	2.49E+05
	HC	(HexNAc)3(Hex)3	242	3.04E+05
	HC	(HexNAc)3(Hex)1(Fuc)1(NeuAc)1	242	3.11E+05
	HC	(HexNAc)1;(Hex)1	55_58	9.11E+05
	HC	(Fuc)1;(Hex)1(Fuc)1	105_106	2.80E+06
	HC	(HexNAc)1(Hex)2	242	3.55E+06
	HC	(HexNAc)1(Hex)1(NeuGc)2;(HexNAc)2(Hex)2	115_117	7.49E+06
	HC	(HexNAc)1;(Hex)1	55_58	8.77E+06
	HC	(HexNAc)2(Hex)2(Fuc)1(NeuGc)1	114	1.65E+07
	LC	(Hex)1(Fuc)1	31	2.03E+05
	LC	(Fuc)1;(Fuc)1	202_203	1.62E+07
Remsima				
	HC	(HexNAc)1	301	5.52E+05
	HC	(HexNAc)1;(Fuc)1	427_429	7.17E+05
	LC	(HexNAc)2(Hex)1	168	1.73E+05
	LC	(HexNAc)2(Hex)1(Fuc)2;(HexNAc)1	197_202	4.93E+07
	LC	(HexNAc)1(Fuc)1;(HexNAc)2(Hex)1(Fuc)1	129_131	2.91E+08

Table 6
continued

		Glycan	Glycosylation position in protein	Peak area
Flixabi				
	HC	(HexNAc)1(Fuc)1	301	4.59E+05
	HC	(HexNAc)1	301	6.15E+05
	HC	(Hex)2	21	1.43E+06
	HC	(Hex)4	427	8.77E+06
	HC	(HexNAc)4(Hex)1	21	5.30E+08
	HC	(HexNAc)1;(HexNAc)2(Hex)1(Fuc)2	21_25	5.30E+08
	HC	(HexNAc)3(Fuc)1;(Hex)1(Fuc)1	21_25	5.83E+08
	HC	(HexNAc)3;(HexNAc)1(Hex)1	21_25	6.28E+08
	HC	(HexNAc)1;(HexNAc)3(Hex)1	21_25	7.70E+08
	LC	(HexNAc)2(Hex)1	168	3.33E+05
	LC	(HexNAc)1	7	4.66E+05
	LC	(HexNAc)1	5	5.72E+05
	LC	(HexNAc)4(Hex)3	162	3.02E+06
	LC	(Hex)1(Fuc)1;(HexNAc)1(Hex)1(NeuAc)2	31_32	7.37E+06
	LC	(HexNAc)1(Fuc)1;(HexNAc)2(Hex)1(Fuc)1	197_202	4.19E+07
	LC	(HexNAc)1;(HexNAc)2(Hex)1(Fuc)2	129_131	1.57E+08
	LC	(HexNAc)2(Hex)1(Fuc)2;(HexNAc)1	129_131	1.57E+08
NIBSC reference				
	HC	(HexNAc)1	301	6.70E+05

HC heavy chain, *LC* light chain

site. Although the approach described here is based on MS spectra generated on Orbitrap Eclipse Tribrid Mass Spectrometer, it is also suitable for data generated on other MS systems, as long as the spectra have sufficient fragments from both peptide backbones and the attached glycans.

4 Notes

1. As PNGase-F is very expensive, prepare only enough volume required for the number of samples while allowing for pipetting losses.
2. To avoid dryness, this practice should be followed throughout the experiment when spinning is involved. Record the spin time required and apply it to the following steps.
3. The method is optimized for a starting material of 50 μg protein. For lower starting material, adjust volume accordingly.
4. Tryptic peptides can be kept at −20°C for long-term storage.
5. As PNGase-F is very expensive, we reduce the starting material to 10 μg, if N-linked de-glycosylation is required.
6. PNGase-F (*N*-Glycosidase F), a recombinant amidase that cleaves between the innermost GlcNAc and asparagine residues of high mannose, hybrid, and complex oligosaccharides, is an effective enzymatic method for removing almost all N-linked oligosaccharides from glycoproteins.
7. Manufacturer's recommended protocol can be found on https://www.neb.com/en/protocols/2014/07/31/pngase-f-protocol?pdf=true&srsltid=AfmBOorb7g02qwus_PH2OmdRiJMa5Q3mtUzo_wDPT7cT8K37Ue7-uMya.
8. OglyZOR is an endoglycosidase (*O*-Glycosidase) that specifically hydrolyses core 1 *O*-glycans on native glycoproteins, and it can be used for glycan analysis, confirmation of *O*-glycan presence and reduction of sample heterogeneity.
9. Manufacturer's recommended protocol for the digestion of *O*-glycosylated proteins with the removal of sialic acids to remove O-linked oligosaccharides from glycoproteins can be found on https://www.genovis.com/wp-content/uploads/INSTRUCTIONS-for-OglyZOR_2000u_reg_17.1.2.pdf.
10. Genovis has since added another three enzymes to its portfolio.
    OpeRATOR (Cat. No. G2-OP1-020), digests proteins carrying core 1 *O*-glycans. https://www.chayon.co.kr/wp-content/uploads/2020/10/Product-protocol-operator-feb-2020.pdf.

ImpaRATOR (Cat. No. G1-IR1-020), cleaves N-terminally of serine or threonine residues modified with mucin-type *O*-glycans, including sialylated species. https://www.genovis.com/smartenzymes/glycan-profiling/imparator/.
OmniGLYZOR (Cat. No. G3-OM6-005), hydrolyzes N- and Mucin-type *O*-glycans. By using OmniGLYZOR, removal of sialic acids using SialEXO® (included) is NOT required for optimal performance. https://www.genovis.com/smartenzymes/glycan-profiling/omniglyzor/.

11. Bioinformatics Solutions Inc. has a standalone PEAKS GlycanFinder version 2.5 software that can (1) identify and quantify intact glycopeptides and released glycans and (2) support both label-free and labeled glycopeptide quantification, such as quantitation analysis of intact glycopeptides with TMT or iTRAQ labeling.
12. Amino acid sequences of Infliximab are as follows:

Heavy Chain
EVKLEESGGGLVQPGGSMKLSCVASGFIFSNHWMNWVRQS-PEKGLEWVAEIRSKSINSATHYAESVKGRFTISRDDSKSAVY-LQMTDLRTEDTGVYYCSRNYYGSTYDYWGQGTTLTVSSAST-KGPSVFPLAPSSKSTSGGTAALGCLVKDYFPEPVTVSWNS-GALTSGVHTFPAVLQSSGLYSLSSVVTVPSSSLGTQTYICN-VNHKPSNTKVDKKVEPKSCDKTHTCPPCPAPELLGGPSV-FLFPPKPKDTLMISRTPEVTCVVVDVSHEDPEVKFNWYVDG-VEVHNAKTKPREEQYNSTYRVVSVLTVLHQDWLNGKEYK-CKVSNKALPAPIEKTISKAKGQPREPQVYTLPPSRDELTKN-QVSLTCLVKGFYPSDIAVEWESNGQPENNYKTTPPVLDS-DGSFFLYSKLTVDKSRWQQGNVFSCSVMHEALHNHYTQK-SLSLSPGK
Light Chain
DILLTQSPAILSVSPGERVSFSCRASQFVGSSIHWYQQRTNG-SPRLLIKYASESMSGIPSRFSGSGSGTDFTLSINTVESEDIADYY-CQQSHSWPFTFGSGTNLEVKRTVAAPSVFIFPPSDEQLKS-GTASVVCLLNNFYPREAKVQWKVDNALQSGNSQESVTE-QDSKDSTYSLSSTLTLSKADYEKHKVYACEVTHQGLSSPVTKS-FNRGEC

Acknowledgments

The project 23FUN06 ProMET has received funding from the European Partnership on Metrology, cofinanced from the European Union's Horizon Europe Research and Innovation Programme and by the Participating States.

References

1. Walsh G (2010) Post-translational modifications of protein biopharmaceuticals. Drug Discov Today 15(1718):773–780. https://doi.org/10.1016/j.drudis.2010.06.009
2. Delobel A (2021) Glycosylation of therapeutic proteins: a critical quality attribute. Methods Mol Biol 2271:1–21. https://doi.org/10.1007/978-1-0716-1241-5_1
3. Burton DR, Dwek RA (2006) Sugar determines antibody activity. Science 313(5787):627–628. https://doi.org/10.1126/science.1131712
4. Culyba EK, Price JL, Hanson SR et al (2011) Protein native-state stabilization by placing aromatic side chains in *N*-glycosylated reverse turns. Science 331(6017):571–575. https://doi.org/10.1126/science.1198461
5. Thompson N, Wakarchuk W (2022) O-glycosylation and its role in therapeutic proteins. Biosci Rep 42(10):BSR20220094. https://doi.org/10.1042/BSR20220094
6. Schwarz F, Aebi M (2011) Mechanisms and principles of N-linked protein glycosylation. Curr Opin Struct Biol 21:576–582. https://doi.org/10.1016/j.sbi.2011.08.005
7. He M, Zhou X, Wang X (2024) Glycosylation: mechanisms, biological functions and clinical implications. Signal Transduct Target Ther 9:194. https://doi.org/10.1038/s41392-024-01886-1
8. Laine RA (1994) A calculation of all possible oligosaccharide isomers both branched and linear yields 1.05 x 10(12) structures for a reducing hexasaccharide: the isomer barrier to development of single-method saccharide sequencing or synthesis systems. Glycobiology 4(6):759–767. https://doi.org/10.1093/glycob/4.6.759
9. Schachner L, Mullen C, Phung W et al (2024) Exposing the molecular heterogeneity of glycosylated biotherapeutics. Nat Commun 15:3259. https://doi.org/10.1038/s41467-024-47693-8
10. Mojumdar A, Yoo HJ, Kim DH et al (2024) Advances in mass spectrometry-based approaches for characterizing monoclonal antibodies: resolving structural complexity and analytical challenges. J Anal Sci Technol 15:23. https://doi.org/10.1186/s40543-024-00437-1
11. Cao W (2024) Advancing mass spectrometry–based glycoproteomic software tools for comprehensive site-specific glycoproteome analysis. Curr Opin Chem Biol 80:102442. https://doi.org/10.1016/j.cbpa.2024.102442
12. Lodge J, Schauer K, Brademan D et al (2020) Top-down characterization of an intact monoclonal antibody using activated ion electron transfer dissociation. Anal Chem 92(15):10246–10251. https://doi.org/10.1021/acs.analchem.0c00705
13. Oates RN, Lieu LB, Srzentić K et al (2024) Characterization of a monoclonal antibody by native and denaturing top-down mass spectrometry. J Am Soc Mass Spectrom 35(9):2197–2208. https://doi.org/10.1021/jasms.4c00224
14. Jensen PH, Karlsson NG, Kolarich D et al (2012) Structural analysis of N- and O-glycans released from glycoproteins. Nat Protoc 7(7):1299–1310. https://doi.org/10.1038/nprot.2012.063
15. Cao L, Qu Y, Zhang Z et al (2016) Intact glycopeptide characterization using mass spectrometry. Expert Rev Proteomics 13(5):513–522. https://doi.org/10.1586/14789450.2016.1172965
16. Sjögren J, Olsson F, Beck A (2016) Rapid and improved characterization of therapeutic antibodies and antibody related products using IdeS digestion and subunit analysis. Analyst 141(11):3114–3125. https://doi.org/10.1039/c6an00071a
17. Senini I, Tengattini S, Rinaldi F et al (2024) Direct glycosylation analysis of intact monoclonal antibodies combining ESI MS of glycoforms and MALDI-in source decay MS of glycan fragments. Commun Chem 7(1):203. https://doi.org/10.1038/s42004-024-01297-x
18. Riley NM, Coon JJ (2018) The role of electron transfer dissociation in modern proteomics. Anal Chem 90(1):40–64. https://doi.org/10.1021/acs.analchem.7b04810
19. Singh C, Zampronio CG, Creese AJ et al (2012) Higher energy collision dissociation (HCD) product ion-triggered electron transfer dissociation (ETD) mass spectrometry for the analysis of N-linked glycoproteins. J Proteome Res 11(9):4517–4525. https://doi.org/10.1021/pr300257c
20. Baba T, Ryumin P, Duchoslav E (2021) Dissociation of biomolecules by an intense low-energy electron beam in a high sensitivity time-of-flight mass spectrometer. J Am Soc Mass Spectrom 32(8):1964–1975. https://doi.org/10.1021/jasms.0c00425
21. Windwarder M, Altmann F (2014) Site-specific analysis of the O-glycosylation of bovine fetuin by electron-transfer dissociation mass spectrometry. J Proteomics 108:258–268. https://doi.org/10.1016/j.jprot.2014.05.022
22. Yu Q, Wang B, Chen Z et al (2017) Electron-transfer/higher-energy collision dissociation (EThcD)-enabled intact glycopeptide/glycoproteome characterization. J Am Soc Mass Spectrom 28(9):1751–1764. https://doi.org/10.1007/s13361-017-1701-4

23. Mao Y, Wang S, Zhao Y et al (2021) Systematic evaluation of fragmentation methods for unlabelled and isobaric mass tag-labelled O-glycopeptides. Anal Chem 93(32):11167–11175. https://doi.org/10.1021/acs.analchem.1c01696
24. Li R, Xia C, Wu S et al (2024) Direct and detailed site-specific glycopeptide characterization by higher-energy electron-activated dissociation tandem mass spectrometry. Anal Chem 96(3):1251–1258. https://doi.org/10.1021/acs.analchem.3c04484
25. Macauslane K, Pegg C, Nouwens A (2024) Electron-activated dissociation and collision-induced dissociation glycopeptide fragmentation for improved glycoproteomics. Anal Chem 96(27):10986–10994. https://doi.org/10.1021/acs.analchem.4c01450
26. Eshghi S, Yang W, Hu Y et al (2016) Classification of tandem mass spectra for identification of N- and O-linked glycopeptides. Sci Rep 6:37189. https://doi.org/10.1038/srep37189
27. Madsen J, Farutin V, Lin Y et al (2018) Data-independent oxonium ion profiling of multi-glycosylated biotherapeutics. MAbs 10(7):968–978. https://doi.org/10.1080/19420862.2018.1494106
28. Polasky DA, Nesvizhskii AI (2023) Recent advances in computational algorithms and software for large-scale glycoproteomics. Curr Opin Chem Biol 72:102238. https://doi.org/10.1016/j.cbpa.2022.102238
29. Cao W, Liu M, Kong S et al (2021) Recent advances in software tools for more generic and precise intact glycopeptide analysis. Mol Cell Proteomics 20:100060. https://doi.org/10.1074/mcp.r120.002090
30. Yang Y, Fang Q (2024) Prediction of glycopeptide fragment mass spectra by deep learning. Nat Commun 15(1):2448. https://doi.org/10.1038/s41467-024-46771-1
31. Bollineni R, Koehler C, Gislefoss R et al (2018) Large-scale intact glycopeptide identification by Mascot database search. Sci Rep 8(1):2117. https://doi.org/10.1038/s41598-018-20331-2
32. Kawahara R, Chernykh A, Alagesan K et al (2021) Community evaluation of glycoproteomics informatics solutions reveals high-performance search strategies for serum glycopeptide analysis. Nat Methods 18(11):1304–1316. https://doi.org/10.1038/s41592-021-01309-x
33. Sun W, Zhang Q, Zhang X et al (2023) Glycopeptide database search and de novo sequencing with PEAKS GlycanFinder enable highly sensitive glycoproteomics. Nat Commun 14(1):4046. https://doi.org/10.1038/s41467-023-39699-5
34. Bern M, Kil YJ, Becker C (2012) Byonic: advanced peptide and protein identification software. Curr Protoc Bioinformatics 12 Dec; Chapter 13:13.20.1–13.20.14. https://doi.org/10.1002/0471250953.bi1320s40
35. Armony G, Brehmer S, Srikumar T et al (2023) The GlycoPaSER prototype as a real-time N-glycopeptide identification tool based on the PaSER parallel computing platform. Int J Mol Sci 44(9):7869. https://doi.org/10.3390/ijms24097869
36. Shen J, Chen Z, Sun S (2022) Identifying intact N-glycopeptides from tandem mass spectrometry data using StrucGP. Biophys Rep 8(5-6):282–300. https://doi.org/10.52601/bpr.2022.220010
37. Zeng W, Cao W, Liu M et al (2021) Precise, fast and comprehensive analysis of intact glycopeptides and modified glycans with pGlyco3. Nat Methods 18(12):1515 1523. https://doi.org/10.1038/s41592-021-01306-0

Chapter 19

Glycoengineering of Monoclonal Antibodies with Homogeneous Glycan Isomers with Asymmetric Antennae

Roger Y. Tam, Grayson Hatfield and Lioudmila Tepliakova

Abstract

Glycosylation is an important post-translational modification that can alter the biological activities of antibodies. In monoclonal antibody biotherapeutic products, glycosylation of the Fc-domain, which is often sensitive to manufacturing conditions, can alter product potency by affecting its interactions with host immune cells. Improving control of antibody glycosylation to produce more chemically defined products and to better understand glycan function is gaining interest, with chemoenzymatic glycan engineering emerging as a powerful technique. This chapter describes methods to obtain purified glycans, including isomers with asymmetric antennae, their enzymatic incorporation into monoclonal antibodies, and a natural killer T cell-based binding assay to determine FcγRIIIA (CD16A) binding affinity.

Keywords Monoclonal antibodies, Glycoengineering, Chemoenzymatic synthesis, Fc-effector activities, Flow cytometry, NK92-CD16A

1 Introduction

Glycosylation is an important post-translational modification (PTM) occurring on numerous proteins that can affect their biological activities, such as, but not limited to, altering their binding to specific targets, immune recognition, and stability to environmental stressors and enzymatic degradation [1, 2]. Monoclonal antibodies (mAbs) comprise a class of biotherapeutics that are used to treat numerous diseases, where glycosylation of the protein backbone plays crucial roles in their safety and potency [3]. Their mechanisms of action vary from drug to drug, and predominantly involves binding of their variable domain to a specific target molecule such as a cell-surface receptor or a molecule in solution [4]. For those that are bound to cell surface receptors, secondary mechanisms of action can involve the mAb Fc domain, where its glycosylation (in the IgG1 subclass of mAbs) can drastically affect its binding to host immune cells to trigger further downstream effector activities to kill the cell it is attached to.

Xu Zhang et al. (eds.), *Protein Post-Translational Modifications*, Methods in Molecular Biology, vol. 3018,
https://doi.org/10.1007/978-1-0716-5166-7_19,

For example, antibody-dependent cellular cytotoxicity (ADCC) is significantly influenced by core glycan fucosylation [5–7] and site-specific galactosylation [8–10], and complement-dependent cytotoxicity (CDC) is correlated to terminal galactosylation [6, 7, 11].

As a PTM, protein glycosylation is non-templated during its biosynthesis and depends on the localized presence of glycan-modifying enzymes (known as glycosyl transferases and glycosidases for enzymes that attach or cleave sugar residues, respectively) in the Golgi bodies and extracellular matrix [12]. From a large-scale manufacturing perspective, this non-templated mechanism of glycosylation is sensitive to cell culturing conditions, resulting in mAbs with heterogeneous glycan structures that can change from batch to batch [13]. For example, pH, temperature, and cell cycles can affect sialylation levels, and manganese concentrations can affect degrees of galactosylation.

Engineering the glycan component can lead to more homogeneous end products [14–17], which may be beneficial in designing mAbs with specific desired effector activities, as well as enabling studying the roles of specific glycan structures, which, from a regulatory perspective, is crucial to better understand how these specific glycan structures contribute to a mAb's safety and efficacy. This latter need is particularly important with the ever-growing field of biosimilars, where manufacturers can use their own cell culture production methods, resulting in glycosylation that may differ compared to the innovator mAb biotherapeutic; it is crucial that regulatory reviewers are able to understand how glycosylation affects product quality. Recent work has shown that glycan isomers can have drastic differences in effector bioactivities [6, 8–11], highlighting the need to also consider these molecular characteristics. These isomers include terminal galactose residues on specific antennae of biantennary glycans, and human- vs. non-human-type glycans, which differ by how the terminal galactose residue is connected to the rest of the glycan [9, 10].

In this chapter, we will describe methods for enzymatic synthesis of site-specific glycan structures onto the Fc domains of monoclonal antibody-based therapeutics. Steps include enzymatic synthesis and purification of site-specific galactosylation, transglycosylation methods, characterization by HPAEC-PAD and intact mass spectrometry, and bioactivity testing by FcγRIIIA binding via NK92-CD16A cell-based flow cytometry assays. Although only the FcγRIIIA bioactivity is described here, the glycoengineered products described herein can be applied to other user-defined bioactivity assays.

2 Materials

2.1 Enzymes, Reagents, and Consumables

1. α-fucosidase (GH29)
2. EndoS–Chitin Binding Domain (EndoS-CBD)

3. EndoS2 D184M (*see* **Note 1**)
4. β-1,4-galactosidase (Sigma, Cat# 10105031001)
5. PNGase F
6. 100% ethanol
7. Diethyl ether
8. Acetone
9. Chitin resin
10. 2-chloro-1,3-dimethylimidazolinium chloride (DMC)
11. Trimethylamine (TEA)
12. Ammonium hydroxide
13. G25 Sephadex resin
14. Protein A resin
15. Bovine serum albumin
16. Ethylenediaminetetraacetic acid disodium salt dihydrate (EDTA)
17. Paraformaldehyde powder
18. Glycine
19. Activated carbon resin
20. Mini-PROTEAN® TGX Stain Free™ protein gels
21. Amicon centrifugal filters (30 kDa MWCO)
22. Buchner funnel
23. Filter paper

2.2 Buffers and Mobile Phases

1. 500 mM potassium chloride (KCl)
2. 500 mM sodium acetate and 50 mM $CaCl_2$, pH 5.7
3. Transglycosylation buffer (TGB): 50 mM Tris, pH 7.3
4. 5 mM $CaCl_2$, pH 5.5
5. 0.84 M 2-chloro-1,3-dimethylimidazolinium chloride (DMC) solution. Weigh 42.83 mg of DMC (0.253 mmol) and add 251.4 μL MQ water for a total volume of 300 μL (*see* **Note 2**)
6. 10% H_2SO_4 (v/v) in ethanol
7. Protein A binding buffer: 100 mM Na_2PO_4, 100 mM NaCl, pH 8.0
8. 50 mM sucrose solution in Protein A binding buffer
9. Protein A elution buffer: 50 mM sodium citrate buffer (pH 3.0)
10. Citrate-buffered saline (130 mM NaCl, 20 mM sodium citrate, pH 5.7)

11. HPAEC-PAD mobile phase A (MP-A): 200 mM NaOH, prepared from 50% w/w NaOH solution
12. HPAEC-PAD mobile phase B (MP-B): 150 mM NaOAc in 200 mM NaOH
13. HPAEC-PAD mobile phase C (MP-C): Milli-Q H_2O
14. HPLC mobile phase D (MP-D): Milli-Q H_2O containing 0.1% trifluoroacetic acid (TFA)
15. HPLC mobile phase E (MP-E): 99% acetonitrile, 1% Milli-Q H_2O containing 0.1% TFA
16. LC-MS mobile phase F (MP-F): 0.1% formic acid in water
17. LC-MS mobile phase G (MP-G): 0.1% formic acid in 80% acetonitrile
18. 1% (10 mg/mL) bovine serum albumin in cold PBS (1% BSA/PBS)
19. 1% BSA/PBS with 2 mM EDTA. To 10 mL of 1% BSA/PBS, add 7.44 mg EDTA
20. 4% PFA/PBS
21. 0.4 M glycine in PBS

2.3 Cells and Media

1. NK92-GFP-CD16A cell lines (ATCC, V variant; F variant).
2. 2 mM folic acid solution: weigh 4.4 mg folic acid (441.404 g/mol) and dissolve in 5.0 mL of 10 mM NaOH to obtain a final folic acid concentration of 2 mM. In a sterile biosafety cabinet and working under aseptic conditions, use 0.22 μm syringe filters to sterile-filter the folic acid solution and store in the dark at 4°C for up to 2 weeks.
3. 20 mM myo-inositol solution: weigh 36.0 mg of myo-inositol (180.16 g/mol) and dissolve in 10.0 mL of Milli-Q water to obtain a final concentration of 20 mM. In a sterile biosafety cabinet and working under aseptic conditions, use 0.22 μm syringe filters to sterile-filter the folic acid solution and store in the dark at 4°C for up to 2 weeks.
4. Stock media for cell culture: to prepare 252.5 mL of the "Stock media," transfer 182 mL MEM α (no nucleotides, no glutamine) to an empty sterile bottle, 31.5 mL heat-inactivated fetal bovine serum (HI-FBS, final concentration = 12.5%), 31.5 mL horse serum (final concentration = 12.5%), 2.5 mL of 200 mM GlutaMAX (final concentration = 2 mM), 2.5 mL of 2 mM folic acid solution (final concentration = 0.02 mM), and 2.5 mL of 20 mM myo-inositol solution (final concentration = 0.2 mM).
5. Complete media for cell culture: to prepare 15 mL of complete media, transfer 15 mL of the above stock media into a new tube, add 30 μL of 50 mM 2-mercaptoethanol (BME) solution (final concentration = 0.1 mM) and 0.3 μL of interleukin-2 (IL-2) to

the solution (specific activity: > 107 U/mg at 0.5 μg/μL; final specific activity: >10 U/μL) (*see* **Note 3**).

2.4 Antibodies

1. Alexa Fluor® 647-conjugated Goat anti-human IgG-F(ab′)2 specific (AffiniPure F(ab′)2 fragment) (Jackson ImmunoResearch, Cat# 109-606-097)
2. APC-CD16 Ab, anti-human, REA423 (Miltenyi Biotec, Cat# 130-113-951)

2.5 Equipment and Software Tools

1. Dionex IC-3000 or 6000 instrument for HPAEC-PAD analysis
2. Dionex CarboPac PA200 IC column (3 mm × 250 mm, 5.5 μm particle size, Cat# 062896)
3. Gold Standard PAD waveform with an AgCl electrode
4. HPLC instrument equipped with UV detector (214 nm) for PGC-HPLC purification
5. HyperCarb PGC column (Thermo Fisher Cat# 35005-159070 A, 10 cm × 150 cm, 5 μm particle size)
6. Orbitrap Fusion Lumos Tribrid mass spectrometer coupled to an Easy-nLC 1200 that was calibrated by infusion prior to analysis with a mixture of caffeine, MRFA, and Ultramark 1621
7. C4 PepMap300 PS-DVB trap column (300 Å × 5 μm × 5 mm)
8. PepSwift monolithic PS-DVB analytical column (200 μm × 25 cm)
9. BD LSR II flow cytometer (FACS Diva, v 8.0.1), equipped with a multi-well plate reader and the following laser/filter sets: blue laser 488 nm (20 mW) with GFP filter set: 505 long pass 525/50 band pass and SSC filter set: 488/10 band pass and FSC; Red Laser 640 nm (40 mW) with APC/AF647 filter set: 660/20 band pass
10. Silica gel TLC paper
11. NanoDrop spectrophotometer
12. Buchi rotary evaporator
13. Bottle-top filter (0.44 μm)
14. LabConco lyophilizer

3 Methods

3.1 Glycan Synthesis

Glycans can be purchased from commercial sources, or isolated from natural sources [18]. While previous methods have reported the isolation of human-type complex glycans [19–21], this section will also include obtaining purified regioisomers with the same molecular weight (i.e., glycans with mono-galactosylation on a specific antenna of biantennary glycans).

3.1.1 SGP Isolation to Obtain Glycopeptides

Initial sialoglycoprotein (SGP) isolated from eggs comprises mixtures of glycans, at approximately a 1:7 ratio of mono-galactosylated at the α3 antenna (G1(3)) and bi-galactosylated glycopeptides (G2) (Fig. 1). These glycans are bound to the SGP peptide (KVA*N*KT, where the Asn (N) residue is glycosylated).

1. Separate egg yolks from egg whites from 48 eggs and transfer the yolks to a 2 L Erlenmeyer flask or beaker.
2. Wash yolks by adding 1 L of 100% ethanol and stir vigorously on a stir plate for 45 min.
3. Let the reaction settle and remove the supernatant by passing mixture through a large Buchner funnel containing filter paper, under vacuum.
4. Transfer any collected precipitate back into the 2 L Erlenmeyer flask or beaker.
5. Repeat Steps 2–4 another two times with 100% ethanol.
6. Repeat Steps 2–4 with 300 mL of diethyl ether.
7. To the remaining solid precipitate, extract SGP by adding 1 L of 35% acetone and stir vigorously for 1 h.
8. Stop stirring and let reaction settle.
9. Pass the supernatant through a large Buchner funnel containing filter paper, under vacuum.
10. Transfer any collected precipitate back into the 2 L Erlenmeyer flask or beaker.
11. Repeat Steps 7–9, stirring overnight and then for 2 h, for a total of three extractions.

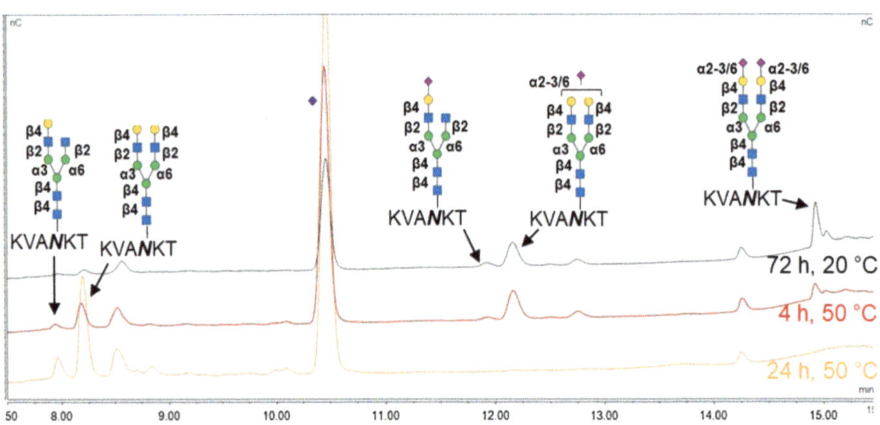

Fig. 1 Representative HPAEC-PAD chromatograms (HPAEC-PAD Method 1) showing the desialylation of SGP glycopeptides isolated from egg yolk using 50 mM HCl

12. After each filtration, combine all 35% acetone fractions.
13. Evaporate acetone using a rotary evaporator and a warmed water bath (If sialylated glycans are desired, do not exceed a water bath temperature of 30°C)
14. Following removal of acetone, filter the solution using a 0.4 μm bottle-top filter to remove precipitates.
15. Concentrate the remaining aqueous solution using a lyophilizer.
16. The resulting solid crude SGP mixture can be stored at −80°C until use. The expected yield is approximately 1.0 g.

3.1.2 Desialylation and Semi-Purification of SGP Glycopeptides

If required, the terminal neuraminic acid residues can be readily cleaved from the glycans using acid treatment as described herein. The crude SGP glycopeptides can be semi-purified by HPLC to remove potential contaminants that may co-elute with the desired glycans in subsequent steps.

1. If sialylated glycans are desired, skip to Step 8 below.
2. To 125 mg of the crude sialylated SGP material obtained in Sect. 3.1.1, Step 16 above, dilute into 2.0 mL with Milli-Q water
3. Add 100 μL of 1 M HCl to obtain a final concentration of 50 mM HCl.
4. Heat overnight at 50°C.
5. The next day, monitor completion of cleavage by HPAEC-PAD analysis using HPAEC-PAD Method 1 (Table 1).
6. Upon reaction completion, neutralize with 2.2 mL of 200 mM NaOH.
7. Confirm neutralization using pH strips.
8. Semi-purify the SGP glycopeptide using an activated carbon column to remove hydrophobic compounds using the following steps.
9. In a fume hood, transfer activated carbon resin into a 5 mL solid-phase extraction tube to a resin height of 1 cm.
10. Wash and equilibrate the resin by sequentially passing through 90% acetonitrile in 0.1% TFA (two column volumes), followed by 20% acetonitrile in 0.1% TFA (one column volume), then 2.5% acetonitrile in 0.1% TFA (three column volumes).
11. Using a pipette, carefully load the neutralized reaction mixture onto the top of the resin in the SPE tube and pass through the resin by gravity.
12. Collect the initial flow-through volume fractions.

Table 1
HPAEC-PAD Method 1

Column:	**Dionex CarboPac PA200 IC column (3 mm × 250 mm, 5.5 μm particle size)**		
Column temp:	**30°C**		
Flow rate:	**0.4 mL/min**		
Representative pump gradients:			
Time (min)	**% MP-A**	**% MP-B**	**% MP-C**
0	20	0	80
1	20	0	80
11	60	20	20
12	20	80	0
13	20	80	0
14	20	0	80
18	20	0	80
Representative elution times:			
Compound		**Retention time (min)**	
G1(3)-SGP		8.0	
G2-SGP		8.2	
G0T		9.4	
G1(6)-T		9.7	
G1(3)-T		9.8	
G2-T		10.1	
Neu_1-G1(3)-SGP		11.9	
Neu_1-G2-SGP		12.2	
Neu_2-G2-SGP		14.9	

13. Pass through 5% acetonitrile in 0.1% TFA (three column volumes), followed by 10% acetonitrile in 0.1% TFA (one column volume), 15% acetonitrile in 0.1% TFA (two column volumes), 25% acetonitrile in 0.1% TFA (three column volumes), and 50% acetonitrile in 0.1% TFA (one column volume) sequentially and collect each fraction.
14. For each collected fraction, perform HPAEC-PAD analysis (HPAEC-PAD Method 1) to locate SGP.
15. Evaporate fractions containing SGP and lyophilize.

16. For fractions containing crude glycan-SGP, purify by PGC-HPLC using HPLC Method 1 (Table 2) (*see* **Note 4**).
17. Collect fractions using a fraction collector and monitor purity by HPAEC-PAD analysis (HPAEC-PAD Method 1; Table 1).
18. Combine fractions containing purified G1(3)-SGP and G2-SGP and lyophilize.
19. Combine impure fractions comprising a mixture of glycan-SGPs and lyophilize for re-purification.

3.1.3 Enzymatic Glycopeptide Cleavage to Obtain Truncated Glycans That Are Monogalactosylated at the α3 Antenna (G1(3)T) and Bi-Galactosylated (G2T)

The galactosylated glycans are enzymatically cleaved from the glycopeptide backbone between the two core *N*-acetylglucosamine (GlcNAc) residues using EndoS enzymes.

1. Mix the bottle of chitin resin well, ensuring no resin is aggregated and stuck on the side of the bottle.

Table 2
HPLC Method 1

Column: HyperCarb PGC column (10 cm × 150 cm, 5 μm particle size)		
Column temp: 40°C		
Detector: UV, 214 nm (for glycans)		
Flow rate: 5 mL/min		
Fractions: 30 s/fraction		
Representative pump gradient:		
Time (min)	**% MP-D**	**% MP-E**
0	99	1
1	94	6
26	85	15
28	50	50
29	50	50
31	99	1
35	99	1
Representative elution times:		
Compound	**Retention time (min)**	
G1(3)-SGP	17.0–17.5	
G2-SGP	17.5–19.5	

2. Using a wide-bore P200 pipette tip, transfer 200 µL of the resin mixture into a conical 1.5 mL tube.
3. Add 1.0 mL of 20 mM Tris/100 mM NaCl, pH 8.5.
4. Close the cap and mix by inverting the tube.
5. Centrifuge for 30 s to pellet the resin at the bottom of the tube.
6. Using a pipette to remove the supernatant only, being careful not to remove any resin.
7. Repeat Steps 3–6 another two times to wash the chitin resin.
8. Following the last wash, add 100 µL 20 mM Tris/100 mM NaCl, pH 8.5, followed by 5 µL of EndoS-CBD.
9. Mix the EndoS-resin slurry well using wide-bore pipette tips and transfer to a solution of SGP (10 mg/mL containing 5 mM $CaCl_2$, pH 5.5).
10. Warm the tube to 37°C until reaction completion (*see* **Note 5**).
11. Monitor completion of cleavage by HPAEC-PAD Method 1 (Table 1 and Fig. 2) (*see* **Note 6**).
12. Following reaction completion, remove the resin by passing the solution through a 1 mL solid-phase extraction column and rinse with Milli-Q water.
13. Lyophilize the solution and store at −80°C until further HPLC purification (see Sect. 3.1.4, Step 6) or use in subsequent reactions.

3.1.4 Enzymatic Cleavage of Bi-Galactosylated Glycan (G2T) to Obtain Glycans That Are Agalactosylated (G0T) and Monogalactosylated at the α6 Antenna (G1(6)T)

1. To prepare glycans containing zero or one terminal galactose residue at the α6 antenna only (i.e., G0T, G1(6)T), dilute bi-galactosyl glycan G2T to 5 mg/mL (1.0 mL) with 150 µL of KCl (500 mM), 150 µL of a buffer containing 500 mM sodium acetate and 50 mM $CaCl_2$ (pH 5.7), and 4.55 mL Milli-Q water.

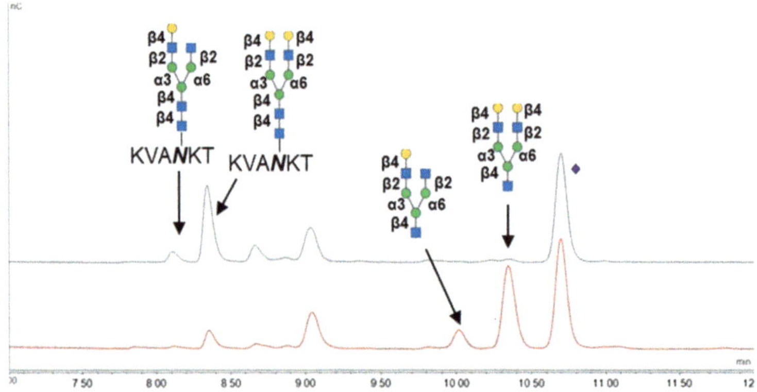

Fig. 2 Representative HPAEC-PAD chromatograms (HPAEC-PAD Method 1) showing cleavage of the glycopeptides into their respective truncated glycans

2. Add 150 μL β-1,4 galactosidase (see Note 7).
3. Incubate reaction at 30°C overnight, or until the desired proportion of G0T to G1(6)T is obtained.
4. Monitor reaction by HPAEC-PAD Method 1 (Table 1 and Fig. 3) (see Note 8).
5. Upon reaction completion, stop reaction by heating reaction to 80°C for 10 min to denature β-galactosidase enzymes.
6. Purify truncated glycans by PGC-HPLC using the following HPLC Method 2 (Table 3) (see Note 9).
7. Collect fractions using a fraction collector or manually, and monitor purity by HPAEC-PAD (HPAEC-PAD Method 1; Table 1).
8. Combine fractions containing purified truncated glycans and lyophilize.
9. Combined impure fractions comprising a mixture of glycans and lyophilize for re-purification.

3.1.5 Oxazoline Synthesis

Convert each purified glycan to their respective oxazoline intermediates for transglycosylation reactions with endoglycosidase enzymes.

1. Prepare HPAEC-PAD using the HPAEC-PAD Method 2 (Table 4) to monitor reaction progress.
2. Prepare a G25 column using a thin long column with a height of approximately 20 cm, with a cotton plug at the bottom to keep resin in the column.
3. Prepare a slurry of G25 Sephadex resin in cold 0.01% ammonium hydroxide.
4. Add slurry into column and let gravity drip.

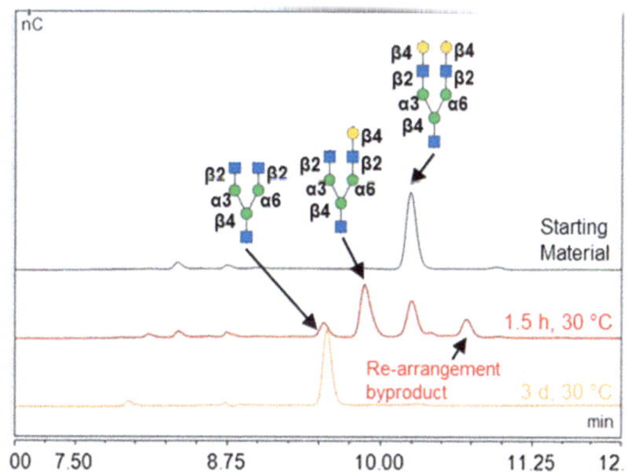

Fig. 3 Representative HPAEC-PAD chromatograms (HPAEC-PAD Method 1) showing the formation α-6 monogalactosylated product (red trace) and the agalactosylated product (orange trace), following treatment with β1-4 galactosidase

Table 3
HPLC Method 2

Column: HyperCarb PGC column (10 cm × 150 cm, 5 μm particle size)		
Column temp: 40°C		
Detector: UV, 214 nm (for glycans)		
Flow rate: 5 mL/min		
Fractions: 30 s/fraction		
Representative pump gradient:		
Time (min)	**% MP-D**	**% MP-E**
0	99	1
1	94	6
42	86.4	13.7
45	25	75
48	25	75
50	99	1
55	99	1
Representative retention times:		
Compound	**Retention time (min): β-anomer**	**Retention time (min): α-anomer**
G0T	17	22
G1(6)-T	26	31.5
G1(3)-T	23	28
G2-T	32	37

5. Equilibrate column with in cold 0.01% ammonium hydroxide (three column volumes) by passing solution on the top and letting drip through. Do not let the top of the resin become dry. Ensure pH of eluent is slightly basic using pH strips (*see* **Note 10**).
6. Cool a 0.6 mL Eppendorf tube on ice.
7. For 2.0 μmol of glycan, add the required amount of glycan into the tube, and then add 76 μL of DMC solution and place on ice.

Table 4
HPAEC-PAD Method 2

Column:	Dionex CarboPac PA200 IC column (3 mm × 250 mm, 5.5 μm particle size, Cat# 062896		
Column temp: 30°C			
Flow rate:	0.4 mL/min		
Representative pump gradient:			
Time (min)	% MP-A	% MP-B	% MP-C
0	40	0	60
7	57	25	18
7.5	40	0	60
10	40	0	60

8. In a chemical fume hood, add 26 μL trimethylamine and place reaction on ice at 4°C.
9. After 30 min, monitor reaction by HPAEC-PAD using Method 2 (Table 4).
10. Once the reaction is confirmed to be completed by HPAEC-PAD, immediately purify by the pre-equilibrated G25 column with the following steps.
11. Place a 96-well polypropylene plate (U-bottom) on ice to cool.
12. Transfer entire reaction on to the top of the resin bed with minimal disruption of the resin bed, and let top of solution pass to the top of the resin.
13. Wash reaction tube with 100 μL 0.01% NH_4OH and add to top of resin bed. Repeat two times.
14. Gently add cold 0.01% NH_4OH to the G25 column and let gravity drip into 96-well plate on ice (*see* **Note 11**).
15. Collect 250–300 μL/well, covering wells throughout to prevent evaporation.
16. To monitor elution of oxazoline, spot 0.5 μL of each fraction onto a piece of silica TLC paper. After all fractions have been spotted (oxazoline typically elutes in fractions 8–13, and triethylamine elutes slightly afterwards), use tweezers to dip the TLC paper into a solution of 10% H_2SO_4 in EtOH in a chemical fume hood (*see* **Note 12**). Dab off excess H_2SO_4/EtOH on the side of the solution jar, and carefully heat TLC paper using a heat gun in a chemical fume hood. Fraction

spots containing glycans will stain a brown/black spot, while those without glycans will be the same off-yellow colour as the background.

17. Confirm purity of fractions using HPAEC-PAD Method 2 (Table 4).
18. Prepare each sample for HPAEC-PAD analysis by diluting in 10–20 mM NaOH (~1/5000 dilution) to ensure oxazoline does not degrade while in sample tube (*see* **Note 13**).
19. Keep 96-well plate in the fridge, covered to prevent evaporation, while waiting for HPAEC-PAD analyses to complete.
20. Combine fractions containing purified oxazolines.
21. Take a quantitative sample and quantify using a glycan standard of known concentration (*see* **Note 14**).
22. Following quantification, aliquot into 100 nmol (or other desired amount) and lyophilize using a centrivap for as little time as possible (*see* **Note 15**).
23. Immediately store at −80°C until use.

3.2 mAb Glycan Remodeling

To facilitate a simpler workflow, the same endoglycosidase is used for both glycan deglycosylation and transglycosylation.

3.2.1 Fc Cleavage (± AlfC)

1. Transfer 1.0 mg monoclonal antibody into an Amicon centrifugal filter (30 kDa MWCO), and buffer exchange with TGB (50 mM Tris, pH 7.3), to a final concentration of 10 mg/mL.
2. Deglycosylate the monoclonal antibody with 10 μg EndoS2 D184M by adding 5 μL (2.0 mg/mL in PBS).
3. For defucosylated glycoforms, add buffer exchanged fucosidase to the reaction. Buffer exchange is performed to remove any glycerol from the enzyme stock solution, and is done by transferring 100 μL enzyme solution into a 10 kDa Amicon tube, followed by spinning and buffer exchange into TGB.
4. Incubate the reaction overnight at 37°C.
5. Monitor completion by SDS-PAGE. To prepare samples, dilute 1 μg of sample with Milli-Q water and reducing loading dye to achieve a final concentration of ~0.25 mg/mL. Heat the gel samples for 7 min at 80°C to reduce samples. Load 0.25–0.45 μg protein/lane into 10% polyacrylamide gels. Run gels at 75 V for 3 min, then 175 V for 50 min (Fig. 4) (*see* **Note 16**).

3.2.2 Fc Transglycosylation

1. Remove cleaved glycans by passing the crude deglycosylated reaction mixture through an Amicon centrifugal filter (30 kDa MWCO) and buffer exchange into cold TGB (*see* **Note 17**).
2. Concentrate back to approximately 10 mg/mL.
3. Warm on a heat block to 25°C.

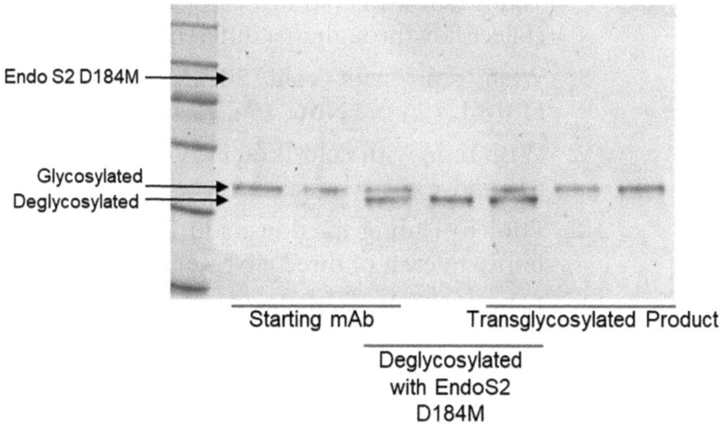

Fig. 4 Representative reducing SDS gel showing the change in migration of the heavy chain following deglycosylation (middle lanes with the lower bands) and transglycosylation (right lanes with the higher bands) reactions. To ensure the change in band migration is consistent through out the gel, a lane comprising both reactants and products is performed, resulting in two bands in the same lane. In the deglycosylated intermediate (middle three lanes), EndoS2 D184M in the reaction is also seen at the top of the gel

4. Prepare a solution of oxazoline by diluting a 200 nmol aliquot of oxazoline with 4.0 μL TGB 7.3, and place on ice until use.
5. Add 1.0 μL of the oxazoline solution (50 nmol) to the mAb reaction mixture containing EndoS2 D184M every 10 min for a total of four additions.
6. After a total reaction time of 35 min (i.e., 5 min after the final addition), monitor the reaction progress using reducing SDS-PAGE as described in Sect. 3.2.1 Step 5.
7. While the gel is running, place the reaction on ice.
8. Upon reaction completion, dilute the reaction with 500 μL of cold Protein A binding buffer (100 mM Na_2PO_4, 100 mM NaCl, pH 8.0), and place in fridge until purification (*see* **Note 18**).

3.2.3 Protein A purification

1. Rinse 200 μL Protein A resin slurry with cold Protein A binding buffer (1.0 mL × 3).
2. Add the resin slurry to the reaction tube and incubate at 4°C with gentle rocking for 5 min.
3. Transfer the reaction/slurry mixture into a 1 mL SPE tube
4. Let solution drain into a collection well/tube by gravity, in the fridge.
5. Pass the collected flowthrough through the resin/SPE Tube again.
6. Repeat steps 4 and 5 once.

7. Wash resin with cold Protein A binding buffer (0.5 mL), and collect flowthrough in a different tube.
8. Wash resin with cold 50 mM sucrose solution in PBS (1.0 mL × 2) (*see* **Note 19**).
9. Wash resin with cold Protein A binding buffer (0.75 mL × 3) to remove any residual sucrose.
10. Prior to eluting fractions, add 200 μL of Protein A binding buffer to each of three next wells.
11. Elute antibody from the Protein A column using 100 μL of Protein A Elution Buffer (50 mM sodium citrate buffer, pH 3.0).
12. Repeat the elution twice and collect the flowthrough into the three wells with 200 μL of Protein A binding buffer, respectively.
13. Monitor approximate concentration of each eluted fraction using a nanodrop spectrophotometer (at Absorbance of 280 nm).
14. Monitor eluted monoclonal antibody using SDS-PAGE (Sect. 3.2.1, Step 5) and combine fractions containing desired Ab product into a single tube.
15. Transfer combined fractions into an Amicon centrifugal filter (30 kDa MWCO) and buffer exchange with Citrate Buffered Saline (130 mM NaCl, 20 mM sodium citrate, pH 5.7) to achieve a final concentrations of approximately 0.3–1.0 mg/mL as quantified by nanodrop (at Absorbance of 280 nm).
16. Store solutions at 4°C.

3.2.4 Released Glycan Characterization Using HPAEC-PAD

1. Characterize remodelled glycans using PNGaseF to release glycans.
2. Purify glycans by activated carbon column.
3. Equilibrate resin with 2.5% acetonitrile in 0.1% TFA.
4. Load sample and wash with three column volumes of 2.5% acetonitrile in 0.1% TFA.
5. Elute glycans with 20% acetonitrile in 0.1% TFA.
6. Filter samples through a 0.2 μm syringe filter, evaporate and lyophilize samples
7. Perform HPAEC-PAD analysis, using HPAEC-PAD Method 1 (Table 1).

3.2.5 Intact Mass Spec Characterization

Characterization by intact mass spectrometry (Table 5) demonstrates homogeneity of the final product, particularly glycosylation of both heavy chains of the Ab dimer, which would otherwise not be detected using the aforementioned HPAEC-PAD analysis.

Table 5
Intact mass spectrometry method for homogeneity characterization

Trap column:	**C4 Pepmap300 PS-DVB trap column (300 A × 5 µm × 5 mm)**	
Separation column:	**PepSwift Monolithic PS-DVB analytical column 200 µm × 25 cm**	
Column temperature: 50°C Flow rate: 1 µL/min		
Representative LC gradient:		
Time (min)	**% MP-F (0.1% formic acid in water)**	**% MP-G (0.1% formic acid in 80% acetonitrile)**
0	100	0
5	50	50
10	50	50
20	10	90
25	10	90

1. Inject 0.5 µL of purified mAb sample into LC-MS system comprising a C4 trap column for desalting with 0.1% formic acid in water (MP-F), and a PepSwift Monolithic PS-DVB analytical column for separation.
2. Analyze eluted proteins with "Intact Protein" mode with Orbitrap Fusion Lumos MS at 15,000 resolutions, from 500 to 4000 m/z with auto maximum injection time, normalized AGC target of 50%, and 10 microscans.
3. Process the generated raw MS data with BiopharmaFinder 3.0. Protein deconvolution was performed with the ReSpect algorithm, using mass output of 130–160 kDa, Model Mass Range of 145,000–150,000, Charge State Range 40–80.

3.3 In Vitro Bioactivity Testing for CD16A (FcγRIIIA) Binding Activity

To determine binding, a cell-based assay can be used, using an NK92 natural killer T cell that overexpresses the CD16A (FcγRIIIA) receptor. Both high- (V176) and low- (F176) affinity CD16A receptor analogues are commercially available. Following binding between the cells and the remodelled mAb-glycoform, fluorescent labelling of the mAb monitored by flow cytometry is used to determine the degree of mAb-glycoform binding.

3.3.1 Cell Culture of NK92-GFP-CD16A Cells

1. Warm Complete media to 37°C.
2. To passage cells, slowly transfer cell suspension from a flask into a 15 mL tube.

3. Centrifuge at 1700 rpm for 3.5 min.
4. Remove media supernatant using a 10 mL strippette.
5. Resuspend cells in 1.0 mL complete media by slowly adding pre-warmed media down the side of the tube.
6. Use a P1000 pipette tip, or Pasteur pipette to resuspend pellet and gently triturate to break up cellular aggregates.
7. Transfer 10 μL of suspended cell solution into a tube, and add 10 μL Trypan blue.
8. Count number of live cells using a hemocytometer or an automated cell counter instrument.
9. Transfer 15 mL complete media into T75 flask.
10. Transfer the appropriate number of live cells for passaging to obtain a final cell density of 6.0×10^4 cells/mL.
11. Incubate at 37°C, 5% CO_2.
12. Passage cells every 3 days.

3.3.2 mAb–NK92-CD16A Binding Assay

1. Pre-cool a 96-deep well plate on ice.
2. Prepare 2.5-fold serial dilutions of antibody glycoforms (80 μL per well), using pre-cooled 1% BSA/PBS.
3. Harvest NK92-CD16A cells and spin down 1700 × rpm for 4 min to obtain pellet.
4. Add 1.0 mL cold 1% BSA/ PBS, centrifuge and remove supernatant.
5. Re-suspend cells in cold 1% BSA/PBS with 2 mM EDTA, and count cells using a haemocytometer or automated cell counter. While counting cells, place cell suspension on ice to keep cool.
6. Mix cell suspension thoroughly and dilute to 6.25×10^5 cells/mL.
7. Transfer 40.0 μL of diluted cell suspension to each well containing antibody, and mix gently with a pipette (2.5×10^4 cells/well).
8. Gently rock deep well plate at 4°C for 30 min.
9. Add 0.5 mL of cold 1% BSA/PBS to each well.
10. Centrifuge at 1700 × rpm (7 min, 4°C) and remove supernatant by inverting plate.
11. To each well containing antibody glycoforms, add 150 μL of a 1/300 dilution of Alexa Fluor-647 Anti-Human IgG (F(ab′)2-specific) in 1% BSA/PBS.
12. As a negative control, also add 150 μL of diluted AlexaFluor-647 Anti-Human IgG (F(ab′)2-specific) to wells in the absence of antibody (i.e., cells only).

13. As a control to directly detect CD16 receptors on NK92 cells, add 150 µL of a 1/50 dilution of APC-CD16 Ab (anti-human, REA423), in 1% BSA/PBS, to a well containing cells only (i.e., without any antibody glycoforms).
14. Incubate with gentle rocking at 4°C for 30 min.
15. Wash cells with 1 mL of cold 1% BSA/PBS with 2 mM EDTA and centrifuge/remove supernatant as above.
16. Re-suspend cells in 135 µL of 1% BSA/PBS and transferred into a shallow 96-well plate containing 45 µL of 4% PFA/PBS.
17. Incubate shallow plates with gentle rocking at room temperature for 30 min.
18. Centrifuge at 200 × *g* (1 min) and remove 140 µL of supernatant.
19. Re-suspend cell pellet in 80 µL of 0.4 M glycine in PBS.

3.3.3 Flow Cytometry Analysis

1. Read plates using a flow cytometer such as a BD LSR II Flow cytometer (FACS Diva, v 8.0.1), equipped with a multi-well plate reader and the following laser/filter sets: Blue laser 488 nm (20 mW) with GFP filter set: 505 Long Pass 525/50 Band Pass and SSC filter set: 488/10 Band Pass and FSC; Red Laser 640 nm (40 mW) with APC/AF647 filter set: 660/20 Band Pass.

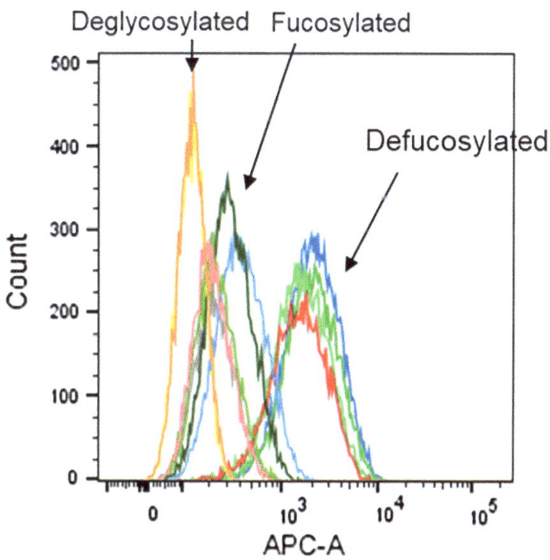

Fig. 5 Representative flow cytometry histogram, showing the degree of binding between NK92-CD16A cells and remodelled glycans at a fixed antibody concentration. The defucosylated glycans show the highest degrees of binding, followed by fucosylated glycans. Within each set of glycoforms, variable galactosylation further affects binding

2. Voltages were optimized in initial experiments and saved as application settings. Application settings were used in FACS DIVA for all voltages in all subsequent experiments to keep measurements consistent between experiments.
3. Use FlowJo (v 10.8.1) for data analysis, and the gating hierarchy is as follows: (i) FSC-A vs. SSC-a, (ii) FSC-H vs. HSC-W, (iii) SSC-H vs SSC-W, (iv) GFP+, (v) AF647±. A representative flow cytometry histogram result is shown in Fig. 5.

4 Notes

1. The sequence for EndoS2 D184M endoglycosidase is derived from amino acids 1–843 of EndoS2 from *S. pyogenes* [22], containing the D184M mutation [17], and a C-terminal decahistidine tag. After cloning into a pET-20b(+) vector, followed by transformation into BL21 (DE3) pLysS *E. coli*, the *E. coli* containing the EndoS2 D184M vector was grown in LB media with ampicillin at 37°C. The enzyme was produced following induction with IPTG and cultured overnight at 16°C. Cells were then pelleted by centrifugation at 4000 × *g* (4°C, 45 min), and the resulting pellet was resuspended in binding buffer (100 mM sodium phosphate/10 mM Tris base buffer, pH 8.0) with 20 mM imidazole. Cells were lysed using multiple freeze/thaw cycles with liquid nitrogen, and the supernatant was purified using Ni-NTA affinity chromatography. Following binding of the protein to the Ni-NTA resin, the resin was washed with six column volumes of binding buffer with 20 mM imidazole, and then Binding Buffer with 60 mM imidazole. EndoS2 D184M was eluted using an elution buffer comprising 100 mM sodium phosphate and 10 mM Tris base buffer with 240 mM imidazole (pH 8.0) for four column volumes into separate fractions. SDS-PAGE was used to confirm protein elution, and purified fractions were combined and buffer exchanged into phosphate-buffered saline (PBS), pH 6.0.
2. Due to the hygroscopic nature of DMC, the following method can be used if working in a glovebox under an inert atmosphere is not available. Place an empty 0.6 mL Eppendorf tube onto an analytical balance and tare the balance. Move the tube into the chemical fume hood. Quickly open the bottle of DMC and use a spatula to transfer a small amount of DMC directly into the tube and immediately close the lid. Fill the the headspace of the DMC bottle with an inert gas such as nitrogen or argon. Clean the spatula with 100% ethanol in the fume hood. Reweigh the tube containing DMC.

3. Prepare aliquots of IL-2 by resuspended vial (100 μg) in 100 μL of 100 mM sterile-filtered HOAc. Dilute with 100 μL of 0.2% BSA in sterile-filtered PBS (final IL-2 concentration = 0.5 μg/μL). Prepare aliquots of 2–5 μL per tube and store at −80°C. After using an aliquot, it can be kept in the fridge and is stable for 4 days.
4. While it is possible to skip this step, the use of PGC-HPLC here helps remove impurities that would otherwise co-elute with glycan peaks in the next section.
5. Binding EndoS-CBD to chitin resin dramatically increased hydrolytic of this enzyme, likely due to avidity effects of localized EndoS bound to the resin which settles at the bottom of the tube.
6. The resulting glycans comprise only 1 GlcNAc residue at the reducing end, compared to conventional full-length glycans which have two reducing GlcNAc residues. Referred to as "truncated" glycans, these will be used to ligate directly onto the antibody that contains a single GlcNAc residue, which then regenerates the native two GlcNAc residues of the native glycan structure.
7. This specific β1–4 galactosidase shows selectivity in cleaving the α3 antenna faster than the α6 antenna, resulting in a monogalactosylated glycan at the α6 antenna. However, prolonging this reaction will end up cleaving the α6 galactose residue as well to form the agalactosylated G0 product.
8. A rearrangement product occurs during cleavage with β -galactosidase, which has a retention time after G2T.
9. With PGC-HPLC, both β-and α-anomers are resolved on the column and are separated during the chromatographic run; however, they will quickly epimerize into an equilibrium mixture following collection in the tubes. Therefore, fractions containing both the β- and α-anomers can be combined together, and will appear as the same peak under HPAEC-PAD analysis.
10. If pH is not basic enough, the oxazoline will hydrolyze as it is being purified or subsequent lyophilization/storage.
11. Do not push the solution through, as this will impair adequate separation from residual triethylamine and DMC.
12. Take extreme precaution with this step, as H_2SO_4 is extremely corrosive and dangerous. Perform these steps in a chemical fume hood, wearing safety glasses, a lab coat, and gloves.
13. Oxazoline peaks elute several minutes prior to the parent glycan peak, using HPAEC-PAD Method 2.

14. Signal intensities between the parent glycan and the oxazoline product are similar by HPAEC-PAD.
15. Do not lyophilize overnight, as the centrivap chamber will heat up and degrade oxazoline; for 100 μL per tube, it takes about 2 h to complete lyophilization, at a vacuum pressure of 0.050 mbar.
16. To ensure accurate analysis of band migration at various glycosylation states, a lane comprising both starting material and final product within the same lane (i.e., a co-spot) should be performed (see Fig. 4).
17. EndoS2 D184M will remain with the mAb at this point and will be used for the next transglycosylation reaction.
18. Purification should occur as soon as possible, but it is possible to store the diluted reaction for one night in the fridge at 4°C.
19. The sucrose wash is important to remove free glycans, which were found to adsorb to the antibody mixture. Inadequate removal of free glycans causes issues when performing glycan analysis with PNGase F, since free glycans will overlap with the cleaved glycans.

Acknowledgments

This work was supported by the Government of Canada through Health Canada.

References

1. Moremen KW, Tiemeyer M, Nairn AV (2012) Vertebrate protein glycosylation: diversity, synthesis and function. Nat Rev Mol Cell Biol 13(7):448–462
2. He M, Zhou X, Wang X (2024) Glycosylation: mechanisms, biological functions and clinical implications. Signal Transduct Target Ther 9(1):194
3. Reusch D, Tejada ML (2015) Fc glycans of therapeutic antibodies as critical quality attributes. Glycobiology 25(12):1325–1334
4. Kothari M, Wanjari A, Acharya S, Karwa V, Chavhan R, Kumar S et al (2024) A comprehensive review of monoclonal antibodies in modern medicine: tracing the evolution of a revolutionary therapeutic approach. Cureus 16(6):e61983
5. Shields RL, Lai J, Keck R, O'Connell LY, Hong K, Meng YG et al (2002) Lack of fucose on human IgG1 N-linked oligosaccharide improves binding to human Fcgamma RIII and antibody-dependent cellular toxicity. J Biol Chem 277(30):26733–26740
6. Wada R, Matsui M, Kawasaki N (2019) Influence of N-glycosylation on effector functions and thermal stability of glycoengineered IgG1 monoclonal antibody with homogeneous glycoforms. MAbs 11(2):350–372
7. Dekkers G, Treffers L, Plomp R, Bentlage AEH, de Boer M, Koeleman CAM et al (2017) Decoding the human immunoglobulin G-glycan repertoire reveals a spectrum of Fc-receptor- and complement-mediated-effector activities. Front Immunol 8:877
8. Aoyama M, Hashii N, Tsukimura W, Osumi K, Harazono A, Tada M et al (2019) Effects of terminal galactose residues in mannose alpha1-6 arm of Fc-glycan on the effector functions of therapeutic monoclonal antibodies. MAbs 11(5):826–836

9. Hatfield G, Tepliakova L, Gingras G, Stalker A, Li X, Aubin Y, Tam RY (2022) Specific location of galactosylation in an afucosylated antiviral monoclonal antibody affects its FcgammaRIIIA binding affinity. Front Immunol 13:972168
10. Hatfield G, Tepliakova L, Tran J, Lu H, Gilbert M, Tam RY (2023) Bivalent non-human gal-alpha1-3-gal glycan epitopes in the Fc region of a monoclonal antibody model can be recognized by anti-Gal-alpha1-3-Gal IgE antibodies. MAbs 15(1):2239405
11. Chung AW, Crispin M, Pritchard L, Robinson H, Gorny MK, Yu X et al (2014) Identification of antibody glycosylation structures that predict monoclonal antibody Fc-effector function. AIDS 28(17):2523–2530
12. Stanley P, Moremen KW, Lewis NE, Taniguchi N, AebiM2022N-glycans. In: Varki A, Cummings RD, Esko JD, Stanley P, Hart GW, Aebi M, (Eds.) Essentials of glycobiology. 4th edn. Cold Spring Harbor Laboratory Press, Cold Spring Harbor (NY), 103–116
13. Hossler P, Khattak SF, Li ZJ (2009) Optimal and consistent protein glycosylation in mammalian cell culture. Glycobiology 19(9):936–949
14. Shin JH, Oh S, Jang MH, Lee SY, Min C, Eu YJ et al (2024) Enhanced efficacy of glycoengineered rice cell-produced trastuzumab. Plant Biotechnol J 22(11):3068–3081
15. Strasser R, Stadlmann J, Schahs M, Stiegler G, Quendler H, Mach L et al (2008) Generation of glyco-engineered Nicotiana benthamiana for the production of monoclonal antibodies with a homogeneous human-like N-glycan structure. Plant Biotechnol J 6(4):392–402
16. Kurogochi M, Mori M, Osumi K, Tojino M, Sugawara S, Takashima S et al (2015) Glycoengineered monoclonal antibodies with homogeneous glycan (M3, G0, G2, and A2) using a chemoenzymatic approach have different affinities for FcgammaRIIIa and variable antibody-dependent cellular cytotoxicity activities. PLoS One 10(7):e0132848
17. Li T, Tong X, Yang Q, Giddens JP, Wang LX (2016) Glycosynthase mutants of endoglycosidase S2 show potent transglycosylation activity and remarkably relaxed substrate specificity for antibody glycosylation remodeling. J Biol Chem 291(32):16508–16518
18. Fairbanks AJ (2018) Synthetic and semi-synthetic approaches to unprotected N-glycan oxazolines. Beilstein J Org Chem 14:416–429
19. Tang F, Wang LX, Huang W (2017) Chemoenzymatic synthesis of glycoengineered IgG antibodies and glycosite-specific antibody-drug conjugates. Nat Protoc 12(8):1702–1721
20. Sun B, Bao W, Tian X, Li M, Liu H, Dong J, Huang W (2014) A simplified procedure for gram-scale production of sialylglycopeptide (SGP) from egg yolks and subsequent semi-synthesis of Man3GlcNAc oxazoline. Carbohydr Res 396:62–69
21. Liu L, Prudden AR, Bosman GP, Boons GJ (2017) Improved isolation and characterization procedure of sialylglycopeptide from egg yolk powder. Carbohydr Res 452:122–128
22. Sjogren J, Struwe WB, Cosgrave EF, Rudd PM, Stervander M, Allhorn M et al (2013) EndoS2 is a unique and conserved enzyme of serotype M49 group A *Streptococcus* that hydrolyses N-linked glycans on IgG and alpha1-acid glycoprotein. Biochem J 455(1):107–118

Chapter 20

Deamidation Analysis of Biotherapeutic Proteins by Capillary Electrophoresis–Mass Spectrometry

Yun Wang and Huixin Lu

Abstract

Deamidation is a common chemical modification that can form spontaneously in solution and alter the chemistry of asparagine and glutamine residues. In biotherapeutics, uncontrolled deamidation poses a strong risk to product safety and efficacy. Traditional methods for assessing deamidation, such as ion-exchange chromatography and peptide mapping, can be limited. Capillary electrophoresis–mass spectrometry (CE–MS), in contrast, is well-suited for separating deamidated variants from their nondeamidated counterparts prior to identification by mass spectrometry (MS). For deamidation analysis, clear separation is critical for unambiguous MS identification as the variants can be shifted in mass by just 0.984 Da. In this chapter, a detailed method for the direct analysis of deamidation in protein biotherapeutics with CE–MS is outlined.

Keywords Deamidation, Biotherapeutic proteins, Capillary electrophoresis, Mass spectrometry, Critical quality attributes

1 Introduction

In proteins, the process of deamidation converts asparagine and glutamine residues into their acidic forms, aspartic/isoaspartic acid and glutamic acid, respectively [1]. As a nonenzymatic and spontaneous process that readily occurs in solution, uncontrolled deamidation of protein therapeutics can significantly impact the consistency and stability of products during routine processing, handling, and storage. The factors that influence deamidation mechanisms and kinetics involve both protein-specific characteristics and environmental conditions [2, 3]. For example, smaller side chains after asparagine residues, and higher temperatures in acidic conditions have been shown to lead to increased deamidation rates. For therapeutic proteins, deamidation in critical structural elements, such as the complementarity-determining regions of antibodies, can alter protein binding kinetics and impair downstream functions [4]. Uncontrolled deamidation can also promote aggregation and compromise protein

Xu Zhang et al. (eds.), *Protein Post-Translational Modifications*, Methods in Molecular Biology, vol. 3018,
https://doi.org/10.1007/978-1-0716-5166-7_20,

structural stability [5, 6]. Given these demonstrated risks, the characterization and quantitation of deamidation variants are important for maintaining product stability and efficacy.

Charge variant analyses are commonly used to profile deamidation variants in protein therapeutics. The introduction of a negative charge lowers the isoelectric point (pI) of deamidated species and allows for high-resolution separation by methods such as isoelectric focusing (IEF), ion-exchange liquid chromatography (IEC), and various capillary electrophoresis (CE) formats. Typically, these formats are coupled to absorbance detectors and, in order to identify peaks as deamidation, fractions need to be collected and digested for peptide mapping by mass spectrometry (MS) [5]. This can be technically challenging due to small fraction volumes and the risk of introducing deamidation artifacts during enzyme digestion. Directly coupling to MS has the potential to overcome these limitations. However, for chromatography separations, IEC running conditions are often incompatible with MS analysis and reversed-phase methods typically lack the resolving power for charge-based separation of biomolecules [7]. Recently, commercial CE–MS systems have become increasingly available for biopharmaceutical applications. Methods, such as capillary zone electrophoresis (CZE), are well-suited for intact charge variant analysis and, when directly coupled to MS, overcome many limitations associated with fraction collection and subsequent enzymatic digest. Compared to traditional charged variant methods, CE-MS offers superior charge-based separation with clear mass identification.

In this chapter, we describe key protocol instructions for the direct analysis of intact deamidation variants in protein therapeutics using CE–MS. The protocol includes steps from sample preparation, system conditioning, MS analysis, and capillary storage. In this method, a volatile, MS-compatible solution is used as the background electrolyte (BGE) with a neutral coated capillary for the CZE separation of charge variants. One advantage of this methodology is the commercial availability of reference standards and protocol instruction. In addition, the characterization of protein therapeutics requires high consistency and robustness, especially for quantitative assays. Therefore, we also discuss specific challenges associated with analyzing protein therapeutics by CE–MS and possible mitigation strategies.

2 Materials

2.1 Reagents and Solutions

1. PepCal mix for calibration (P/N 5045759, SCIEX) in 10% acetic acid (*see* **Note 1**).
2. Three-Protein Test Mix for capillary performance (P/N 477436, Sciex) containing Cytochrome C, Lysozyme, and Ribonuclease A (*see* **Note 2**).

3. Pierce Intact Protein Standard Mix with six-proteins for mass accuracy (optional; *see* **Note 3**).
4. pI 4.1 marker to evaluate spray stability (P/N A58481, Sciex) (optional).
5. 20 mM sodium phosphate buffer pH 8.0: Dissolve 2.840 g of disodium hydrogen phosphate in 0.8 L water. Titrate with phosphoric acid or sodium hydroxide to pH 8.0, as necessary (*see* **Notes 4 and 5**). Then, add water to 1 L. Minimize the amount of acid or base added for the best MS results.
6. 20 mM Tris–HCl buffer pH 8.0: Dissolve 2.422 g of Tris base in 0.8 L water. Titrate with hydrochloric acid or sodium hydroxide to pH 8.0, as necessary. Then, add water to 1 L. Minimize the amount of acid or base added for the best MS results.
7. 50 mM ammonium acetate pH 3.1 as CE–MS BGE (*see* **Note 6**): Prepare 50 mM ammonium acetate by adding 666 μL of 7.5 M ammonium acetate stock with 100 mL water. Prepare 50 mM acetic acid by diluting 286 μL glacial acetic acid in 100 mL water. Adjust the pH to 3.1 ± 0.1 by titrating the 50 mM acetic acid with the 50 mM ammonium acetate. Store at 4°C and monitor pH periodically.
8. 10 mM ammonium acetate pH 3.1 as MS sample buffer: Dilute the CE–MS BGE in water as needed. For example, 10 mL CE–MS BGE in 40 mL water.
9. 0.1 M HCl for capillary cleaning.
10. Water (LC–MS grade).
11. Methanol (LC–MS grade).
12. Interferon α2b CRS (EDQM cat no. I0320301).

2.2 Consumables

1. 0.5 mL Amicon Ultra centrifugal filter with appropriate molecular weight cutoff (MWCO) to retain the protein of interest (e.g., 3–10 kDa MWCO for 19 kDa protein).
2. NanoVials for CE–MS samples (P/N 5043467, Sciex).
3. CESI vials for CE–MS buffers (P/N B11648, Sciex).
4. CESI vial caps (P/N B24699, Sciex).
5. 0.2 mL skirted tubes for sample vials in CESI vials (P/N: T325-12N).
6. Neutral OptiMS Surface Cartridge (P/N: B07368, Sciex): 30 μm I.D. × 91 cm total length.

2.3 Equipment

1. Table-top centrifuge capable of 15,000 × *g*.
2. Stir plate.
3. pH meter.

4. TripleTOF 6600+ mass spectrometer (Sciex) with OptiFlow interface (Sciex).
5. NanoSpray Ion Source (Sciex).
6. CESI 8000+ capillary electrophoresis (Sciex).

2.4 Software

1. Analyst TF 1.8.1 for control of TripleTOF 6600+: a software patch is required to reduce curtain gas to 4–5 psi (*see* **Note 7**).
2. 32 Karat for control of CESI 8000+.
3. PeakView 2.2 for electropherogram analysis (*see* **Note 8**).

3 Methods

3.1 Accelerated Degradation for Deamidated Variants

1. Rinse a centrifugal filter by adding 500 μL water and spinning for 5 min at 15,000 × *g*. Remove any remaining liquid above and below the filter. Repeat twice.
2. Rinse the centrifugal filter with 20 mM sodium phosphate buffer or 20 mM Tris–HCl buffer for accelerated degradation by adding 500 μL buffer and spinning for 5 min at 15,000 × *g*. Remove any remaining liquid above and below the filter. Repeat twice.
3. Add 0.25 mg of an IFN α2b CRS sample to the filter and top up with sodium phosphate or Tris–HCl buffer. If the sample volume is too high for one filter, add a portion of the sample to the filter and spin at 15,000 × *g* for 5–20 min or until volume is reduced by half. Repeat until the sample is completely concentrated into the centrifugal filter (*see* **Note 9**). Adjust spin intervals as necessary during sample concentration.
4. Perform a buffer exchange by spinning down the sample at 15,000 × *g* for 5–20 min, discarding the filtrate, and replenishing with buffer (*see* **Note 9**). If the sample cannot be brought down to the minimum filter volume, record the approximate volume of the sample and the amount of buffer added to calculate the extent of sample matrix dilution.
5. Repeat buffer exchange twice more or until sample matrix components have been sufficiently reduced.
6. Transfer the sample to a low-protein-binding microtube. Note the approximate volume.
7. Add buffer to bring volume to 500 μL for 0.5 mg/mL or adjust as needed (*see* **Note 10**).
8. Remove aggregates by an additional spin at 15,000 × *g* for 10 min in the microtube. Transfer the supernatant to a new microtube.

9. Incubate the sample at 37°C. Deamidation should be detectable around the second week of incubation (*see* **Note 11**). Samples can be used for method optimization or as a reference for impurities.

3.2 MS Configuration and Calibration

Calibration should be regularly performed to monitor instrument performance. For CE–MS, calibrant can be directly infused through a capillary and the NanoSpray Ion Source, which allows for automatic calibration if needed (*see* **Note 12**). The following is an example of how to perform calibration through a dedicated capillary.

1. Install a capillary for calibration (*see* **Note 13**).
2. Prepare PepCal calibrant in a CESI nanovial by adding 19 μL 10% acetic acid and 1 μL PepCal mix. Lower amounts of acetic acid can also be used if a contaminant peak at 537 m/z is observed (*see* **Note 14**).
3. Use direct control or program a method to rinse the separation capillary in the forward direction with methanol and water at 100 psi for 5 min each.
4. Perform an injection of PepCal using direct control for 180 s at 100 psi. Ensure “For Capillary Fill” is checked.
5. Continue direct injection of PepCal at 20 psi with “For Capillary Fill” checked (*see* **Note 15**).
6. On Analyst, ensure that the curtain gas is set to 5 psi (*see* **Note 16**). Set the following method parameters for MS calibration: Polarity to “Positive”; Scan type to “TOF MS”; TOF Masses from 400 to 1200 m/z; Ion Source Gas 1 and 2, and Temperature (TEM) to 0; Declustering Potential (DP) to 80 V; Collision Energy (CE) to 10 V.
7. Set the ion spray voltage floating (ISVF) to 1800 V, acquire for 30 s, and recalibrate. If no signal is observed, increase the ISVF in 100 V increments until 2400 V. If still no signal is observed, increase the temperature.
8. Repeat Steps 5–7 for MS/MS calibration at high resolution and high sensitivity. Apply the same settings as above except: Product of 758.91 m/z and TOF Masses from 100 to 1500 m/z with Collision Energy at 40 V.
9. For capillary storage, rinse the separation capillary in the forward direction with methanol and water at 100 psi for 5 min each.
10. Dry the separation capillary with a 5 psi vacuum for 15–20 min using an empty vial.

3.3 Capillary Preparation

1. Install a Neutral OptiMS Surface Cartridge onto the CESI 8000+.

2. Place the sprayer tip in a 50 mL tube containing approximately 5 mL of water—enough to submerge the tip of the sprayer tip.
3. Hydrate the capillary overnight with the method steps outlined in Table 1.
4. After capillary hydration, place the capillary sprayer tip onto the NanoSpray holder.
5. Continue to electrical conditioning with the method steps outlined in Table 2.
6. Monitor the capillary tip during forward and reverse pressure to ensure an active flow.
7. Monitor the current during the 30 kV electrical conditioning step to ensure a stable signal (*see* **Note 17** and Fig. 1).
8. (Optional) Evaluate spray stability with a pI 4.1 marker (*see* **Note 18**).

3.4 Performance Evaluation

Due to variability in capillary performance over time, it is recommended to perform a separation with a protein standard with Three-Protein Test Mix to assess resolution prior to use. For neutral capillaries, this assessment can confirm if the coating is intact. Optionally, MS performance and accuracy of the subsequent deconvolution for large intact proteins can be confirmed with the Pierce Intact Protein Standard Mix. This mix contains six proteins with a wide distribution of molecular weights and m/z.

3.4.1 Evaluation with Three-Protein Test Mix

1. If opening a new Three-Protein Test Mix, reconstitute in 2.0 mL of water. Aliquot and store frozen (−20°C) for long-term use.

Table 1
CESI method for overnight hydration of neutral capillary

Event	Value	Duration (min)	Direction[a]	Note
Rinse	100 psi	5 min	Forward	0.1 M HCl separation line fill
Rinse	100 psi	10 min	Forward	BGE separation line fill
Rinse	75 psi	5 min	Reverse	Water CLC fill
Rinse	100 psi	30 min	Forward	Water separation line fill
Rinse	5 psi	300 min Alternating overnight	Forward	Water separation line fill
Rinse	75 psi	5 min Alternating overnight	Reverse	Water CLC fill

[a]Forward rinses for separation line. Reverse rinses for conductive liquid capillary (CLC)

Table 2
CESI method for electrical conditioning of neutral capillary

Event	Value	Duration (min)	Direction	Note
Separate—pressure	100 psi	3 min	Reverse	BGE separation line fill
Separate—pressure	100 psi	5 min	Forward	BGE CLC fill
Separate—voltage	30 kV 5 psi	60 min	1 min ramp, normal polarity, both pressure directions	Electrical conditioning
Separate—voltage	1 kV 5 psi	5 min	5 min ramp, normal polarity, both pressure directions	Ramp down

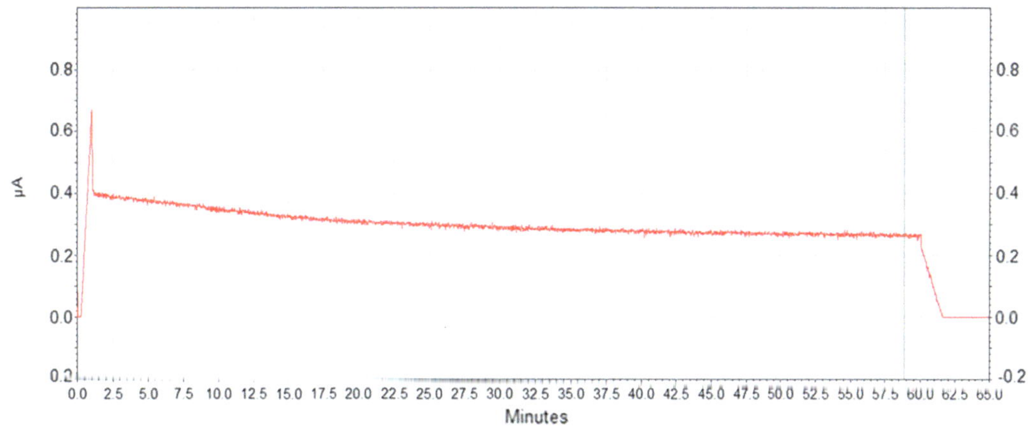

Fig. 1 Example steady current during electrical conditioning

2. Dilute 20 μL of reconstituted Three-Protein Test Mix in 80 μL of the MS sample buffer. Store on ice in the short-term when not used.
3. Add the prepared standards to the sample vials.
4. Prepare one vial of 0.1 M HCl for the inlet tray.
5. Prepare three vials of CE–MS BGE (50 mM ammonium acetate pH 3.1) for the inlet and outlet trays. To prevent the capillary ends from drying when in the home position, add one vial of CE–MS BGE at A1 of both inlet and outlet trays. These can

be used for separation. The last CE–MS BGE vial can be used for rinses.

6. Prepare two vials of water for the inlet and outlet. These will be used for water dips.
7. Continue analysis with the method steps in Table 3. Assign positions to the solutions prepared in Sect. 3.4.1 Steps 4–6, accordingly.
8. MS acquisition parameters include: Polarity to "Positive"; Scan type to "TOF MS"; TOF Masses from 500 to 3000 m/z; Ion Source Gas 1 and 2, and Temperature (TEM) to 0; Declustering Potential (DP) to 100 V; Collision Energy (CE) to 10 V.
9. Figure 2 shows an example electropherogram of the Three-Protein Test Mix. Resolution between each peak should

Table 3
CESI method for three-protein test mix analysis

Event	Value	Duration	Direction	Note
Rinse	100 psi	3 min	Forward	0.1 M HCl separation line fill
Rinse	75 psi	3 min	Reverse	BGE CLC fill
Rinse	100 psi	10 min	Forward	BGE separation line fill
Inject—pressure	2.5 psi	15.0 s	Forward	Hydrodynamic sample injection
Wait		0 min		Water dip
Inject—pressure	2.5 psi	10.0 s	Forward	Hydrodynamic buffer plug injection
Separate—voltage	30 kV 0.5 psi	10 min	1 min ramp, normal polarity, both pressure directions	Separation at low flow
Relay on				For MS acquisition[a]
Separate—voltage	30 kV 1.5 psi	10 min	1 min ramp, normal polarity, both pressure directions	Separation at high flow
Separate—voltage	1.0 kV 50.0 psi	5 min	5 min ramp, normal polarity, both pressure directions	Ramp down

[a]Can be started concurrently with or a few minutes after the separation begins

be qualitatively similar to the example. Cytochrome C and Lysozyme should be well-separated.

3.4.2 Evaluation with Intact Protein Standard Mix

1. If opening a new Pierce Intact Protein Standard Mix (76 μg), reconstitute in 50 μL water. Aliquot and store at 4°C for use within 7 days.
2. Add 5 μL of the reconstituted Intact Protein Standard mix to 15 μL of the MS sample buffer.
3. Prepare one vial of 0.1 M HCl for the inlet tray.
4. Prepare three vials of CE–MS BGE (50 mM ammonium acetate pH 3.1) for the inlet and outlet trays. To prevent the capillary ends from drying when in the home position, add one vial of CE–MS BGE at A1 of both inlet and outlet trays. These can be used for separation. The last CE–MS BGE vial can be used for rinses.
5. Prepare two vials of water for the inlet and outlet. These will be used for water dips.
6. Continue analysis with the method steps in Table 4. Assign positions to the solutions prepared in Sect. 3.4.2 Steps 12–14, accordingly.
7. MS acquisition parameters include: Polarity to "Positive"; Scan type to "TOF MS"; TOF Masses from 500 to 3000 m/z; Ion Source Gas 1 and 2, and Temperature (TEM) to 0; Declustering Potential (DP) to 100 V; Collision Energy (CE) to 10 V.

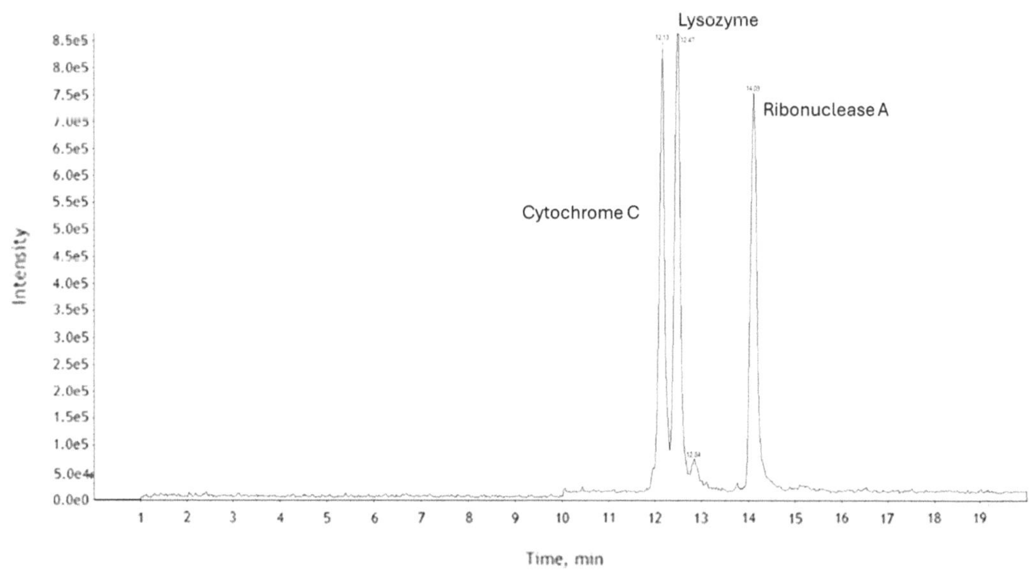

Fig. 2 Example spectrum of three-protein test mix

Table 4
CESI method for Pierce Intact Protein Standard Mix analysis

Event	Value	Duration	Direction	Note
Rinse	100 psi	3 min	Forward	0.1 M HCl separation line fill
Rinse	75 psi	3 min	Reverse	BGE CLC fill
Rinse	100 psi	10 min	Forward	BGE separation line fill
Inject—pressure	2.5 psi	15.0 s	Forward	Hydrodynamic sample injection
Wait		0 min		Water dip
Inject—pressure	2.5 psi	10.0 s	Forward	Hydrodynamic buffer plug injection
Separate—voltage	30 kV 0.5 psi	35 min	1 min ramp, normal polarity, both pressure directions	Separation at low flow
Relay on				For MS acquisition[a]
Separate—voltage	1.0 kV 50.0 psi	5 min	5 min ramp, normal polarity, both pressure directions	Ramp down

[a]Can be started concurrently with or a few minutes after separation begins

8. Figure 3 shows an example electropherogram of the Pierce Intact Protein Standard Mix. Theoretical average masses are provided in Table 5. Five proteins should be detectable (Bovine Carbonic Anhydrase II, *Escherichia coli* Exo Klenow, Human IGF-I LR3, chimeric *Streptococcus* Protein AG, and *Streptococcus dysgalactiae* Protein G).

3.5 Sample Preparation

This method has been applied to therapeutic protein drug products or drug substances in albumin-free formulations. Large albumin peaks can interfere with peaks of interest. Removing albumin peaks will require additional separation steps not described in this protocol.

1. Rinse a centrifugal filter by adding 500 μL water and spinning for 5 min at 15,000 × *g*. Remove any remaining liquid above and below the filter. Repeat twice.
2. Rinse the centrifugal filter with sample buffer by adding 500 μ L buffer and spinning for 5 min at 15,000 × *g*. Remove any remaining liquid above and below the filter. Repeat twice.

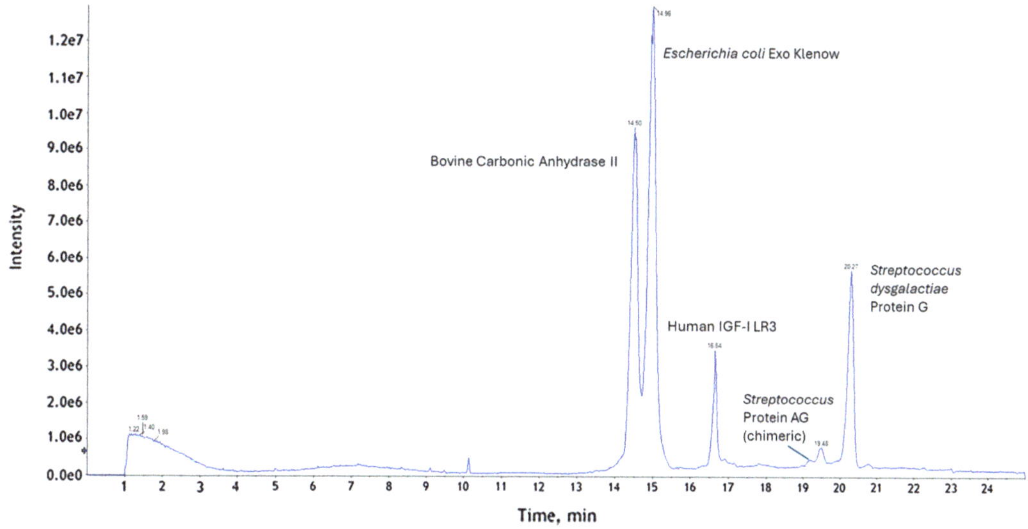

Fig. 3 Example spectrum of Pierce Intact Protein Standard Mix

Table 5
Theoretical masses for Pierce Intact Protein Standard Mix

Protein name	Protein accession	Theoretical avg mass (Da)
Human IGF-I LR3	P05019 (40-118)	9111.47
Streptococcus dysgalactiae protein G	P06654(223-413)	21,442.61
Bovine carbonic anhydrase II	P00921	28,981.29
Streptococcus protein AG (chimeric)	P02976, P19909	50,459.74
Escherichia coli Exo Klenow	P00582 (324-928)	68,001.15

3. Add 400 μL of sample buffer and 100 μL of the protein sample (0.5 mg/mL) to the filter.
4. Perform a buffer exchange by spinning down the sample at 15,000 × *g* for 5–20 min or until the volume is ~100 μL, discarding the filtrate, and replenishing with buffer to ~500 μL.
5. Repeat buffer exchange twice more.
6. Transfer the sample to a low-protein-binding microtube. The volume should be ~500 μL.
7. Add buffer to bring volume to 500 μL for 0.1 mg/mL or adjust as needed (*see* **Note 10**).
8. Remove aggregates by an additional spin at 15,000 × *g* for 10 min in the microtube. Transfer the supernatant to a new microtube being careful not to disturb any pellet.

3.6 CE–MS

1. Prepare one vial of 0.1 M HCl for the inlet tray.
2. Prepare three vials of CE–MS BGE (50 mM ammonium acetate pH 3.1) for inlet and outlet tray (*see* **Note 19**). To prevent the capillary ends from drying when in the home position, add one vial of CE–MS BGE at A1 of both inlet and outlet trays. These can be used for separation. The last CE–MS BGE vial can be used for rinses.
3. Prepare two vials of water for both inlet and outlet trays. These will be used for water dips.
4. Continue analysis with the method steps on the CESI 8000+ outlined in Table 6. Assign positions to the solutions prepared in Sect. 3.6 Steps 1–3, accordingly. The cleaning method can be separated from the injection and separation steps.
5. On the Analyst, ensure that the curtain gas is set to 5 psi (*see* **Note 16**). Set the following method parameters for MS calibration: Polarity to "Positive"; Scan type to "TOF MS"; TOF Masses from 500 to 3000 m/z; Ion Source Gas 1 and 2, and Temperature (TEM) to 0; Declustering Potential (DP)

Table 6
CESI method for sample analysis

Event	Value	Duration	Direction	Note
Rinse	100 psi	3 min	Forward	Cleaning: 0.1 M HCl separation line fill
Rinse	75 psi	3 min	Reverse	Cleaning: BGE CLC fill
Rinse	100 psi	15 min	Forward	Cleaning: BGE separation line fill
Inject—pressure	2.5 psi	15.0 s	Forward	Hydrodynamic sample injection
Wait		0 min		Water dip
Inject—pressure	2.5 psi	10.0 s	Forward	Hydrodynamic buffer plug injection
Separate—voltage	30 kV 1.0 psi	25 min[a]	1 min ramp, normal polarity, pressure both directions	Separation
Relay on				For MS acquisition[b]
Separate—voltage	1 kV 50 psi	5 min	5 min ramp, normal polarity, forward	Ramp down

[a]Adjust separation duration accordingly [b]Can be started concurrently with or a few minutes after separation begins

to 100 V; Collision Energy (CE) to 10 V. The "Time bins to sum" is an MS acquisition parameter that impacts resolution and sensitivity. This should be modified with the protein size, with larger proteins requiring higher time bins to sum (>50). For a 19 kDa protein, our lab has tested 12 time bins to sum.

6. An additional script for Intact Protein Mode (IPM) should be enabled for analysis of proteins with >6 kDa. To confirm that this is enabled in Analyst, go to Script > IntactProteinMode and click "IPM Enable." An additional option can be checked to "Decrease CEM by 100 V." This option will decrease the CEM voltage by 100 V in TOF-MS mode. In our experience, this has affected the accuracy of deconvoluted masses and, with the 1 Da difference associated with deamidation, this can significantly impact PTM identification. As a result, the impact of this option should be separately evaluated in each lab. In our lab, this option is unchecked (IPM must be disabled first before the option can be modifiable) and IPM is enabled.
7. Set the ISVF to the voltage required for stable spray, if tested. Otherwise, start at 1800 V. Increase in 100 V intervals if spray is not observed during the sequence.
8. After analysis, clean the capillary with the method parameters in Table 7.
9. Store the capillary at 4°C.

3.7 Analysis

1. Open the .wiff file in PeakView 2.2.
2. Select a peak area for deconvolution (*see* **Note 20**). Double-click the area to display the summed m/z spectrum. Alternatively, double-click a single point to view the m/z of that point.
3. With the m/z spectrum selected, go to Bio Tool Kit > Reconstruct Protein for deconvolution.
4. Reconstruction options can be modified according to the characteristics of the m/z spectrum (start and stop m/z) and the protein size (start and stop mass). Two parameters, the "Input spectrum isotope resolution" and the "step mass," will affect

Table 7
CESI method for capillary storage

Event	Value	Duration	Direction	Note
Rinse—pressure	100 psi	5 min	Forward	0.1 M HCl separation line fill
Rinse—pressure	100 psi	10 min	Forward	BGE separation line fill
Rinse—pressure	75 psi	5 min	Reverse	Water CLC fill
Rinse—pressure	100 psi	20 min	Forward	Water separation line fill

the resulting deconvolution. The "Input spectrum isotope resolution" setting refers to the resolution of the m/z spectra. Deconvolution with low resolution compared to high resolution can result in mass differences >1.0 Da. Similarly, lower step masses will result in higher resolution for closely spaced deconvoluted peaks but also increase processing time. As a result, it is recommended to deconvolute all peaks using the same resolution and step mass values for comparisons within the same spectrum.

5. Mass spectra and peak tables can be exported for further analysis and processing.

4 Notes

1. An alternative option is the X500B ESI positive calibration solution (P/N 5049910). Ideally, the calibrant will cover the same m/z range as the ions of interest.
2. The Three-Protein Test Mix is only used for verifying the capillary separation performance. The deconvoluted mass of each protein in the mix has not been verified.
3. Pierce Intact Protein Standard Mix can be used to verify the mass accuracy after deconvolution. The separation resolution of each protein peak has not been verified.
4. Phosphate conditions have been shown to promote deamidation. The optimal phosphate concentration can be evaluated.
5. For the best MS results, minimize the amount of phosphate and sodium added. For better precision of components in the final buffer solution, disodium hydrogen phosphate and sodium dihydrogen phosphate solutions can be prepared at the same concentration and combined without pH titration [8].
6. Separation performance in CE–MS is sensitive to changes in pH and just a shift in 0.1 pH units can affect resolution. As a result, the pH of the BGE should be monitored over time to assess any impact on method performance.
7. High curtain gas will cause the capillary to vibrate during analysis and interfere with signal acquisition. Since the lowest allowable curtain gas in Analyst is 10 psi, a software patch is needed to lower the curtain gas to 4–5 psi. Contact a Sciex customer service representative for further detail.
8. Depending on the analysis objectives, other analytical software may also be used to process the results. PeakView provides deconvolution algorithms for identifying peak masses.

Generating plot overlays and integrations with exported .csv files of mass spectra can be performed in R.

9. Proteins can aggregate with rapid concentration. To mitigate aggregation, monitor the sample in the liquid above the filter and centrifuge in shorter time intervals.
10. The final protein concentration can be confirmed by UV-Vis spectrophotometry or a colorimetric assay.
11. Deamidation can be monitored by ion-exchange chromatography in addition to CE–MS.
12. Alternatively, calibrant can also be directly infused by syringe through the DuoSpray Ion Source. However, this requires that the source be replaced for calibration. Automatic calibration within a sample sequence cannot be performed with this source.
13. This capillary can be dedicated for calibration or the cartridge to be used for analysis. A Silica Surface OptiMS Cartridge can be used for calibration.
14. If a strong 537 m/z peak occurs, continue to rinse the capillary with PepCal mix until the signal dissipates. If the signal persists, the CLC should be conditioned with a series of reverse NaOH rinses followed by HCl, water, and 10% acetic acid. For further details, consult the CESI 8000 Plus High Performance Separation-ESI Module User Manual.
15. Large droplets at the end of the capillary can interfere with signal acquisition. Gently tap the capillary platform to remove large drops before starting signal acquisition.
16. The setting for curtain gas in the Manual Tuning screen will remain at 10 psi. However, the current curtain gas value can be located in the "MassSpec TripleTOF6600 Detailed Status" screen, which is accessible by clicking the bottom-right icon 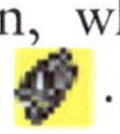.
17. Uneven current across the capillary could be an indication of capillary blockage or deposits in the capillary. If capillary current is not steady after cleaning and continued conditioning, it may be necessary to replace the capillary.
18. Spray stability under the voltage parameters used in sample separation can be evaluated by monitoring the acquisition of directly infused pI 4.1 marker diluted in CE–MS BGE. This can be prepared by diluting 2 μL of the pI 4.1 marker with 100 μL of CE–MS BGE. In brief, fill the separation line with sample and CLC line with CESI-MS BGE. Under direct control, apply the desired voltage and pressure parameters used in the method, including ramp time. Acquire with an ISVF of 0 to observe no signal. To assess the minimum voltage, start with an ISVF of 1500 V and ramp up in 100 V increments

until continuous spray is established. Ions can be extracted and monitored for stability. For additional detail, consult the Neutral OptiMS Cartridge Instruction Guide from Sciex.

19. To prevent buffer depletion when performing several injections within the same sequence (i.e., >5–10 injections), multiple vials of BGE can be prepared for separation. After a predetermined number of injections, the BGE vial can be automatically changed to a new vial either by creating a new method or setting the increment in the method parameters.

20. If the separation resolution between native and deamidated species is low, there may be significant overlap between peaks. Since the mass difference between the two species is only 1 Da, it may be necessary to sum over a narrower region within each peak or select a single point for each peak in order to avoid overlapping regions between peaks.

Acknowledgments

We would like to thank Michel Girard and Simon Sauvé for their guidance in the development of this protocol. We would also like to thank Tobias Bader and Stewart Bigelow for their technical contributions to this work. This work was supported by the Government of Canada through Health Canada.

References

1. Manning MC, Patel K, Borchardt RT (1989) Stability of protein pharmaceuticals. Pharm Res 6(11):903–918
2. Tyler-Cross R, Schirch V (1991) Effects of amino acid sequenc800e, buffers, and ionic strength on the rate and mechanism of deamidation of asparagine residues in small peptides. J Biol Chem 266(33):22549–22556
3. Pace AL, Wong RL, Zhang YT, Kao YH, Wang YJ (2013) Asparagine deamidation dependence on buffer type, pH, and temperature. J Pharm Sci 102(6):1712–1723
4. Schmid I, Bonnington L, Gerl M, Bomans K, Thaller AL, Wagner K et al. (2018) Assessment of susceptible chemical modification sites of trastuzumab and endogenous human immunoglobulins at physiological conditions. Commun Biol 1(1):28
5. Zhang YT, Hu J, Pace AL, Wong R, Wang YJ, Kao YH (2014) Characterization of asparagine 330 deamidation in an Fc-fragment of IgG1 using cation exchange chromatography and peptide mapping. J Chromatogr B Analyt Technol Biomed Life Sci 965:65–71
6. Alam ME, Barnett GV, Slaney TR, Starr CG, Das TK, Tessier PM (2019) Deamidation can compromise antibody colloidal stability and enhance aggregation in a pH-dependent manner. Mol Pharm 16(5):1939–1949
7. Badgett MJ, Boyes B, Orlando R (2017) The separation and quantitation of peptides with and without oxidation of methionine and deamidation of asparagine using hydrophilic interaction liquid chromatography with mass spectrometry (HILIC-MS). J Am Soc Mass Spectrom 28(5):818–826
8. Ganesh K, Soumen R, Ravichandran Y, Janarthanan. (2016) Dynamic approach to predict pH profiles of biologically relevant buffers. Biochem Biophys Rep 9:121–127

Index

Xu Zhang et al. (eds.), Methods in Molecular Biology, Methods in Molecular Biology, vol. 3018,
https://doi.org/10.1007/978-1-0716-5166-7,

MIX
Papier aus verantwortungsvollen Quellen
Paper from responsible sources
FSC® C105338

If you have any concerns about our products,
you can contact us on
ProductSafety@springernature.com

In case Publisher is established outside the EU,
the EU authorized representative is:
Springer Nature Customer Service Center GmbH
Europaplatz 3, 69115 Heidelberg, Germany

Printed by Libri Plureos GmbH
in Hamburg, Germany